Challenges of Policing Democracies

Challenges of Policing Democracies

A World Perspective

Edited by

Dilip K. Das

*State University of New York
at Plattsburgh, NY, USA*

and

Otwin Marenin

*Washington State University
Pullman, WA, USA*

Gordon and Breach Publishers

Australia Canada France Germany India
Japan Luxembourg Malaysia The Netherlands
Russia Singapore Switzerland

Copyright © 2000 OPA (Overseas Publishers Association) N.V. Published by license under the Gordon and Breach Publishers imprint.

All rights reserved.

No part of this book may be reproduced or utilized in any form or by any means, electronic or mechanical, including photocopying and recording, or by any information storage or retrieval system, without permission in writing from the publisher. Printed in Singapore.

Amsteldijk 166
1st Floor
1079 LH Amsterdam
The Netherlands

British Library Cataloguing in Publication Data

Challenges of policing democracies : a world perspective
 1. Police power – Political aspects 2. Police administration
 – Political aspects 3. Police – Public opinion – Political
 aspects 4. Democracy
 I. Das, Dilip K., 1941–   II. Marenin, Otwin
 362.2'3

ISBN 90-5700-558-1

CONTENTS

List of Figures	vii
List of Tables	ix
Foreword	xi
Preface	xv

I   Challenges of Policing Democracies

1	Challenges of Policing Democracies: A World Perspective *Dilip K. Das*	3
2	Policing in Democratic Societies: An Historical Overview *Peter C. Kratcoski, with the assistance of Wojciech Cebulak*	23

II   Country Perspectives

3	Challenges of Policing Democracies: The Croatian Experience *Sanja Kutnjak Ivkovich*	45
4	Estonian Police *Ando Leps*	87
5	The Challenges of Policing Democracy in Hungary *Istvan Szikinger*	115
6	The Challenges of Policing Democracy in Poland *Emil W. Pływaczewski*	143
7	Challenges of Policing Democracies: The Russian Experience *Yakov Gilinskiy*	173
8	Challenges Facing Democratic Policing in South Africa *Jeffrey Lever and Elrena van der Spuy*	195
9	Challenges of Policing Democracies: The Case of Austria *Maximilian Edelbacher and Gilbert Norden*	215
10	The Challenges of Policing Democracy: The British Experience *Gregory J. Durston*	243
11	Challenges of Policing Democracies: The Dutch Experience *Dann van de Meeberg and Alexis A. Aronowitz*	285

III Reflections on Challenges of Policing Democracies

12 Democracy, Democratization, Democratic Policing   311
    Otwin Marenin

IV Country Studies

13 Policing Macedonia   335
    Dime Gurev

14 Policing Democracy: The Slovenian Experience   343
    Jernej Videtic

15 The Impact of Human Rights on Identity Checks by the Police:
    A Swiss Perspective   353
    Laurent Walpen

Index   361

FIGURES

3.1	Police Education and Training in the Republic of Croatia	56
3.2	Organization of the Ministry of the Interior of the Republic of Croatia, at National Headquarters	60
9.1	Frequency of Conflicts Between "Official Regulations" and "Citizen-Friendliness" within the "Police" and "Gendarmerie"	223
9.2	Number of Registered Criminal Acts, Number of Solved Crimes, and Clearance Rate in Austria, 1975–1995	225
9.3	Total Convicted, as well as Share of Foreigners and Youth among Convicted, in Austria 1975–1995	227
11.1	Dual Authority over the Dutch Police	287

TABLES

1.1	Levels of Democracy of the Participating Countries	3
3.1	Crime Rates in some Central and Eastern European Countries 1993–1994	68
3.2	Indicators of Crimes Known to the Police and Indicators of Other Important Security Issues 1991–1996	69
4.1	Number of Population (thousands)	88
4.2	Gross Domestic Product, Crime and Crime Detection in 1989–1994	89
4.3	Gross National Product per Head of Population in 1991 (in US Dollars)	89
4.4	Crime Rates and Trends	96
4.5	Distribution of Crimes	96
4.6	Trends in the Commission of Murder	97
4.7	Trends in Theft	97
4.8	Trends in Juvenile Crime	98
4.9	Net Balance in Migration	99
4.10	Ethnic Composition of Population	100
4.11	Ethnic Composition of Criminals	101
4.12	Composition of Police Personnel	102
4.13	Background of Personnel	103
4.14	Schooling and Age of Personnel	103
6.1	Number of Persons Stopped at the Border	153
6.2	Foreigners Suspected of Committing Crimes in Poland in the Years 1976–1994	155
7.1	Trends in Crime in Russia, 1985–1998	181
7.2	Drug Related Crimes in Russia, 1987–1995	185
9.1	The Rank of the "Police"/"Gendarmerie" within the Hierarchy of Institutions, according to the Degree of Confidence Expressed by the Population (by Survey Results)	224
11.1	Crime in the Netherlands	289
11.2	Overview of Ethnic Origin of Detainees in 1993	295

FOREWORD

This volume is based on a series of papers presented at a symposium held in May 1995 at the Oñati Sociology of Law Institute in Spain. Participants from fifteen countries, mostly European, met to discuss perceived global challenges of policing democratic societies during a decade of dramatic geo-political change.

Nowhere was this change more evident than in Eastern Europe—the region represented most extensively at the symposium by participants from Estonia, Hungary, Poland and the Russian Federation. From the Balkans, still reeling from the bloody conflict arising from the disintegration of the former Yugoslavia, came representatives from Macedonia and Slovenia, while the strife torn regions of the Middle East had a single representative from Israel. The continents of South America and Africa were represented singly—the former by a participant from Chile and the latter by a participant from South Africa. Both Chile and South Africa remain nations emerging only recently from long periods of rule by brutal and repressive regimes.

To complete the mix of nations participating at the Oñati symposium were representatives from Austria, Switzerland, The Netherlands and the United Kingdom—all European countries with established democracies but still possessing a rich diversity of legal, political, cultural and allied traditions. In the case of three of these countries—Austria, The Netherlands and the United Kingdom—these traditions have the additional and shared overlay provided by their respective membership in the European Union.

In his introduction to this book, Professor Dilip Das has referred to the conceptual dilemmas encountered in any international gathering of this type at which the participants possess such varied backgrounds and experience regarding the topic under review. The two principal concepts being considered—policing and democracy—contain sufficient room even within a single society for extensive dialogue and debate about their content and meaning, let alone among the sixteen representatives at Oñati. Not surprisingly, no common understanding seems to have been reached about the definition of these concepts at Oñati, but the discussion flowing from the symposium, which is now reflected in this book, provides valuable insight and guidance for anyone seeking to advance policing within the context of democratic principles.

Considerable attention has already been given, principally at the level of the United Nations, for the development of common norms and standards of law enforcement including the use of force, the handling of juveniles, the prosecution of suspects and the treatment of offenders. The translation of these norms and standards into specific policing practices and procedures adopted by each of the member states of the United Nations remains a matter of ongoing concern. The Oñati symposium papers identify many of the difficulties being encountered along the road to reform in the area of law enforcement. Professor Das has expanded these policing reform dilemmas under three principal headings—organizational, operational and professional. A similar approach is adopted here.

The organizational challenges for policing which confront nations in a state of transition to democracy continue to be formidable. These challenges are not confined usually to settling upon and putting in place a desired structural model for policing. Frequently, they extend to the design and implementation of an entirely new justice system. Policing reform can be quite meaningless if it leaves untouched former criminal laws and procedures used to maintain a non-democratic regime in power. Sweeping away an entire justice system of this type and replacing it with one which both espouses and operates according to democratic principles is a complex and time consuming task. In South Africa, for example, it is a task which has been encompassed as part of a process of drafting and passing into law a new constitution for the country, and the entrenchment within that constitution of a bill of rights. The process continues with public hearings before a National Truth and Reconciliation Commission to allow those who committed crimes in the past to seek atonement for their misdeeds—and reintegration into South African society. This Commission has already dealt with many cases involving current members of the South African police responsible for wrongdoing under the former apartheid regime.

Operational concerns also place significant obstacles in the path of policing reforms within nations experiencing radical change, such as South Africa. For example, the significant growth in organized crime which afflicts contemporary South Africa, and to an even greater extent the Eastern European nations, is well documented. So too is the troubling growth in violence associated with much of this criminal activity. There have been major and related shifts of populations across national boundaries as people seek to flee from the brutality and hardships of their own societies to safer and more prosperous havens. Much of this movement of people has occurred in Europe, exerting great stress upon the humanitarian and liberal principles espoused by Western European democracies and leading to calls for more repressive policing measures to combat illegal migration. Similar developments have occurred in the United States where the surge of "economic" migrants from Latin America and elsewhere continues to be a significant problem for United States law enforcement officials.

Many professional challenges to policing reform were also identified by the Oñati symposium participants. Foremost among these were various forms of police deviance, including the excessive use of force and corruption. Such deviance is certainly not limited to police in nations undergoing a transition to democracy. The pressures and temptations to behave in violent and corrupt ways are, however, no doubt more compelling and explicable in a society where in the past the police were required to act as the repressive bastion for non-democratic and corrupt regimes. Quashing such police misconduct and replacing it with acceptable and lawful behavior requires much more than the introduction of a code of ethics and like provisions. Much of the deviance exhibited by the police mirrors that which continues to be displayed by ruling elites in the new "democratized" societies they now serve. Rooting out this deeply entrenched corruption remains a serious problem and a significant threat to the long term viability of many of the supposedly democratic governments established over recent years in Eastern Europe and other parts of the world.

Despite the presence of these serious, sustained and widespread challenges to the implementation of new ideals and practices in the policing of many emerging democracies, the general message which arises from the Oñati symposium papers contained in this book is both optimistic and encouraging. Less than a decade ago it would have been impossible to contemplate, let alone organize, a gathering like this to discuss a topic which touches upon one of the rawest nerves in any society, namely the nature and quality of policing. The paradigm shift which has occurred in the global balance of power following the end of the Cold War has now fostered an environment which permitted representatives of the thirteen nations participating in the Oñati symposium to engage in an open and frank dialogue about matters which would previously have trespassed immediately into barred and secret affairs of state.

Further dialogue of this type must be an essential ingredient of the change process required to bring about lasting police reform in countries such as those represented at Oñati. It should also be recognized that such reform cannot be delivered overnight. A timely reminder of this fact, and of the need for patience and an understanding of the difficulties involved, is contained in this book. The last great era of international policing reform, at least from the perspective of the common law, took place near the beginning of the nineteenth century with the establishment of the London Metropolitan Police Force by Sir Robert Peel. From this development spread the concept of a permanent and publicly funded system of policing to service community needs and operate with community support and control. That concept required more than a century to be adapted and implemented across the broad reaches of the then British Empire. Hopefully, the new democracies of this and the next century will require less time to embrace democratic policing. The present book promises to assist in facilitating and hastening the process of reform.

Duncan Chappell

PREFACE

A book is the product of many contributors. Most of the papers which now constitute this volume were first presented at the second International Police Executive Symposium at the School of Sociology of Law at Oñati, Spain in May 1995. The conference was supported by a grant from that school and was attended by police officials and academics from thirteen countries. The organizing framework for each chapter—types of challenges to democratic policing and the policy responses undertaken by governments and the police—was developed at the conference and is summarized and elaborated in Das' introductory chapter in this volume.

The Oñati meeting was the second in what has become a series of international symposia organized by the International Police Executive Symposium (IPES), an organization founded by Dilip Das in 1994 to encourage international, cross-cultural and cross-occupational dialogues among police officials, academics and policy makers by bringing them together in intensive four-day workshops in interesting parts of the world, normally during the first week in June of each year. Here they can discuss specific issues, policy proposals and developments in comparative and international policing. The first IPES annual meeting was hosted by the Ministry of Justice and the Police in Geneva in 1994, on the theme of "Police Challenges and Strategies." After the Oñati conference, successive symposia dealt with "Organized Crime" (hosted by Kanagawa University in Yokohama, 1996), "International Police Cooperation" (hosted by the Austrian Federal Police in Vienna, 1997), and "Crime Prevention" (hosted by the Dutch Ministry of Justice and EUROPOL in The Hague, 1998). The theme of the symposia for 1999, hosted by the police of Andhra Pradesh (a state of the Indian Union), was "Public Order Policing."

As mentioned, the papers in this volume were initially written by participants to the Oñati symposium. However, all the papers were subsequently rewritten to fit the organizing scheme developed by the symposium participants in order to highlight policing challenges and responses in each country. In some cases revisions were done by the original authors; in other cases we and the original authors asked recognized experts to join as co-authors or become sole authors; in one case we specifically commissioned a paper on a country which we felt would be of interest as a case study (Croatia), and finally a concluding reflections chapter by Marenin was added. Some of the original papers which could not be rewritten extensively are included in the Appendix. They are included there mainly because their organization was not as complete in following the "Challenges..." organizing framework as were the other chapters. Our goal was to make the organization and discussion of challenges and responses to democratic policing as consistent, comparative and informative as possible.

We would like to thank the contributors to this volume. The book is their work. We only proposed and suggested and sat back to watch the fruits of their labors roll in. We wish to thank Kirsty Mackay, Lauren Orenstein and Louise Timko whose unstinting support (including some delicious breakfasts and dinners) sustained this work, kept us on track and, by providing the promise of a final product, motivated us to pursue what sometimes felt like a tedious and endless task. We hope our efforts, as well as those of the contributors, repay their support, commitment and faith. We also want to thank Pam Robertson, who supervised the final editorial work on this project. Without the editors the book would not have achieved the professional look it now has. Mark Simon deserves our thanks for his continued support. We are grateful to Duncan Chappell for the "Foreword" which he graciously wrote for this volume, despite his frighteningly busy schedule in moving from one challenging assignment to another in various parts of the world.

We hope that this work will encourage a dialogue among practitioners, policy makers and academics in the fields of policing, criminal justice and related topics. This is one of the most important objectives of the International Police Executive Symposium.

Section I
Challenges of Policing Democracies

1

Challenges of Policing Democracies: A World Perspective

DILIP K. DAS

INTRODUCTION

The Second International Police Executive Symposium (Oñati, May 1995) was organized on the theme of "Challenges of Policing Democracies: A World Perspective." It was attended by police leaders, academics and justice professionals from thirteen countries[†]. Among them there were six emerging democracies, four established democracies, and three mixed democracies. The countries placed in the first category were Estonia, Hungary, Macedonia, Poland, the Russian Federation, and Slovenia. Included in the second category were Austria, the United Kingdom, the Netherlands, and Switzerland. Chile, Israel, and South Africa were incorporated in the third category (see Table 1.1). Chapters on Israel, Macedonia and Chile, however, are not included in this book.

For the purpose of the symposium, the most relevant characteristics of the emerging democracies were that they had just recently thrown off decades of authoritarian rule and were attempting to adopt democratic political cultures and institutions. They were struggling to survive economically and all of them were passing through varying degrees of anomie (see Merton, 1938; 1957) as they strove to adjust to unprecedented developments. They were the young democracies. The police in these countries were considered by the established democracies to be in need of modern equipment and training (Lintner, 1994). The established democracies were

Table 1.1. Levels of democracy of the participating countries

Established	Mixed	Emerging
Austria	Chile	Estonia
United Kingdom	South Africa	Hungary
The Netherlands		Macedonia
		Poland
		The Russian Federation
		Slovenia

the stable western European countries, highly affluent and willing to export police technology to the emerging democracies. The countries in the category of mixed democracies had democratically elected governments. They were economically better off in their own regions. But they were characterized internally by volatile politics, ethnic violence (for example, Israel and its relations with the Arabs) and residues of recent authoritarian regimes (Chile and South Africa) which had unabashedly restricted democratic rights (see McCormack, 1990). The police had a rather severe image in these countries. Chile's police, for example, were part of the military junta and were charged with the extermination of opposition leaders.

Keeping in mind that the police were regarded as anathema or a potential threat in a democracy (Alderson, 1979; Alderson and Stead, 1973; Goldstein, 1977; Skolnick, 1975), the objectives of the symposium were established as follows: (1) to appreciate at first hand what the police in emerging democracies regarded as challenges in operating within the newly democratic political environment, the established democracies and the mixed democracies were invited to present their contemporary experiences of these challenges); (2) to explore the similarities and the differences of the challenges, if any, from one category of democratic societies to another; and (3) to discuss the responses and the remedies adopted by various countries at different levels of democratic achievement.

PERSPECTIVES ON THE CONCEPT OF DEMOCRATIC POLICING

In order to facilitate a meaningful exchange of views on the theme, the participants were exhorted to spell out their concepts of democratic policing. It was recognized that such concepts were likely to be colored by culture (Triandis, 1994). Shelley (1995) referred to numerous conceptual problems she encountered in a similar gathering held in Budapest because the participants from Western Europe and those from Central and Eastern Europe came from vastly different societies. In the symposium under review, there were participants from Asia, Africa, and South America; so it was recognized that they were likely to analyze the challenges of policing and crime in very different ways. They were using similar words, but these denoted different concepts. It is easy to agree with Shelley that the participants from established, emerging and mixed democracies were perhaps not using the same intellectual arsenal to interpret the experiences of their police.

The Israeli participant's perspective was that the police were militaristic, armed and a coercive instrument of control at the disposal of the government (see Bittner, 1970). They could be utilized for policing democracies but they could not be democratic. His point of view was based on his experience in Israel, where the police were heavily armed, disproportionately engaged in security and protection roles, and only minimally involved in service tasks (see Friedman, 1986). A similar position was adopted by the Estonian participant who argued that the police could not be more democratic than the society they came from. Estonia was just emerging as a democracy but the police had been traditionally militaristic. Yet judged by the fact that they treated all criminals alike (Russians, Estonians and other groups) and that religion did not play a role in police policies, policing in this country should be viewed as democratic. The Hungarian participant stated that the police who worked with the community were democratic if they worked in accordance with popular wishes. Those engaged predominantly in law enforcement tasks could be democratic if proper and adequate supervision existed to ensure their conformity to democratic principles.

The Austrian participant viewed democratic police as people-oriented police—as giving citizens a role in crime prevention and control. It was recognized that, in the

Austrian concept of the state, there was an inherent conflict between state power (i.e., police power) and civil society (i.e., people power). Accordingly, police decision-making was required to be transparent. The Austrian view, based on experience with a sinister tradition of secret police and foreign occupation (see Emerson, 1968), was also shared by South Africa, Slovenia and Poland. The Polish participant also mentioned that democratic policing demanded giving the local people a role. The South African participant commented that decentralization, community policing, civilian control, and commitment to the minimum use of force were hallmarks of democratic policing his country wished to usher in.

The Chilean, the Dutch and the Macedonian participants opined that democratic policing called for the police to be regulated by law. The Swiss view was nearer that of Chile, the Netherlands and Macedonia: democratic policing meant that all police activities must be regulated in accordance with law. The police practicing such a style were totally accessible to the public and maintained maximum transparency. In the United Kingdom adherence to the rule of law and accountability to the public were considered essential elements of democratic policing. According to the Russian participant, democratic policing denoted the protection of the public by the police and the avoidance of abuse of power. This could be achieved by adherence to the rule of law. However, laws must be based on democratic principles in tune with the international standards of human rights as well as due process.

The Dutch participant also said that because the Dutch police operated in a democratic environment, they had considerable individual rights including the right to unionize and strike. The participant from Israel advocated that in a democracy the police should maintain institutional structures and procedures to learn, enforce, and reward compliance with civil and human rights (M. Amir, 1995).

In brief, in policing democracies the values and norms the police were expected to demonstrate were concerns for

(1) Rule of law;
(2) Accountability to the public;
(3) Transparency of decision making;
(4) Popular participation in policing;
(5) Minimum use of force;
(6) Creating an organization that facilitates learning of civil and human rights; and
(7) Internal democracy in the organization.

CHALLENGES

The challenges to democratic policing were identified by the participants from the various countries. These challenges can be broadly discussed under organizational, operational, and professional categories. Within the organizational category are the internal administrative changes in police agencies. The operational category will include challenges to the police in adhering to democratic values in their work and activities in the field. Discussed within the professional category are the problems and developments affecting the professionalism of police officers.

Organizational Issues

As the old regimes were disintegrating and democratic edifices were erected, the police, closely associated with the pre-democratic regimes, were the target of those who sought to remove the evils evidenced by the old structure. This meant getting rid of the old guards in the police, accomplished according to the situations prevailing in various countries.

Young Personnel and Old Culture

In Estonia most of the top officials who were also Russian-speaking either were forced to leave or left voluntarily. As a result, the police force turned out to be overwhelmingly dominated by younger officers who had little education. In the best European tradition, education was a characteristic of the older officers. In the pre-democratic days some had held, for example, law degrees but no lawyers came to the police now, as prestige, pay, and prospects of promotion were bleak. They could find better amenities elsewhere, including in organized crime operations. Unprecedented developments in the police and the country as a whole also caused frequent changes of the Ministers of the Interior, some of whom were uneducated, making challenges of policing democracies even more difficult.

The number of those who had been police since the Soviet days were not inconsiderable, and as a result the old culture of communist policing, based on autocracy and total disregard of human rights, did not disappear. In Hungary the police leaders of the pre-democratic days—who had been well qualified for their positions—were replaced by lower-level police officers as these were considered to be less contaminated by the values of the older system. But it really did not ensure a departure from the older style, values and methods. In the Russian Federation younger officers were about 55 percent and in Poland about 40 percent of the officer corps.

In the mixed democracies the exodus of former officers was not considerable, and the older culture continued. In South Africa the police were facing enormous challenges in transformation to the democratic norms due to the resistance of many old officers who believed that the community service orientation of the police would not be adequate to handle police problems. Israel's police claimed that they had to be violent or they would fail to keep a balance between ruthless efficiency and human rights because the threat from outside was real, despite the political changes in the Palestinian territory. In Chile the militarism of the police appeared to have remained fully intact. After the fall of the Pinochet regime—with its atrocious human rights record, for which the police must share the responsibility—the newly elected president of Chile retained members of the discredited junta to remain as commanders of the air force and the police (*The Europa World Year Book, 1994*).

Centralizing Trends

In the wake of the turmoil resulting from political upheavals, there was a movement toward consolidation of police services in the emerging and mixed democracies. Greater nationalization, which meant centralization of police authority at the hands of the national government, was reported from South Africa, Slovenia, Macedonia, Estonia, Hungary, and Poland. Incidentally, this period of consolidation and consequent nationalization of police services in these countries coincided with a streamlining, restructuring and centralization of the police organizations in established democracies such as Austria, the United Kingdom and the Netherlands. Even in Switzerland, which took considerable pride in cantonal autonomy, the growing influence of the Association of the Cantonal Chiefs of Police resulted in the twenty-six local police services paying more and more attention to concerted action in the interest of national security.

In Austria the Federal Gendarmerie were regrouping widely scattered smaller posts in order to create larger and wider jurisdiction for better consolidation and coordination of police activities in the rural areas (see Das, 1994). In the Netherlands the police were restructured into a smaller number of regional services with a coordinating national agency endowed with several central responsibilities and a national role. In the United Kingdom, too, the police were becoming increasingly nationalized with a nationwide information network, the growing centralizing influence of the Chief Officers' Association, greater authority of the Home Secretary over the police including control over the Public Authority and non-tenured chief constables (Forrester, 1995; Reiner, 1991). The police were also authorized a greater use of arms, and protective armor against potential attacks on police officers became compulsory. In the American concept of policing such tendencies towards centralized consolidation of police authority would be considered challenges to democratic policing (see Bayley, 1985; Berkeley, 1969; Fosdick, 1975).

In brief, organizational developments were considered challenges to the adherence of democratic standards and values by the police in different societies:

(1) Youth, inexperience, and inadequate training and education characterized the new police in the former communist countries.
(2) Continuation of the past culture of authoritarianism and lack of respect for human rights was ensured by retention of a considerable number of officers from the previous regimes.
(3) Greater centralization marked the police organizations of several established democracies as well as those of other countries.
(4) The role and philosophy of the police in the emerging democracies did not change.

Operational Issues

Crime and migration were cited as the most challenging issues for democratic policing. The police felt that they were expected to protect their societies at all cost from a whole host of evils connected with these matters, such as fear of crime, victimization, ethnic tension, and prejudice, among others.

Crime

Newly emerging crimes and rising crime rates created undue pressure on the police to do something quickly and posed consequent challenges to democratic police methods. Public pressure was capable of driving the police to look for expedient short cuts, which have the potential to violate accepted civil liberties (Martin & Romano, 1992) even in established democracies.

Increases in crime, particularly the rise in organized crime, were alarming to all emerging democracies (Estonia, Hungary, Macedonia, Poland, the Russian Federation and Slovenia). In mixed democracies, like South Africa, organized crime registered a noticeable increase; with the end of apartheid and changes in the government, many more opportunities for white collar and organized crime had emerged. Austria and the Netherlands, the established democracies, also reported a rise in organized crime. In Slovenia commercial crime increased, as measured by the amount of money involved. Compared to the figures of white collar crime in 1989, Hungary experienced a 725 percent rise in this category of crime by 1994. The rise in this type of crime led Poland to open an economic crime section to handle white collar crime, and Macedonia as well as the Russian Federation were doing the same. It was the Estonian experience that when older white collar criminals were arrested, newer ones appeared on the scene. It was explained that such crimes were manifestations of capitalist greed (see Coleman, 1985). White collar crime, particularly in the area of banking, was a major problem for the police in Geneva.

Emerging democracies like Hungary, Macedonia, and Poland were transit areas for drugs transported to more affluent western neighbors. Hungary had well-developed manufacturing facilities for morphine in the pharmaceutical sector and this posed a threat for abuse of this product. In Poland the large scale production of amphetamine was a menace. A higher number of drug crimes were reported by South Africa. There were deteriorating developments in drugs and drug-related crimes in Israel. Austria and the Netherlands experienced an increase in drugs and drug-related crimes.

Crimes were becoming increasingly more brutal. In Estonia, for the first time in its history, there was one murder a day in 1994. Even in Macedonia, otherwise a peaceful country, there were three hired murders in 1994, an unprecedented development. Austria, Hungary, Poland, the Russian Federation and Slovenia experienced an increase in criminal brutality. An American type of racketeering—squaring accounts, premeditated murders, grievous bodily harm, and missing-persons believed to have been killed—were indicators of violence in crime. Intergang violence was widespread (Hungary, the Russian Federation, Austria and South Africa). In the Russian Federation there were organized mafia-type violent criminal groups and criminals had easier access to weapons. Criminals as victims of violence at the hands of fellow criminals were testimony to the growing use of violence in the underworld (Hungary and the Netherlands). In Hungary and Poland organized crime gave rise to hitherto unknown murders for pecuniary gains, rivalries and revenge. There were also execution-type killings. Political extremism and political violence were clearly evident in some societies irrespective of their level of democracy (Austria and South Africa). All the countries observed more brutality in crimes committed by juveniles. In Austria there were violent youth gangs and in the Netherlands more violence in crimes was exhibited by Moroccan and Turkish youths. Juvenile criminals incarcerated in the Russian Federation were 181,479 out of the total incarcerated population of 1,742,855. This was the highest rate of juvenile incarceration in the whole of Europe (Gilinskiy, 1994).

Migration

In this period of radical political changes, unprecedented social disturbances (including the ethnic conflicts in the former Yugoslavia) and great economic upheavals, migration, which included illegal immigrants, refugees and asylum-seekers, was a multi-faceted challenge to the police in the newly emerging democracies as well as to their counterparts in the mixed and the established democracies. The police were called upon to take care of the problems of citizen fear, hostility and tension generated by the interactions of citizens and migrants.

The police were aware that they should follow international human rights standards as contained in such documents as the *Convention Relating to the Status of Stateless Persons* (1960) and the *Convention Relating to the Status of Refugees* (1954). But migrants were perceived to be responsible for the general increase in disorder and criminal activity. Such perceptions, however unscientific, created popular uncertainty and fear (Austria, The Netherlands, South Africa, Hungary, Macedonia, and Poland). It was mentioned, for example, that Hungarians are convinced that organized crime is the result of a non-Hungarian mentality (Szikinger, 1996), or, in other countries, that crime must be caused by foreigners—Russians, Chechnyans, Ukrainians, Bulgarians or Serbs (Nemeth, 1996). In the Russian Federation, younger people among migrants, particularly among some groups (Azerbaijan migrants, for example) had shown criminal tendencies. Migrants became a police problem as they were resented and attacked by the indigenous poor and nationalistic groups for their perceived impact on conditions of scarcity in the country to which they had fled. After deportation migrants reentered the same country which created a nuisance for the border police (South Africa). Migration was a challenge to democratic policing. As politics took precedence, the compassionate treatment demanded by international norms became a casualty to reality.

Professional Issues

Symposium participants were of the opinion that a lack of individual professionalism could be a real challenge to democratic policing. Violence, corruption, politicization and stereotypes about crimes and criminals understood as aspects of police subculture by several participants were discussed as indicators of an absence of professionalism. But the concept and reality of a police subculture was either not understood or denied by a number of participants (Austria, Chile, Macedonia, the Netherlands, Switzerland and Poland). Others (Estonia, Hungary, Israel, the Russian Federation, and South Africa) were, however, painfully aware of its existence.

Police Subculture and Violence

Police violence was a challenge to varying degrees in all democracies. In 1994, there were one hundred thirty cases of police violence in Austria. Although an Amnesty International Report referred to the existence of extensive brutality, police investigations did not confirm the majority of the alleged cases. However, some amount of brutality in the police did exist, and it was a threat to individual professionalism. While there were instances of police violence in Estonia, Hungary and Macedonia, it was in South Africa where a virulent subculture of police violence prevailed. Several officers there were sent to prison for violence, and cases of brutality constituted a sizable part of 9,000 cases of police misconduct investigated in 1994. In Israel the police had a violent image and were required to be tough, as the threat from the Arabs was considered very serious. In the United Kingdom police violence was reportedly mostly confined to misdirected enthusiasm. There were cases of abuse of power during investigation; but violence against the police had resulted in regulations requiring the wearing of protective armor. Talks were also going on regarding arming the police officers in the United Kingdom. There was a report of police violence in Geneva (Switzerland) circulated by Amnesty International. However, none of these allegations of the use of excessive force was found to be true during enquiries.

According to the Austrian analysis, police violence was situational. In handling mass demonstrations, for example, police officers were under stress and prone to use violence. Macedonia also experienced more police violence in handling demonstrations involving Albanians. In Switzerland use of force was generally due to strong resistance of arrestees at the time of the arrests. In Hungary, police violence was directed to specific groups. Gypsies, for example, bitterly resented that the police mistreated them. Reports of rough police behavior with the members of the general public were more frequent from the Waarmostraat, the sailors' quarters and red light district, in Amsterdam. The South African police perception that the criminal was the enemy of the society and that the police were expected to treat him as such generated police violence towards so-called undesirable people. The Estonian participant maintained that police deviance, which included homicide, suicide, and other violent acts, reflected social deviance.

Corruption

Seven cases of police corruption were reported in Austria in 1994 and it was feared that the number would increase in the future because of the corruptive influence of expanding organized crime. Adequate salaries helped prevent corruption among police officers in the United Kingdom, although it created a mentality of an occupation army (Baldwin, 1961) as officers lived away from the inner city where they worked. In 1994 there were twenty cases of police corruption in the Netherlands. Since the pay of police personnel in Switzerland was rather high, punishment against such behavior was extremely harsh and as there was no tradition of corruption in public service, the cases of police corruption were rare. For the established democracies (Austria, the Netherlands, Switzerland and the United Kingdom) police corruption was not a challenge to democratic policing as such.

Corruption in Estonia was a legacy from communisan when bribery, misappropriation and other forms of illegal financial gains involving people in high places were not investigated. The extent of Estonian police corruption was not known except that there were both meat-eaters and grass-eaters among the corrupt police officers in Estonia (see Knapp Commission Report, 1972). According to the Russian participant, the corruption in all powerful organs of the state and among general administrators at the highest levels and the police, which bedeviled the former regime, seemed to have become worse following the violent upheavals caused by the demise of the communist government. The Russian militia was unwilling to combat organized crime and drugs. Thousands of cases of corruption were investigated annually in Poland. It was a measure of the seriousness of the problem that in 1994 as many as sixty officers were dismissed because of corruption, two cases ended up in courts, and officers were incarcerated. In Hungary cases of police corruption were among 620 such cases involving all public servants. In 1994 there were thirty cases of corruption among 100 cases of police misconduct investigated in Macedonia. Corruption was not uncommon in the South African police and to that extent the police were a reflection of the society in general. The public in South Africa bought their way out of law violations. Annually a few cases of police corruption took place in Chile. In Israel the police were violent but not corrupt. But this violent police face the danger of becoming corrupt as organized crime spread its tentacles in the country.

Politicization

Politicization of the police was somewhat common across all levels of democracies but its extent and consequent impact on professionalism varied (see Reiner, 1992).

Police politicization, though prohibited by the new constitution in Hungary, was not eliminated from the highest level of the police hierarchy. In the Russian Federation older officers at high levels continued from the communist days. The government was able to manipulate them to alter crime statistics for political purposes. Hungary and Estonia regretted the lack of a legal culture in their societies, blurring the distinction between politics and law.

In Israel the police wanted to protect the status quo and were very conservative. Politically, the police were on the right. In South Africa the police suffered from the stigma of being the protectors of an illegitimate regime and defenders of the apartheid policies. In that country the police also had to handle violent political crimes routinely, and this responsibility had a tendency to make them appear a political tool. Politically, the stock of the police in Chile was low. Thus, the pariah status of the police in South Africa and Chile was not conducive to their capacity for maintaining democratic neutrality and professional autonomy in the new political order. In the United Kingdom there was an erosion of the non-political and ministerial status of the police. Political vulnerability was a reality as the chief constables were no longer tenured. In Austria there was an emphasis on political compatibility between politicians in power and police executives working in their jurisdictions. In Switzerland (Geneva), too, the political views and preferences of the minister directing the police could influence decision making by chiefs of police, particularly affecting assignment, promotion and other personnel matters. Even in those societies (Austria, the Netherlands, Switzerland and the United Kingdom) with a strong legal culture, the police did not conform to the rule of law because they were called upon to handle politicized issues.

Stereotyping

Austrian police stereotypes consist of seeing the migrant as criminal although the reality differed. Street crimes, such as burglaries, robberies and the importation of drugs did show disproportionate involvement of migrants and foreign elements. But white-collar crimes, the drug trade and consumption at the street level and terrorist deeds of left-wing activists overwhelmingly involved Austrians. The Dutch police criminal stereotype was a foreigner, involved in drugs and habituated to committing small crimes. In Estonia, non-Estonians were handled by the police as criminal suspects. They were regarded as criminally more active than the general population. It was part of Macedonian police folklore that criminals were from two strata of society: the richest and the poorest. The Gypsies, the poorest criminals, were not as serious a threat to democratic values as the richest criminals, such as the money launderers. The police in Hungary could not conceive that the middle and upper classes were as prone to criminal propensities as the underclasses. The police also protected the deviant among themselves. In the Russian Federation the criminal stereotype was that the people from the erstwhile south—from the Caucasian and Asiatic regions like Georgia and Tazakistan—were criminals. Homeless people were viewed as criminals.

Migrants became objects of moral panic (Cohen, 1980) in a number of societies. They became identified as the dangerous class. The police (Austria, Israel, and Hungary) who identified themselves with the social strata which stood for conservative social values were affected by the sense of nervousness and fear that gripped their societies. In all levels of democracies, the police were not immune from the prejudice and fear members of the public had regarding certain ethnic groups (see DeKeseredy & Schwartz, 1996).

RESPONSES AND REMEDIES

The diverse challenges emanating from the organizational, operational and professional issues have been tackled in various ways by countries at various levels of democratic development. Their responses and remedies can be discussed under these headings: (1) International Cooperation, (2) Legal Resources, (3) Harnessing the Media, and (4) Administrative Measures. Again, it will be noted that the utilization of these ways and means has varied from country to country and from one level of democracy to another.

International Cooperation

International police cooperation is rapidly expanding as a result of a variety of policy objectives pursued by various countries. Some countries are actively engaged in expanding international police cooperation (Austria and Poland) in pursuance of their national policies. As a neutral country, Austria, a bridge between the eastern and western nations before the fall of communism, took the initiative to establish the Middle European Police Academy where police officers from the neighboring emerging democracies are trained with Austrian officers to improve their professional skills. By an act of parliament, Poland declared international cooperation one of the most important tasks of the police (Stanczyk, 1996). Switzerland has traditionally maintained contacts with the police of other countries through liaison officers.

The need for international police cooperation is nowhere felt more keenly than in the sphere of organized crime, white-collar crime, and illegal migration. The realization of the global nature of modern crimes and the need for developing an international response is widely recognized (Austria, Chile, Estonia, Hungary, Macedonia, the Netherlands, Slovenia, South Africa, and Switzerland). Police officers of some countries (The Netherlands) are working in foreign nations to track criminal elements who operated in their own countries. They work in close collaboration with the police of the host nations. Dutch police officers work in Singapore, Thailand, the Netherlands Antilles, the Russian Federation, Spain, France (Interpol), Venezuela, the United States, and Brazil. Influences among police officers involved in this process of cooperation tended to be mutual; the expanding network of international police exchanges was not just in one direction.

United States police officers work in all types of democracies represented in the symposium. The Netherlands reported that US police influenced Dutch police more than any other foreign police agency because of an extensive FBI network. However, US influence exerted in the course of developing a global strategy against organized crime (particularly drug-related crimes) does not always strengthen democratic norms and values as the attitude and values of American police are seen by the police in some countries (Austria, Hungary, Switzerland and the Netherlands) to be heavily affected by crime-fighting zeal. According to the Hungarian participant, the police in his country felt considerably exposed to influence from their American counterparts who would like to "use" (Szikinger, 1996) them in the fight against international crimes.

Emerging and mixed democracies (Estonia, Hungary, Macedonia, Poland, South Africa and Israel) receive opportunities for training from established democracies like France, Germany, the Netherlands, the United Kingdom and the United States of America. Through the FBI Academy and American universities there has been considerable influences in the nature of the training and education of police in the established democracies of Europe (Austria, the Netherlands, and Switzerland).

Advice and counsel by the police of the western democracies to the mixed and emerging democracies was also noted. Macedonia received advice from the German police in developing their strategy against the Albanian demonstrators. Germans and the British held several seminars on the general theme of the policing of democracies. The Russian Federation received some organizational ideas from the West.

Aid, by giving modern police technology, has become a standard response from the established democracies to the other countries. Hungary received vehicles from France, Germany and Switzerland. Macedonia, too, received such technological aid from established democracies. The Russian Federation, particularly in the major urban centers, has received cars and electronic equipment from the established democracies.

Legal Resources

All groups representing different levels of democratic achievement reported how newer laws were enacted or proposed to give the police sources of authority and legitimate tools of action. Laws were the standard resources for problems of reforming the police organization (organizational challenges) for dealing with new types of crimes (operational challenges), and for tackling police deviance (professional challenges).

Laws were used in different types of democracies to bring about organizational change. In 1992 Estonia passed a law for reforming the police organization. In 1994 a new police act was passed in Hungary for democratizing the police organization, including simpler complaint procedures against the police. In the Netherlands, the police underwent major reform in accordance with a new law. According to this law, the police force in the Netherlands was restructured into twenty-five regional police agencies and a national police agency. In the Russian Federation, the KGB had been made, by a presidential decree, into a federal security service, but no definite arrangement for controlling it had been set in place. Formally, the Parliament controlled it, but actually it was under the President's control. In South Africa, a new constitution was adopted recently. In accordance with the new national constitution, such public institutions as the police were being restructured. As a vast country with much diversity, the laws reforming the police organization in South Africa were designed to be flexible, with emphasis on decentralization.

Another way the emerging democracies tried to prevent misuse of police power was to separate intelligence units from the police (Hungary, Macedonia and Slovenia). Laws were enacted to achieve this. Intelligence was handled now at the ministerial level, at a higher political level, and not by the police as had been the practice in former times. New laws were enacted in Macedonia and Slovenia to separate politics from police work in order to eliminate police politicization. Laws were passed in Slovenia restricting bugging and wiretapping.

A number of societies made new laws streamlining, clarifying and strengthening police authority in the interest of protecting civil rights and to make the police more effective. Austria's new *Police Security Act* (1991) was, however, too demanding and as a result the street personnel were not adhering to it strictly. Further, a debate was raging in that country regarding whether police powers should be increased. In Hungary criminal procedures were being modified by law in order to make them conform to western standards. In Macedonia a new law was being enacted for the purpose of dealing with organized crime and racketeering.

Police powers in the Netherlands were being increased, authorizing the police to undertake building surveillance and the confiscation of illegally gained wealth. New laws were being passed, granting the Dutch police more power to investigate group crimes, including organized crime. This legislation would make it a crime even to prepare to create organized crime groups. Recently, in the Russian Federation, there had been partial legal reforms. For example, a new penal code was enacted. In Switzerland the laws against organized crime and laundering of money have been fortified with harsher penalties. New laws have been established requiring that a banker must inform the police if s/he thinks the money sent to that bank was being laundered. In the United Kingdom the new police act has given the police stronger power in the handling of civil disorders. Poland also had new laws safeguarding human rights and democracy.

Yet can new laws effectively change cultural traditions unless there are supportive social changes as well (Theobold, 1989)? Laws now grant free legal service to suspects in South Africa. Detainees in police cells in that country are allowed to contact members of their families. Detainees' rights were safeguarded by new laws in Switzerland and Austria. In the United Kingdom new legislation in the areas of child abuse, family violence, and victims' rights had resulted in changes in the role of the police. Chile's experience was that law could be misused if the police used it as a pretext for legality. The more committed to the profession the officers were, the more likely that they would change and adhere to the new laws. The participant from Chile also mentioned that social changes were capable of positively creating an impulse for change for the police. Several participants (Hungary, Slovenia, South Africa, Israel, and Austria) implied that "You cannot teach an old dog new tricks." Police administrators and officers who were indoctrinated under the old system would find it difficult to change.

It was mentioned by the Estonian and the Russian participants that corruption in high places was not investigated during the former regime. That tradition did not die the day communist regimes collapsed. Indeed, now there was a law in Estonia which had created a commission to deal with police corruption and complaints. But citizens were dissatisfied with the decisions in their cases. The 1994 Police Act of Hungary did not protect human rights of offenders because of the rapid deterioration in the law and order situation in the country.

In the Russian Federation there was a new constitution safeguarding human rights. But President Yeltsin had been issuing directives to fight the rising tide of violent crime and these directives were not consistent with the human rights standards incorporated in the constitution. Historically, the Russian Federation had been a non-democratic country and the country's transformation to democracy through law was rather challenging. Russians liked the government to exercise power and authority (see Gibson, 1994). Both the participants from Estonia and Hungary regretted the lack of legal culture in their countries. They also added that former leaders did not bother too much about which actions were legal.

The Media and the Police

It was recognized by the participants that in a democratic society the media could be an important ally (as well as an adversary) of the police in reaching out to the people with new plans, programs and ideas. In one emerging democracy, Macedonia, the media have helped the police. According to the Macedonian participant, in his emerging democracy—where the media were controlled by the government—reporting was accurate. The police could expect help from the media in meeting the challenges of policing such as organized crime, migration and other issues discussed above. In the mixed democracy of South Africa the police view of the media was the same as that of Macedonia. The media were helpful to the police in their plans for seeking public support for meeting the challenges of policing. In Switzerland the police found an open door to the media very useful for cooperation (see Das, 1995).

Estonia has sought to moderate negative media attitudes toward the police. The police were publicly known to be not doing well and the media reflected that general feeling. According to the Hungarian participant, the media were politicized, critical of the police and negative in reporting police work. The police tried to manipulate the media with their monopoly over crime news. In the Russian Federation some sectors of the media were supportive of the police and the latter supported such journalists, but some journalists were critical. In the Netherlands the police felt helpless as investigative reporters published reports of crime and the police. In such journalistic accounts, police secrets were sacrificed and that was considered harmful for police investigations and related matters. In Austria the police could manipulate a crisis in order to help themselves through the media. In that country the media were under the influence of the conservatives who generated fear of crime through exaggerated reports about the rise of crime and by depicting the police as incompetent. The police might be provoked to respond undemocratically to such exposure by launching a so-called war on crime. In that country the conservatives could also make the media put pressure on the parliament which, in turn, increased the heat on the police. A similar view was expressed by the participant from Israel. In the United Kingdom, too, the police found the media capable of causing fear of crime and moral panic (See DeKeseredy & Schwartz, 1996).

Administrative Measures

Participants emphasized the need for greater professionalism to be achieved through better training, education and other organizational steps. For some (Chile, Estonia and Poland), professionalism was equated with having developed a knowledge base of policing, including crime investigation techniques, the use of sophisticated communication equipment, or becoming very specialized in police work. For others (Austria, the United Kingdom, South Africa and Hungary) professionalism implied an individual commitment to duty, adhering to a code of ethics, using one's discretion in decision making and assuming responsibility for decisions, and developing a police capacity for contributing to a better quality of life.

According to the participant from Chile, the status of police work must be enhanced so that young people would view it as a first choice for a profession rather than the last. This could be done by paying more attention to matters involving human dignity. According to the participant from Israel, more emphasis should be placed on the police having broad-based education with an emphasis on human rights rather than on the techniques of policing and the acquisition of appropriate equipment.

The Netherlands participant maintained that the police organizational system in his country was suitably modified to permit the police to develop closeness with the people and also to increase their crime-fighting capacity. In that country the recent reorganization of the police allowed more discretion and local autonomy which was helpful in being people-oriented. The Dutch police became more specialized, particularly those units dealing with organized crime and drugs. This was an effective approach to crime. The Swiss participant felt that the police in a democratic society needed to establish communications and working relationships with political and all other segments of the citizenry. It was pointed out by the participant from the United Kingdom that the police became more professional in the sense of having knowledge of and using sophisticated equipment and crime-solving techniques. They were drafting a code of ethics that would emphasize service.

Controlling police deviance was considered important. In Chile police corruption was prevented through harsh punishments. Citizens' complaints were handled by the police administration. In the case of high-ranking police officials, complaints could go directly to the ministry or the court. As mentioned elsewhere, a new law against corruption was passed recently in Estonia which led to the establishment of a commission to deal with corruption complaints. There was a very strict policy against police corruption in Hungary. Complaints of corruption could be made not only against a police officer but also against the head of his/her agency. In Israel citizens had access to a civilian forum for lodging complaints against police deviance including violence, but the first step in the process was completed by the police. Disciplinary action against corrupt police officers in Macedonia was very severe. Abuse of power—which was more frequent than corruption in Macedonia—could be reported directly to the Ministry of the Interior. In the Netherlands police leaders were encouraged to take a more forceful position in cases of abuse of police power. Citizens had a right to file complaints against the police before a police review commission. There was also a complaints commission that existed outside of the police organization. In Poland citizen complaints were first checked at the police division level. They were then sent a the second, prosecutorial level for a decision on the appropriate action. The ombudsman also dealt with some categories of complaints against the police. Slovenia had established a mechanism to handle citizen complaints against the police. Complaints came to the Bureau of Complaints within the Ministry of the Interior. In South Africa complaints against police corruption and deviance were brought before an independent judge. In the United Kingdom citizens' complaints were first investigated by the police. More civil suits against police misconduct had been filed recently (see Harrison, 1987).

There were utterances from the participants that the zeal for fighting the new waves of crime, the crusade for stopping the tide of migration and the drive for professionalism should be balanced by thoughtful, innovative, change-oriented and intellectually alive problem-solving approaches (Goldstein, 1990; Skolnick & Bayley, 1986). It was recognized that interactions between police and citizens in proactive and cooperative ways has led to crime prevention and control and provided protection against police violence, corruption and deviance (Trojanowicz & Bucqueroux, 1990).

CONCLUSION

It is noted that there is no consensus as to what specifically constitutes democratic policing. The concept does not seem to depend on the length of a country's democratic experience, nor does the nature of the challenges faced by the police in different societies depend on the level of democracy. Having said that, it must be acknowledged that perhaps the level of democracy makes a difference in the degree of the seriousness of some of the challenges and the availability of responses to confront them. There was a vast difference in the degree of affluence among the countries at the various levels of democracies. Their history, tradition, demographic composition, political complexities and social stability were vastly dissimilar. These factors are, no doubt, responsible for variances noted in regard to the challenges of policing democracies as well as responses and the remedies for the same.

Multi-dimensionality in the Concept of Democratic Policing

Irrespective of their democratic level, some countries (Chile, Macedonia, The Netherlands and Switzerland) emphasized the rule of law as the fundamental basis of democratic policing. It is difficult to attribute a unique reason for their thinking so. It must be presumed that the rule of law is an obvious element of democratic policing and it is understandable why it is mentioned by several countries.

Other characteristics mentioned by contributors are based on their specific cultures. Austria's argument that democratic policing must embrace orientation to the publican and transparent decision making is understandable and results from the fact that since the days of the monarchy Austrian police have been ruler-oriented and highly centralized. Secret police have been an important part of policing (Das, 1994). Civilian control, community policing, and minimum use of force, the qualities stressed by South Africa, are understandable with reference to the repressive police practices in support of apartheid only a few years back. Polish references to local participation in democratic policing can be understood within the context of the recent reform in Poland giving local people a say in the working of the national police organization. Equally easy to understand is the stress from the Russian Federation that democratic policing should be rooted in the rule of law but laws must be based on democratic principles in tune with universal standards of human rights. This definition must be read with other statements made by the Russian participant to the effect that although there was a new constitution in the country enshrining the principle of the rule of law, there were many presidential edicts creating laws in violation of constitutional provisions.

Differences and Similarities of the Challenges

There were some differences in challenges to policing democracies but there were similarities as well across different levels of democratic countries. Organizationally, the emerging democracies were facing unprecedented challenges because many of the educated, well-trained and experienced officers left the police or were fired. They were replaced by younger and less-qualified personnel which created some chaos. But the break with the past was only partial as those officers who continued were also considerable in number. So a police culture acquired under the dictatorial regimes continued. Continuation of the old culture was also a challenge in the mixed democracies. Chile's police remained highly militaristic, centralized and nationally controlled without a serious purge. That was also the problem in South Africa. Following the accord with Palestinians the perceived need for violence and brutal efficiency on the part of the police of Israel did not change. Ironically, even in the established democracies political climate, economic necessity and new waves of crime led to greater consolidation and consequent centralization of police services. There has been almost a cultural shift in the tradition of the police in the United Kingdom. From being an unarmed Bobby with her/his humble calling and attached to a local police institution, the British police officer has moved in the direction of a more sophisticated, "modern" idea: an efficient, highly-paid professional shielded with protective armor and working for an increasingly centralized organization (Holdaway, 1993). While centralization itself may not be undemocratic (see Bayley, 1985) it does inject a greater dose of bureaucratic control.

Challenges from crime and migration, the operational issues, in all levels of democracies, can be broadly traced to common causes: freedom, turmoil and disturbances associated with tremendous changes. The collapse of border restrictions, freer movement of people and greater freedom of information was global and universal. So new developments in the field of crime and criminality as well as large-scale transborder migration affected numerous countries. Organized crime, gangs and other types of criminal activities had been in a low key in the former communist regimes as the police were feared, lawbreakers were subjected to harsh treatment and guns were difficult to procure (President's Commission, 1986). Also, given the absence of governments accountable to people, it would have been naive to expect communist governments to give an accurate picture of organized and violent crime (see Hirst, 1986; Szikinger, 1996).

Professional challenges like violence, corruption, politicization, stereotyping and moral panics differed only in degree across the democracies. Police violence, albeit situationally generated, was considered a challenge everywhere. While it was termed as misplaced enthusiasm (Forrester, 1995) shown at the time of interrogation in the United Kingdom, South Africa talked of a subculture of police violence aggravated by political violence in the country. Police violence seems to be universal as the ultimate resource of the police is their capacity for coercion (Gaines, Kappeler & Vaughn, 1994) and since violence against the police is always a possibility (More, 1992). Compared to violence, corruption was not a major challenge in any of the established democracies or Israel, a mixed democracy. None of the police organizations represented at the symposium were free from moral panics generated by stereotyping (see Shusta, Levine, Harris & Wong, 1995).

Differences in Responses

Although there were several common responses to the challenges of policing democracies, it is obvious that the extent and nature of utilization of the responses, as well as results achieved, differed among various levels of democracies.

International police cooperation, sought to strengthen the inadequate operational and professional capacity of the emerging and mixed democracies, is not utilized for the same purpose by the established democracies. The latter make use of international contacts for obtaining criminal intelligence and for placing their own officers in other locales to track criminals wanted by them. The established democracies are basically the donor nations and the others are recipients. Bigger and more powerful nations are even capable of exporting some not-so-attractive features of their policing (e.g., riot techniques taught by Germans to Macedonians and bitterly resented by Albanians who bear the brunt of these imported tactics) (see Das, 1993).

Law has been used extensively in emerging, mixed and established democracies to bring about organizational, operational and professional improvement. It is noted in the emerging democracies that there is often a gap between law as it is written and law as it is carried out (Tomasic, 1987). In the Russian Federation presidential decrees are supplanting the laws incorporated in the constitution. The story of law's failure is not much different in Hungary, Estonia or South Africa. Established democracies like The Netherlands, Switzerland and the United Kingdom display a sense of confidence in law's ability to achieve the purpose it is intended for. Comparatively, Austria, with its more complex history, has had some difficulties in implementing the new human rights oriented police laws. There is an acknowledgment of the difference in making and implementation of laws everywhere (Pinkele & Louthan, 1985).

Police organizations across the various levels of democracy recognize the positive as well as the negative impact of the media in the sphere of moral panics, fear of crime, public image of the police, and the like. Press relations, hence, are an important problem as well as an opportunity for the police (Waddington, 1994; also see

Black, 1989; Scheingold, 1991). Only two countries, namely a mixed democracy (South Africa) and an emerging democracy (Macedonia), consider the media an ally. They are able to count on media support in presenting police plans, programs and agenda to the public. In the Russian Federation, where the media are owned partially by the government, the police receive mixed support and, in turn, the police treat the media as friend or enemy depending upon their predilections. The media image of the police is disparaging in Estonia and Hungary; the Netherlands finds media interference with police work impeding investigations; and Austria's experience is that the media dominated by conservative ownership attempt to define the police role and their attitude to problems.

Across the various levels of democracy, police deviance is controlled by sanctions from civilian review boards, ombudsmen, courts and police departments themselves. However, innovative methods are being tried not only in regard to controlling police deviance but also to strengthen police crime control capacity by developing partnerships with citizens, political representatives, and other public agencies (see Farmer, 1984; and Fogel, 1994). Alliances with the public, which to become established require patience, being in the right place and luck (Carter, 1992), are more in evidence in the established democracies. Some mixed democracies like South Africa and Israel are pursuing their own versions of involvement with the public. There are police-directed citizen patrols at night in the cities of Israel. South Africa is practicing community policing with official recognition of local customary laws. In the emerging democracies the partnership with the public has not been so innovative (see Williams & Serrins, 1993) although there is a recognition of the need for greater relationships with citizens at the grassroots level as evidenced by Poland's legal provision for local input in policing.

In the ultimate analysis each society has to find out what is feasible for it to do to organize the police on democratic standards and values (see Monk, 1996). It must be realized that professional police standards (the rule of law, accountability, and transparency of decision-making, etc.) are to a large extent universal. However, the police everywhere function within cultural limits and constraints as well as economic realities.

NOTE

† The symposium on the theme of Challenges of Policing Democracies, held at the Institute of the Sociology of Law in Oñati (Spain) on May 17–20, 1995, was attended by Prof. Menachem Amir (Institute of Criminology, The Hebrew University, Jerusalem, Israel); Press Officer Emile Berthod (Geneva Police, Switzerland); Hofrat Mag. Maximilian Edelbacher (Chief, Major Crime Bureau, Federal Police, Vienna, Austria); Lieutenant-Colonel Albertus Wynand Eksteen (Efficiency Services, South African Police, Pretoria, South Africa); Dr. Janos Fehervary (Ministry of the Interior, Vienna, Austria); Superintendent Paul Forrester (Police Staff College, Bramshill, United Kingdom); Colonel Nelson Godoy Barrientos (*Carabineros de Chile*, Embassy of Chile, Madrid, Spain); Dime Gurev (Ministry of the Interior, Skopje, Macedonia); Police Inspector Andrezj Koweszko (Chief, National Central Bureau of Interpol, Warsaw, Poland); Prof. Peter Kratcoski (Official Reporter, Kent State University, Kent, Ohio, USA); Dr. Ando Leps (Member of Parliament and Councillor to the Central Investigation Bureau, Tallinn, Estonia); Marko Pogorevc (Ministry of the Interior, Ljubljana, Slovenia); General Jerzy Stanczyk (Chief Commander, Polish State Police, Warsaw, Poland); Dr. Istvan Szikinger (Director, Legal Policy Institute, Budapest, Hungary); Daan Van De Meeberg (Deputy Chief Constable, National Police Agency, Driebergen, the Netherlands); and Jernej Videtic (Ministry of the Interior, Ljubljana, Slovenia).

Peter Kratcoski made an invaluable contribution to the symposium by his accurate and thorough documentation of the views expressed by various participants during the symposium. He was ably helped by his gracious and efficient wife, Lucille. Their contribution to this Executive Summary through their copious notes is gratefully acknowledged.

REFERENCES

Alderson, J. (1979). *Policing freedom: A commentary on the dilemmas of policing in western democracies.* Estoven, Plymouth: Macdonald and Evans.

Alderson, J. C., & Stead, P. J. (1973). *The police we deserve.* London: Wolfe.

Amir, M. Police in Israel. Unpublished paper.

Baldwin, J. (1961). *Nobody knows my name.* New York: Dial Press.

Bayley, D. H. (1985). *Patterns of policing: A comparative international analysis.* New Brunswick, NJ: Rutgers University Press.

Berkeley, G. E. (1969). *The democratic policeman.* Boston: Beacon Press.

Bittner, E. (1970). *The functions of the police in modern society.* Washington, DC: U.S. Government Printing Office.

Black, D. (1989). *Sociological justice.* New York: Oxford University Press.

Carter, D. L. (1992). Community alliance. In L. Hoover (Ed.), *Police management.* Washington DC: Police Executive Research Forum.

Cohen, S. (1980). *Folk devils and moral panics.* Oxford: Basil Blackwell.

Coleman, J. W. (1985). *The criminal elite.* New York: St. Martin Press.

Consortium of Social Sciences Association (1994). *Democratic transformation: Africa, Latin America and Russia. A congressional breakfast seminar.* Washington, DC: Author.

Das, D. K. (1993). *Policing in six countries around the world.* Chicago: Office of International Criminal Justice.

——. (1994). Can police work with people? A View from Austria. *The Police Journal, 68* (4), 334–346.

——. (1995). Police challenges and strategies: The executive summary of International Police Executive Symposium, Geneva, Switzerland, 1994. *Police Studies, 18* (2).

DeKeseredy, W. S. & Schwartz, M. D. (1996). *Criminology.* Belmont, CA: Wadsworth Publishing Company.

The Europa World Year Book 1994 (1994). London: European Publications Limited.

Emerson, D. (1968). *Metternich and the political police.* The Hague: Martinus Nijhoff.

Farmer, D. (1984). *Crime control.* New York: Plenum.

Fogel, D. (1994). *Policing in Central and Eastern Europe.* Forssa, Finland: Kirjapaino Oy.

Forrester, P. (1995, May). Paper presented to the Symposium on Policing and Democracy, Oñati, Spain.

Fosdick, R. B. (1975). *European police systems.* New York: Century Company. Original work published 1915

Friedman, R. R. (1986). The neighborhood police officer and social service agencies in Israel: A working model for cooperation. *Police Studies,* 175–183.

Gaines, L., Kappeler, V. & Vaughn, J. (Eds.) (1994). *Policing in America.* Cincinnati: Anderson.

Gibson, J. L. (1994). Russia. In Consortium of Social Sciences Association, *Democratic transformatio: Africa, Latin America and Russia. A congressional breakfast seminar.* Washington DC: CSSA.

Gilinskiy, Y. (1994). Crime in Russia. Unpublished paper.

Goldstein, H. (1977) *Policing a free society.* Cambridge, MA: Ballinger.

——. (1990). *Problem-oriented policing.* New York: McGraw-Hill Publishing Company.

Harrison, J. (1987). *Police misconduct: Legal remedies.* London: Legal Action Group.

Hirst, P. Q. (1986). *Law, socialism and democracy.* London: Allen and Unwin.

Holdaway, S. (1993). Modernity, rationality and the baguette: Cooperation and the management of policing in Europe. *European Journal on Criminal Policy and Research, 1*(4), 53–70.

Knapp Commission (1972). *The Knapp Commission report.* New York City Commission to Investigate Allegations of Police Corruption and the City's Anti-Corruption Procedures. New York: George Brazillier.

Lintner, E. (1994). Drug trafficking and organized crime in Central and Eastern Europe. In *10th international conference on democracy challenged and put to the test—The problem of combating terrorism, drugs, and organized crime*. Munich: Hans Seitel Foundation.

Martin, J. M. & Romano, A. T. (1992). *Multinational crime: Terrorism, espionage, drug and arms trafficking*. Newbury Park, CA: Sage.

McCormack, T. (Ed.). (1990). *Studies in communications, 1990: A research annual: Censorship and libel, the chilling effect*. Greenwich, CT: JAI Press.

Merton, R. (1938). Social structure and anomie. *American Sociological Review, 3*, 672–682.

———. (1957). *Social theory and social structure*. New York: The Free Press.

Monk, R. C. (Ed.). (1996). *Taking sides: Clashing views on controversial issues on crime and criminology*. Guildford: Dushkin Publishing Group/Brown & Benchmark Publishers.

More, H. W. (1992). *Special topics in policing*. Cincinnati: Anderson.

Nemeth, Z. (1996) Organized crime in Hungary. Unpublished paper.

Pinkele, C. F. & Louthan, W. C. (1985). *Discretion, justice and democracy: A public policy perspective*. Ames, IA: The Iowa University Press.

President's Commission on Organized Crime (1986). *The impact: Organized crime today*. Washington, DC: U.S. Government Printing Office.

Reiner, R. (1991). *Chief constables: Bobbies, bosses or bureaucrats*. New York: Oxford University Press.

———. (1992). *Politics of the police*. (2nd ed.) Toronto: University of Toronto Press.

Scheingold, S. A. (1991). *The Politics of street crime: Criminal process and cultural obsession*. Philadelphia: Temple University Press.

Shelley, L. (1995). Concluding remarks. In , L. Shelley & J. Vigh (Eds.), *Social changes, crime and police* (pp. 195–201). Chur, Switzerland: Harwood Academic Publishers.

Shusta, R. M., Levine, D. R, Harris, P. R. & Wong, H. Z. (1995). *Multicultural law enforcement: Strategies for peacekeeping in a diverse society*. Englewood Cliffs, NJ: Prentice-Hall.

Skolnick, J. H. (1975). *Justice without trial: Law enforcement in a democratic society*. New York: John Wiley.

Skolnick, J. H., & Bayley, D. H. (1986). *The new blue line*. New York: The Free Press.

Stanczyk, J. (1996). Police in Poland. Unpublished paper.

Szikinger, I. (1996). Police in Hungary. Unpublished paper.

Theobold, R. (1989). *Corruption, development and underdevelopment*. Durham, NC: Duke University Press.

Tomasic, R. (1987). *The sociology of law*. Beverly Hills, CA: Sage.

Triandis, H. C. (1994). *Culture and social behavior*. New York: McGraw-Hill, Inc.

Trojanowicz, R., & Bucqueroux, B. (1990). *Community policing: A contemporary perspective*. Cincinnati: Anderson.

Waddington, P. A. J. (1994). *Liberty and order: Public order policing in capital city*. London: UCL Press.

Williams, J. L., & Serrins, A. S. (1993). The Russian militia: An organization in transition. *Police Studies, 16* (4), 124–128.

2

Policing in Democratic Societies: An Historical Overview

PETER C. KRATCOSKI,
with the assistance of **WOJCIECH CEBULAK**

The purpose of this chapter is to present a general overview of policing in various countries that have established democratic forms of government. Terrill (1995, p. 485) notes that "police agencies in industrialized countries throughout the world tend to follow one of three basic organizational models: centralized, coordinated, and fragmented."

The organizational model followed by a country is often the consequence of specific cultural and political influences, and it is not likely that one can predict the form of policing that will emerge. As Terrill observed, there are many factors that will influence the organizational model for policing agencies in any given country.

The political ideology of the country at the time when the police agencies were developing is a key consideration, as is the fact that the political ideology may change over time. Social and cultural factors also must be considered. If a country is in a period of rapid growth, with a heterogeneous population, coupled with what appear to be uncontrollable crime problems, the citizens, even in a democratic society, may be more willing to vest considerable power in the hands of the police than might be the case if the country's population were homogeneous and crime problems seemed controllable. There may be cultural or social characteristics of a particular country that lead inevitably to suspicion toward and distrust for the police. In these instances, the political ideology would tend to lead to decentralization of police powers, because the citizens demand this. Obviously, there are numerous examples of incidents in which the demands of the citizens are not addressed and the police establishment comes under the control of those in political power. In certain countries, the characteristics of the population are so varied that one organizational model is not adequate to serve the interests of persons throughout the country, and a fragmented model is adopted.

Even in those countries where a democratic political tradition has existed for centuries, and the basic police organizational structure has been established for a long period of time, police organizations must constantly adjust to new demands and must modify their operations to meet these demands. For example, the English model, which was enacted in 1829 with the Metropolitan Police Act, served as the basic model for the development of police organizations in many countries, including the United States. However, changes were made in the English model's organizational structure at various times, with the most recent changes occurring in 1995 (Forrester, 1995).

In this chapter, the origins of policing in various democratic countries will be considered. Specific attention will be given to the political, socio-cultural, and economic factors that were important during policing's development process. If the police organizational model changed significantly from its original form, those factors that appeared to have been responsible for the change will be discussed.

PERSPECTIVES ON THE CONCEPTS OF DEMOCRATIC POLICING AND DEMOCRACY

The concept "democracy," in terms of government, implies that the citizens themselves are the ultimate source of power. But in most democratic societies the people do not govern directly. Instead, they elect persons to represent them in the formation of government policies and decisions. Yet the people retain the final authority for government decision making, in the sense that they are responsible for electing the officials who legislate, implement, and enforce laws.

At the Symposium on Challenges of Policing Democracies, held in Oñati, Spain, in 1995, participants were asked to give their perspectives on democratic policing. It was noted by the representative from Israel that policing and democratic police are two different phenomena. That is, police organizations are not usually structured democratically, even though they are charged with enforcing the laws of democratic societies. The major distinction is that in a democratic society the police are ultimately under the control of the people, since they are the ones who elect those who appoint or choose the persons who are responsible for establishing the laws that the police enforce, while within the police organization itself the organizational structure tends to be authoritarian. The common view expressed at the symposium was that "democratic policing meant that all police activities must be regulated in accordance with law. The police practicing such a style were totally accessible to the public and maintained maximum transparency" (Das, 1995, pp. 5–6).

In a totalitarian society, the police often are utilized as an instrument of those in control to assist in maintaining this control, and there is generally evidence of abuse of such power. In fact, those police who commit abuses of power toward the citizens may also be subject to arbitrary decisions pertaining to their own well being or their authority as police.

As we noted earlier, the concept "democratic policing" may be a misnomer when applied to the internal functioning of police departments. Police departments traditionally have been structured along a military model. The departments are developed to reflect a military style of organization and hold values and philosophies very similar to those of the military. Decision making is concentrated at the top levels, and there is very little room for individual decision making, with the exception of the immediate decisions officers must make when in the process of performing their duties. The structure of a traditionally organized department is in the form of a pyramid, with patrol officers at the bottom, their immediate supervisors accountable to another level of supervisory personnel immediately above, and various levels of administration, such as commanders of divisions and bureaus added, with the chief of police at the top of the pyramid. In democratic societies, for example in the US, the chief of police is typically not the ultimate authority within the police department at the local level, since the chief's appointments and policies are subject to review by a safety director, the mayor, the city council, or by a police administrative agency. We see, then, that policing a democracy does not require that the police be organized democratically.

Police Professionalism in a Democratic Society

There is growing acknowledgement that police departments organized and administered along rigid, bureaucratic, militaristic lines are not conducive to effective policing in a complex, democratic society. In order to be effective, officers today must understand the law and its dynamics, interact effectively and fairly with the public, have the patience and training to handle the routine matters that make up much of the job's activity, and exercise their own judgment wisely in situations that demand police discretion. Professionalism, as it pertains to police, involves the internalization and acceptance of the identifying characteristics of the profession. These include having a code of ethics, a defined body of knowledge, continued education in the field, accepted standards for entrance into the occupation, and a commitment to service. While being proficient in the use of policing techniques and being knowledgeable about the equipment needed to perform effectively would be included, the essential factor in professionalism is that the professional officer, regardless of rank, has some degree of autonomy in decision making. This is usually termed "police discretion."

In *Black's Law Dictionary*, discretion is defined in the following way:

> when applied to public functionaries, discretion means a power or right conferred upon them by law of acting officially in certain circumstances, according to the dictates of their own judgment and conscience, uncontrolled by the judgment or conscience of others (Black, 1991, p. 323).

Theoretically, the code of ethics for police officers, regardless of their positions, should include commitment to operating under the rule of law. An officer of a lower rank would not be required to follow a command made by a superior, if the lower ranking officer were aware that to do so would be a law violation. At the symposium, it appeared that all of the representatives from the participating countries, regardless of the level of democracy achieved there or the length of time that democracy had been established, agreed that in policing democracies the values and norms of the police had to include: the rule of law, accountability to the public, transparency of decision making, popular participation in policing, minimum use of force, creating an organization that facilitated learning of civil and human rights, and internal democracy in the organization (Das, 1995, pp. 6–7).

Community Involvement in Police Work

The extent to which the citizens of a country believe that they have a right or even a duty to become involved in the law enforcement functions of that society is closely linked to the culture of the country. Culture in a very broad sense refers to the way of life of a society. A culture embodies norms, values, language, beliefs, traditions, history, expectations, and behavior patterns. Culture is learned through the process of socialization. This process of learning begins at birth and continues throughout life.

The first transmitters of culture would be immediate family members, followed by the schools, the community, and economic and political institutions. Cultures are not static. Traditions, values, and expectations of behavior change, and people must adapt to these changes.

Richards (1996) notes that the concept of "community" is difficult to apply in an objective manner because the term has so many different meanings and reference points. Perhaps the most general usage implies people living in a particular geographic area, normally a small locality, who share common interests (p. 3). He observes that policing in England has traditionally been community based. In tracing the history of community involvement in maintaining social control in England, he notes that the Anglo Saxons who settled in England in the sixth century brought their own culture with them, including laws and methods of enforcing them. He states that,

> according to Anglo Saxon custom, if someone broke the law it was not just a crime against the victim, but against the whole community, with the responsibility for law and order being vested in the citizens themselves. In this regard, all males between the ages of 12 and 60 were responsible for this duty. They were organized in groups (called tythings) of about ten families, and each member of the group was held responsible for the behaviour of the others. If one member of the tything committed a crime, the others had to apprehend that person and bring him before the court. If they failed to do so then all members of the tything were punished—usually by way of a fine. These principles are embedded in Common Law to this day, and responsible, in part, for the evolution of the Constable (p. 6).

In regard to the police, citizens are socialized to perceive them as friends, protectors, helpers, or persons to be feared, depending on their specific learning experiences. Within a single society, perceptions of the police vary, according to socialization. For example, research has shown that many African-Americans living in inner city areas in the United States have quite different perceptions of the police than persons living in other areas of the cities or in suburban areas.

The degree to which the citizens are expected to assist with police work will vary by country, depending on the traditions of the culture of that society. Even in countries with long-established democratic traditions, the relationship between the police and the public has changed and continues to change. When the United States became independent from Great Britain in the eighteenth century, there was citizen fear of having a strong centralized police force, and there were strong efforts to prevent the police from becoming too powerful. Thus, much of the policing fell on the shoulders of the citizens. Eventually, the needs of the country became more complex because of industrialization and urbanization, and the role of citizens in day-to-day police activity diminished. In the late twentieth century, there have been attempts to increase public cooperation and involvement with the police, through the introduction of community policing, crime watch programs, and other programs that require extensive police-citizen cooperation and interaction. This occurred because of dissatisfaction with police services, hostility toward the police on the part of some groups because of alleged discrimination, and a general perception that the police had distanced themselves from the public and lost sight of their function of serving the public.

The People's Republic of China, which does not have a democratic political structure, nevertheless has integrated the citizens into the very fabric of policing. Since rights are viewed as belonging to the community, the citizens are actively involved in punishing and rehabilitating those who violate the rights of others. In the People's Republic of China, the police organizations are under the direct control of the central government through the Ministry of Public Security, which controls the centralized armed police who are in charge of the nation's security and also supervises local police by making policy and giving guidance to the public security organizations that have been established in twenty-two provinces, five autonomous regions, and three municipalities (Bracey, 1996, p. 226). Strong citizen involvement in law enforcement occurs through the use of public security committees. The members of these committees are elected, and they provide police-citizen liaison, carry out patrol and other crime prevention measures, are involved in mediation of disputes, and supervise offenders who return to the community after imprisonment.

There are more than one million public security committees (Dai and Huang, 1993, pp. 140–41). Although the police are required to listen to the citizens and accept their suggestions and criticism, in the final analysis the police and government officials set policies by determining what constitutes the common good. However, the security committee members perform important functions. They provide information to the police on the local level that can be very helpful in solving crimes. They also mediate disputes informally and divert many cases from formal handling. Although the People's Republic of China can directly document the very substantial numbers of citizens involved in maintaining social control in their communities, various writers on policing in China (Bracey, 1996, p. 230; Yang, 1994, p. 58) are quick to point out that the matters of establishing policies and performing the more specialized police work are functions of the Ministry of Justice, the Ministry of Public Security, and the local police departments that come under their jurisdictions. In addition, many of the matters handled by the public security committees are not treated as criminal violations, and while the end result may be diversion from the criminal justice system, an avoidance or circumventing of the due process rights of those accused can also occur (Bracey, 1996, p. 231).

In the following section of the paper, an attempt will be made to trace the development of the police in various countries and to illustrate how they fit into the models of policing developed by Terrill (1995). Before pursuing this task, a description of the three models will be given.

THREE MODELS OF POLICE ORGANIZATIONAL STRUCTURES

The Centralized Model

This organizational model can be found in both democratic and non-democratic countries, but it tends to be more closely associated with the non-democratic countries. According to Terrill (1995), the centralized model has its origins in ancient Rome. The central government establishes a police force that is imposed on the population. Terrill notes that

> since the eighteenth century, this model has been the creation of national governments for the general welfare of citizens, and it has been national governments that centrally administer, supervise, and coordinate their police. In such a model, the police are perceived as representatives of the state, and perceive themselves as such (pp. 485–86).

Terrill gives France, Japan, and the former Soviet Union as examples of countries following this model.

The Coordinated Model

The coordinated model, in contrast, involves considerable sharing of power by the central government and local communities. Certain specified police functions are given to the centralized police force, and others belong to the local police. This model has been closely associated with England and Wales and with numerous other countries, including Canada and Australia, which have adopted variations of the English system.

The Fragmented Model

The fragmented model of police organization is exemplified by that developed in the United States, and its development "is attributed directly to the federated nature of the political system of the country" (Terrill, 1995, p. 487). In the United States, a person is a citizen of the country as well as a citizen of a state and of a community. In the fragmented model, the centralized government is responsible for the enforcement of laws that have national application, the state has responsibilities for enforcement of laws specific to that state, and the local government (county or municipality) formulates laws that are applicable in that jurisdiction.

CASE STUDIES OF VARYING POLICE ORGANIZATIONS

Using the model developed by Terrill (1995), the organizational structures of various police departments can be categorized in terms of the locus of control and the underlying assumptions about the role the police play in relationship to the governance of the country. In this section of this chapter, the different models of police organization will be illustrated through case studies. For the centralized model, Finland will be used. Policing in Great Britain illustrates the coordinated model. The United States has been chosen to illustrate the fragmented model.

The major purpose in presenting the case studies is to trace the origins and development of the police systems of these countries. A detailed explanation of the police functions and the various units developed to perform these functions will not be given. Rather, the material is designed to illustrate the major historical, cultural, or legislative actions that helped shape the evolution of these police organizations.

Policing in Finland: The Centralized Model

The origins of the Finnish police can be traced to the Middle Ages. During the years when Finland was part of the kingdom of Sweden, two separate forms of policing were used for the towns and for the rural areas. A law governing town policing, enacted in 1350, made the town councils responsible for maintaining public safety and order, and the town councils were supervised by the king's bailiff (Laitinen, 1994, p. 25). By the seventeenth century, as the crown consolidated and increased its authority, administration and supervision of the towns were transferred to provincial governors appointed by the king. Finland became an independent country in 1917. It is a parliamentary democracy with an elected president.

Currently, the police in Finland operate on three levels—the security police, mobile police, and central criminal police. All of these come under the Ministry of the Interior. The security police investigate offenses against the state or against law and order in the community, and deal with crimes that are a threat to public safety. The mobile police are the national police reserves, while the central criminal police investigate crimes of a complex nature. A provincial administration board in each province is responsible for the administration of the local police, who perform routine police operations within the communities. The two types of local police include the town police, headed by police chiefs, and the rural police, headed by district sheriffs. Criminal investigations may be initiated by the local police, but they are usually conducted in cooperation with the central criminal police (Laitinen, 1994, p. 124).

The English Police as an Exemplification of the Coordinated Model

Historically, the origins of policing in many democratic countries can be traced to adaptations of the English model. The English system of law enforcement was adopted by other English speaking countries and strongly influenced the American system. The history of policing in England began with local communities largely responsible for policing themselves. Tithing, described earlier in this chapter, was one of the earliest forms of social control used during the Anglo Saxon era. The statutes or ordinances enacted then were called dooms. During the medieval period, the sheriff, as the local representative of the king, was responsible for supervising the tithings. Following the Norman Conquest, in 1066, the period of feudalism ensued. A manor system, with serfs working land owned by a lord and receiving a certain level of physical and economic security in return, predominated.

The constable became the principal policing officer during this period, and this position was retained in the English system long after feudalism declined. From 1285 to the middle of the fourteenth century, a system of justices of the peace and parish constables was established, which remained in effect until 1829. The constables were appointed by the local justices and had responsibility for patrolling during the daylight hours. In the evening, night watchmen lighted the street lamps and patrolled the streets. Under the system of "hue and cry," constables or night watchmen could call out for assistance in apprehending fleeing criminals or for other emergencies. The Justice of the Peace Act of 1327 established these officials as formal peace officers, and a similar act of 1361 formally recognized their authority to administer justice at the local level as representatives of the king.

By the nineteenth century, the justice of the peace and constable system was inadequate to keep order and dispense justice in large metropolitan cities such as London. The night watchmen and constables proved to be inept in many instances, and merchants hired their own "thief catchers," who pursued criminals or brought back stolen property for fees. In 1829, the Metropolitan Police Act provided for a uniformed police force of officers who would not only react to committed crimes, but patrol the city of London with the aim of preventing criminal activity and promoting the public welfare. The Act provided for a single police force that would cover an area extending approximately seven miles from the center of the city. Richardson (1995, p. 556) observed that

> policing involved keeping city streets clean as well as the good order and discipline of its residents. The presence of a police officer might deter residents from airmailing their garbage into the streets, as well as keep street walkers from plying their trade.

The new police departments were structured to gain the support and cooperation of the citizens. There was some lack of acceptance of the police, particularly by the lower classes, because the police were perceived as instruments of the more well-to-do citizens. To prevent police corruption, officers were hired to work in neighborhoods other than those in which they resided. To reduce political influences on the police, they were not given the right to vote (Richardson, 1995).

The Metropolitan Corporations Act of 1835 required all new towns in England and Wales to establish police forces, and the County Police Act of 1839 helped in the development of police for the fifty-six counties in England and Wales. The central government had the power to regulate officer selection, training, and pay schedules. The County and Bureau Police Act of 1856 required the establishment of a police force for each county and borough, which were to be placed under the direction of the Home Secretary.

The Police Act of 1964 reaffirmed the principle of local participation in police development. As Stead (1985, p. 103) observed,

> the Police Act of 1964 marks a decisive stage in police development. It reaffirmed the principle of local participation in police governance. It gave Parliament a whole range of subjects on which it can question the Home Secretary and it makes possible the consolidations of the 1960s.

While the basic organizational structure and philosophy of British policing remained fairly consistent since its beginnings, there have been various stages of reorganization and the need to create new units occurred as social conditions changed and new demands on police developed.

Currently, police forces in England and Wales are structured in the following manner. Each force has a Chief Constable, who is responsible for the police operations of that force. The Metropolitan and City of London forces are each headed by a Commissioner. Each force has associated with it a Police Authority. These are composed of politicians from the local community and magistrates, who are lower court judges. The force for London has a separate and distinct authority. In accordance with the Police Acts of 1964 and 1976, the local authorities have the responsibility to maintain policing in their areas. Each force appoints a police committee to oversee policing issues.

The Home Secretary (Secretary of State for Home Affairs) is responsible for developing and maintaining a national policy on policing, and the central government finances more than half of all the costs for each local force. The chief constable for each force is appointed by the police authority for that area. It is the responsibility of the police authority to provide an adequate police force, and consult with the chief constable on local issues. The authority also has the power to remove a chief constable, with the approval of the Home Office, and the authority is made aware of disciplinary actions taken against officers (Walker & Richards, 1995, p. 43).

Policing in the United States: The Fragmented Model

Although many aspects of policing in the United States can be traced directly to English antecedents, the geographic, political, and cultural factors unique to the American experience strongly influenced the structure and operation of police agencies in the United States. Terrill classified U.S. policing as following a fragmented model, because policing functions are performed by the federal, state, and local governments. The citizens strongly desired local autonomy and management, and this is reflected in the styles of policing that developed.

The New England colonies imitated the English system most closely. In this area, the population was quite homogeneous, and a strong moral and religious orientation was common. The English constable/justice of the peace system, established here and in the rural areas of the northeastern United States, still serves these areas. In the larger cities on the Eastern seaboard, the rapid growth of the cities and the influx of immigrants created a need for more professional police forces. In 1845, New York City instituted its own department, modeled in many ways on the London force. Originally, the officers did not wear uniforms or carry firearms. An important way in which they differed from the London police was that they were allowed to be active in politics and were appointed to their positions or removed by the political party in power. Also, officers tended to be of the same ethnic backgrounds as the citizens they protected and to reside in the areas of the city where they worked. The city of Boston developed a force with more centralized control and less political influence on the officers. The officers were drawn from rural areas of New England rather than from the city, station houses connected by telegraph wires were built, and the officers wore uniforms (Johnson, 1988, p. 178).

In the southern United States, the early police forces were most concerned with protecting the property and rights of trading companies, and a more military style of policing with harsh punishments for those convicted of economic types of crimes, such as stealing, was used. After more settlers arrived and communities grew up, the more traditional styles of English policing were introduced. However, the institution of slavery had a very strong influence on policing in the South. Slave patrols, recruited from the population of white men between ages 18 and 50, were required to patrol roads and inns, searching for runaway slaves or those violating curfew (Johnson, 1988, p. 182).

On a plantation, the master had the power to deal with law violations, particularly if the violations were committed by slaves. There was constant fear of slave uprisings, and in 1806, the city of Charleston, South Carolina, created an armed city guard that patrolled the streets after dark to protect the citizens from criminals and to control blacks (Johnson, 1988, p. 182). As urbanization occurred, the informal law enforcement groups died out, and police departments organized in a quasi-military style were developed.

The westward movement and frontier violence and crime presented other law enforcement problems. Initially, law and order maintenance involved practical activities for survival. In the organized groups traveling west or establishing settlements, those who owned property or needed protection banded together, and group consensus set the standards for conduct. When some citizens felt that those in authority were not performing effectively, they took the law into their own hands through vigilantism, apprehending or punishing offenders without waiting for due process rights to be applied.

The differences in policing and law enforcement practices dictated by sectionalism caused the development of the fragmented, widely varying styles of policing that exist today in the United States at the local and state levels. The police departments established in the nineteenth century were closely tied to the political divisions of the areas they served, and had a decentralized, neighborhood-based organization. Police patrolled on foot and performed both order maintenance and service functions. An important aspect of policing was keeping the prominent citizens and the powerful politicians satisfied.

In the twentieth century, particularly during the 1930s, there were efforts to reduce the amount of political control of the police and eliminate corrupt practices that existed in the large cities. By the 1950s, a professional model of police work emerged. This involved a centralized organizational design, specialization, a bureaucratic structure, and a rather formal and impersonal style of communication between police and the citizens. However, it soon became evident that an impersonal, formalized style of policing was not totally effective. The growing crime problems in the 1970s and the 1980s and citizen dissatisfaction with and distrust of the police, especially in urban minority areas, in many instances caused American policing to turn toward a community-policing, problem-solving orientation, with greater involvement of citizens in the policing process and better communication and cooperation between police and the citizens. In the 1990s, community policing has gained wide acceptance, with large cities and smaller communities using foot or bicycle patrol, police neighborhood mini-stations, citizen crime watches, and similar programs.

THE STRUCTURE AND FUNCTIONS OF POLICING IN COLONIZED COUNTRIES

During the past several decades, we have seen numerous countries gain their independence. Often, the peoples of these countries had been controlled by administrators from foreign countries for hundreds of years. For countries moving from a totalitarian form of government to a democratic form, the transition is not always smooth and, in fact, may never be completely accomplished. In some cases, the new government may be as oppressive as the old one, and the structure and functions of the police do not change appreciably.

Clinard and Abbott (1973) note that when a colonized territory becomes an independent country and the administration of the country changes hands, many of the political, social, and economic institutions remain intact, even though they are now under new political and administrative powers (p. 216). With regard to the police, they state that, during the transition from colonial rule to independent rule (a time period often characterized by disorder), the police organization may remain virtually intact, because the new political administrators must rely on the existing police to maintain law and order during this period.

While the police establishment is generally considered to be very important in helping the new independent country establish a "rule of law" foundation, it is observed by Clinard and Abbott (1973, p. 216) that

> at present police do not contribute adequately to the national development of countries previously under colonial rule, largely because they have been greatly affected, and to some extent handicapped, by the ideologies and police practices of the colonial administrations.

By being tied to the colonial administration for their livelihood, police administrators were forced to consider the protection of the property and persons of the colonial businesses and the political establishment as being their primary concern. For example, before India became independent the police establishment there employed a more strict law-and-order approach and more repressive police practices than those utilized in Great Britain, even though both police organizations were grounded in the same tradition.

Although independence generally results in dramatic changes in the political systems of developing countries, it may have limited effects on the structure and functioning of the police administration. The police continue to perform the same functions as before, but now they are serving a new "master." At times, the threat of revolution and citizen turmoil in countries that have recently become independent produces an atmosphere quite similar to that which existed before independence.

The police find that they are asked to perform the same function as before, that is, keeping the people under control.

As newly independent countries continue to develop and become more economically secure, the functions of the police tend to become more outward directed. For example, the police may assume a military function by being given the responsibility for protecting the country's borders. Matters of immigration, illegal trafficking of goods, and management of prisons may become police functions.

As noted earlier, the organizational structure of a police force and the functions performed by that force are subject to change. For police in developing countries, the new practices, values and traditions that eventually emerge after independence are integrated with those that were imposed on them by the colonizing power before the country became independent. A new set of police operations, values, and traditions, grounded in the old but having a distinctive nature because of new situations and experiences, will emerge.

Case Study: Policing in Ghana

During centuries of colonization, European countries acquired territories in North America, South America, Asia, and Africa. When these countries—including England, the Netherlands, Spain, Portugal, and France—set up plans of governance for their colonies, the political structure tended to imitate that of the mother country. In some instances local inhabitants were allowed to continue practices from their own cultural heritage, but in most cases a new government and legal system were imposed.

The British colonial empire was the most far-flung and endured for the longest period of time. Although most of the British colonies have now gained their independence, the institutions of the new countries tend to retain many of the features put into operation by the English.

Arthur and Marenin (1996), in tracing the evolution of policing in Ghana, West Africa, note that, beginning in the fifteenth century, Portuguese, Spanish, Dutch and English settlers came to the area. The British gradually gained control, and Ghana was governed as a British colony until it became independent in 1957. Before the outside settlers arrived, Ghana had a well-organized system for order maintenance. The tribal chiefs were responsible for establishing and maintaining order, aided by members of their own tribes who helped them enforce rules and judge the penalties for rule violators. According to Arthur and Marenin, "tribal elders, chiefs, and their ruling councils derived their powers of social control from religion, which permeated the fabric of precolonial society in Ghana" (p. 165). This policing was decentralized, with the powers of the chiefs and those they appointed being limited to their own communities, and implemented for security and protection.

When the British began to gain control of this territory in the sixteenth century, they integrated the existing system of local social control into their governance plans. The coordinated model of police organization best illustrates colonial policing. As Arthur and Marenin observed, "the local-authority police forces enforced traditional laws and directives or ordinances of the chiefs, as long as these ordinances did not threaten British political or economic interests" (p. 167).

However, to safeguard British economic interests, maritime police and railway police were also established, and several large British companies brought in their own private policing units. Gradually, police in Ghana came to resemble more and more closely the British police in the England. In 1914, a centralized force, called the

Gold Coast Police Force, was established for the entire country (p. 168). This included Escort Police, who were responsible for protecting the interests of the mining and financial companies, and General Police, who performed traditional law enforcement duties.

Even after the centralized force was implemented, the local forces, associated with the tribal chiefs, continued to function, particularly in rural areas, until they were abolished in 1963 (p. 169). The Escort Police were also disbanded.

After independence, the centralized police force in Ghana faced identification dilemmas. Should they ally themselves with the political leaders, remain neutral, or be the champions of ordinary citizens? During the more than forty years of Ghana's independence, the country experienced all of the problems that newly emerging democracies have been facing more recently. These include internal dissension, a troubled economy, and rising crime rates. The police must tread a thin line between maintaining the political status quo and responding to the needs of citizens for justice and protection. They are trying to establish their independence as an agency that performs traditional police functions and are also attempting to limit their identification with or alliance with the military.

POLICING IN EMERGING DEMOCRACIES

As noted earlier in this chapter, an underlying principle of a democratic society is the rule of law. This implies that the people inherently have certain rights, and it is the obligation of the government in a democracy to assure that these rights are protected. Several authors (Berman, 1992; Iakovlev, 1990; Juviler, 1990) make a distinction between the terms "democracy" and "democratization" (Terrill, 1997, p. 299) to illustrate the difference between countries that have established a government that adheres to the rule of law and follows the principle of government for the people and thus qualify as "democracies," and countries that are in transition from a totalitarian, monarchial, or military controlled form of government and are now aspiring to become democracies through the process of "democratization." According to Terrill (1997, p. 300), the most important aspect of democratization is the understanding that the rule of law is crucial to a democracy and that the government must adhere to the rule of law, even though it may be difficult.

The statements and writings of many of the representatives from the newly developed democracies of Eastern Europe who attended the Symposium on Challenges of Policing Democracies might lead one to conclude that these countries, although politically structured as democracies, may still be in the democratization process.

Representatives of Hungary, Poland, Slovenia, Russia, and Estonia mentioned that the movement toward democracy resulted in internal changes for the police. Specifically, many of the top police officials were forced to leave because they had been associated with the old regime. A decline in the quality of the police departments resulted, with new recruits being less educated, inexperienced, and poorly trained. They also mentioned that some of the old high-level police officials did maintain their positions, and that "the old culture of the communist policing, based on autocracy and total disregard for human rights, did not disappear" (Das, 1995, p. 8).

For several of the emerging democracies, the police authority remained highly centralized, just as it was before the movement toward democracy. It is ironic that in many of these countries the police were greatly disliked, if not hated outright by the ordinary citizens, but nevertheless the consensus was that a strong, centralized police force was necessary to maintain law and order and to combat the serious crime problems of the country, including organized crime, white collar crime, violent crime, smuggling, black marketeering, and illegal immigration. It was reported that some citizens were even willing to give up their civil rights and allow the police "carte blanche" power to control crime and protect the citizens, using whatever means were necessary.

The Case of Poland: An Emerging Democracy

Background

Poland is similar to most of the other emerging democracies of Eastern Europe in the sense that the culture, values, and traditions of the country evolved over hundreds of years and, because of this, the people kept a clear sense of their own identity during the years when they were under the rule of a foreign power.

Poland's history as a distinct country covers almost a thousand years. According to Sigler and Springer (1996, p. 292),

> Poland has moved from a dynastic monarchy to a democracy. In the process it has been governed as a constitutional monarchy, a conquered state, a communist state, and a transitional democracy.

Poland gained some status as a national state in the tenth century when it came under the protection of the Roman Catholic Church. Throughout much of the history of Poland, the Catholic Church has been very closely involved with the government of the country. By the fourteenth century, Poland was a coherent state (Knoll, 1972, p. 293). In the 1600s, Poland developed a constitutional monarchy, with the king elected from the ranks of the nobles rather than obtaining his position through heredity. The *Seym* (Polish Parliament) was the governing instrument for Poland for almost two centuries, and as a result of various struggles that occurred between nobles, peasants, and townspeople, various reforms were instituted and eventually a written constitution was adopted in 1791 (Ludwikowski & Fox, 1993).

The written constitution was in effect for only one year. Then some of the nobles, eliciting assistance from Russia, overthrew the constitution and reestablished the old system. According to Ludwikowski and Fox (1993, p. 294),

> even though Poland was under foreign rule for much of the time during the 18th, 19th and 20th centuries, the democratic principles embodied in the 1791 constitution were also embodied in the 1871 constitution and in the constitution of 1921, the constitution adopted by Poland after it gained independence after World War I.

Poland's independence from foreign domination was rather short-lived. Following World War II, the country became dominated by the USSR. As with any country in which the government and laws are imposed on the people, there was some degree of compliance with the laws, some resistance, and some circumvention. Since the people of Poland had had some experience with being governed by a constitution, the ideals of democracy were never completely forgotten, even during the times when the rule of outsiders was most oppressive.

Post-World War II Police in Poland: A Brief Historical Outline†

Majer (1994) reports that on July 21, 1944, a decree was announced declaring the establishment of the Polish National Liberation Committee (*Polski Komitet Wyzwolenia Narodowego*) with its thirteen departments, one of which was the Department of Public Security. The Committee announced a manifesto a day later which proclaimed the establishment of Citizens' Militia (*Milicja Obywatelska*) whose task it was to uphold order and security. The former State Police (from the period between the two World Wars) was officially dissolved. Under the influence of communist forces (which were themselves highly influenced by ideas in force in the Soviet Union), it was determined that the Citizens' Militia, as an armed agency upholding public order, should constitute part of the governmental Department of Public Security. This philosophy placed the Citizens' Militia directly within the framework of the apparatus of political power. The position of subordination of the Citizens' Militia to the Department of Public Security ended with a decree of December 7, 1954, which disbanded the Department and in its place established two separate agencies, the Department of Internal Affairs (*Ministerstwo Spraw Wewnetrznych*) and the Public Security Committee (*Komitet do Spraw Bezpieczenstwa Publicznego*) attached to the Council of Ministers.

Another milestone in the development of post-World War II police in Poland was, as Majer reports, the decree of December 21, 1955 on the organization and scope of functions of the Citizens' Militia, which was in force (with some amendments) until 1983. The importance of the decree lay in the substantial widening of the scope of functions performed by the Citizens' Militia. The functions covered the upholding of public safety, protection of public peace, law and order, protection of citizens' lives and health, protection of socialized and private property, prevention of crime, crime fighting, conducting investigations and other necessary measures in criminal cases, enforcement of traffic laws, maintaining law and order in various types of public places, issuing gun permits, and other functions.

According to Majer, another set of changes occurred at the end of 1956, when the Citizens' Militia lost its independence from security agencies. The law of November 13, 1956, dissolved the Public Security Committee attached to the Council of Ministers. A new Security Service (*Sluzba Bezpieczenstwa*) was established and strong structural and organizational ties were created between the Citizens' Militia and the new Security Service. This state of affairs lasted without major changes until 1990. During that period, public perception identifying the Citizens' Militia with the Security Service and vice versa was a major obstacle for Militia officers in effectively performing their functions.

Majer also points out that the end of the 1950s saw a number of positive changes, among them the establishment of regional criminalistics laboratories which was an important step in the creation of a branch of criminalistics within the Citizens' Militia. That in turn improved substantially the agency's ability to solve crimes. This positive trend in acquiring modern investigative technologies continued through the 1970s and a number of successes were scored by the Militia in that period, e.g., the apprehension in January 1972 of Zdzislaw Marchwicki (known popularly as the "vampire of the Silesia region") who murdered fifteen women in the 1964–1970 period.

During the 1960s and 1970s a number of organizational changes were introduced in the Citizens' Militia, designed to enable the force to focus on crime-related issues and to transfer many non-criminal obligations to other state agencies. In 1972, the Council of Ministers established the Internal Affairs Academy (*Akademia Spraw Wewnetrznych*). In 1974, further legal measures were introduced in the field of training and job requirements. Among them, high school education was announced as a prerequisite for appointment as a Citizens' Militia officer.

During the 1980s, relations between society and the Citizens' Militia deteriorated visibly. Among the important factors was the fact that Militia officers took part in the enforcement of Martial Law which was declared on December 13, 1981. Literature on the subject acknowledges that no research was ever done on the attitude of the Militia on the political conflict in Poland during 1980–1981, but it is fair to say that there was quite a substantial divergence of views among the officers themselves, reflecting the bigger picture of society at large.

After Tadeusz Mazowiecki took office as prime minister in August 1989 (the first non-communist prime minister after World War II), a number of laws were passed in 1990 which, among other things, disbanded the Citizens' Militia and the Security Service and established a new State Protection Agency (*Urzad Ochrony Panstwa*) and a new Police. The establishment of the latter was clearly an attempt to make a link to the historical State Police of the pre-World War II period, and not just in name.

The newly established police force began its operation under very unfavorable conditions. The main problem was (and continues to be) a rapid increase in crime rates which, as has been demonstrated by research, tends to accompany socialism-to-capitalism transitions such as the one Poland has been going through since 1989. As Cebulak (1996, p. 77) observed,

> to the extent that crime is a product of socio-political change, crime rates are bound to increase much more during a socialism-to-capitalism transition than during a capitalism-to-socialism transition. Some inherent traits of socialism-to-capitalism transitions explain why crime rates increase much more during those transitions than during capitalism-to-socialism ones.

The 1990s have seen dramatic increases in all crime rates, not just violent crime. This rise has been accompanied by increased brutality by criminals, increased and new types of organized crime, and the internationalization of the crime problem. Unfortunately, faced with all these challenges, police have not managed to secure the necessary personnel increases. Compared with the standards of some other European countries, Polish police fall short by about 30,000 officers.

Another issue in the post-1989 period was the question of former Citizens' Militia officers continuing to work for the new post-communist police of the new, capitalist Poland. Faced with the new atmosphere of "verification procedures" and "political correctness," many officers resigned on their own. A number of experts have pointed out, however, the absurdity of eliminating highly trained officers and have insisted that political past should be clearly separated from professional expertise. As a result of these and other practices, about 30,000 completely new employees, as of 1993, had been hired and required training from scratch.

When it comes to police efficiency, even though by official data there have been improvements since 1992 (e.g., according to Majer, Polish police currently solve nine out of ten reported murder cases), data do not match the perception among the public which is increasingly alarmed by rising violence and deteriorating public safety. The police themselves are aware of the problem but frequently find themselves powerless in the face of inadequate staffing, very serious financial hardships, and inadequate, partially obsolete, or otherwise deficient laws. Financial hardships are so severe that there were cases in the mid-1990s of officers appearing on television and asking the viewers to send in money with which the police could later buy gasoline for their patrol cars. Partly as a response to some of these problems, and in the face of increasing levels and complexity of organized crime, a law was passed by Parliament on July 21, 1995 (Law..., 1995) empowering police to use traditionally "Western" police methods such as police provocation and controlled delivery. Under the new regulations, police are able to pose undercover, for instance as drug dealers or offerers of bribes.

Police are also able to use electronic surveillance and censor mail, but all the above-mentioned measures may only be used with regard to certain specific kinds of crime which the law describes. Among them are: deprivation of liberty for ransom; illegal manufacturing, possession, or trafficking in weapons, ammunition, explosives, intoxicating or psychoactive drugs, nuclear or radioactive materials; giving or accepting bribes of enormous value in connection with performing a public office function; counterfeiting or forgery of money, and some other crimes.

The law of July 21, 1995, also provides for the establishment of a "court police" whose function is to provide order maintenance and law enforcement on court premises. It also provides that police and the Border Guard shall receive a specified percentage of monies going into the State Treasury as a result of detection and solving of crime by these two law enforcement agencies.

The new measures are designed to increase police efficiency, rebuild public trust and confidence in the police, and bring the force more in tune with contemporary world trends in the field. Only time will tell, however, to what extent the law on paper can and will be translated into real progress in day-to-day policing. The unprecedented rise in crime rates which has continued since 1989, as well as very serious financial and staffing problems, constitute without a doubt one of the most serious challenges which Polish law enforcement has ever faced.

Political Policing in Russia

As in the past, the population of Russia today is made of a large number of distinct national and racial groups. Terrill (1997, p. 292) states that

> Russia is a federation consisting of six categories of administrative units. These include: 21 republics, six territories, 49 provinces, two federal cities, one autonomous province and 10 autonomous regions. Among these administrative units, the republics have the greatest claim to self-government. Although Russians comprise more than 80 percent of the country's population, there are some 126 nationalities with distinct racial, linguistic and religious preferences.

Before the revolution of 1917, Russia was predominantly rural. Its peasant society was held together by the common link of Russian Orthodox Church traditions, a deep love of country, and the presence of totalitarian authority. According to Terrill (1997, p. 294), after the break-up of the USSR in 1991, changing the economic system from a socialist to a free-market economy has been the focus of attention. There have been few changes in the Russian criminal justice system.

Before the 1917 Revolution, Russian policing followed a centralized policing model. When the Union of Soviet Socialist Republics was formed, a new, centralized police force was organized to combat counter-revolution. The Extraordinary Commission for Combating Counter-Revolution and Sabotage (*Cheka*) was entrusted with this task. As noted by Terrill (1997, p. 308), the Russian Tsar Ivan had created such a political police force in the sixteenth century. The modern *Cheka* relied heavily on the local law enforcement officials for information and support. *Cheka* came to be regarded as too repressive, and it was disbanded in 1922.

The United State Political Administration (*OGPU*) was created in 1924. This administrative unit was responsible for both internal security (railways/waterways) and external security (borders). The *OGPU* was at the height of its power during the rule of Stalin, with its most noteworthy accomplishment being the elimination of all open or suspected opposition to Stalin. The establishment of an effective borderguard, the development of an internal passport system, centralizing the administration of the corrective labor camps, and the eradication of peasant opposition to the efforts to collectivize farms all occurred during this period (Terrill, 1997, p. 309). Although this state security force was totally independent and separate from other law enforcement units and owed allegiance only to the central administration in Moscow, the local police agencies nevertheless often joined forces with the *OGPU* agents when important matters pertaining to national security required a concerted police effort.

Following World War II, a reorganization of government in the USSR. was completed. The new Ministry of Internal Affairs (*MVD*) assumed responsibility for the militia and the Ministry of State Security (*MGB*) had responsibility for state security matters and regular law enforcement functions (Terrill, 1997, p. 309) The notorious Committee for State Security (*KGB*) was formed in 1954. This centralized policing unit, headquartered in Moscow, maintained the responsibility for state security until the collapse of the Soviet Union. The *KGB*, under the control of the community party, was authorized to operate outside the law when national interests demanded such action. Although its operations were more extensive, the goals of the *KGB* were not radically different from those of *Cheka*, the first police political security organization formed after the Revolution.

POLICING AND DEMOCRACY

In this chapter, the historical roots of the policing systems in various countries were traced, for the purpose of illustrating the manner in which the culture, political structure, and specific political, economic, or historical events led to the adoption of a particular type of police organizational structure.

During the Symposium on Challenges of Policing Democracies the representatives from the various countries agreed that the adoption of a democratic form of government does not guarantee that a country will become a democracy or that the police will adopt a new organizational structure or embrace a different set of values. The cornerstone of a democracy is the rule of law, and political leaders must recognize the need for a separate police organization to enforce the law and an independent judicial system to assure that the actions of the police are in accordance with the letter of the law.

The research reported here on the origins and development of the police in various countries illustrates that establishment of a specific form of government does not automatically guarantee a just society or the rule of law. In addition, it can be shown that even countries born out of revolutions motivated by the desire to return political power to the people experienced discrepancies between what was promised by leaders and what actually occurred. Those responsible for overthrowing dictators often used their newly acquired power to attempt to control the police and courts in manners very similar to those used by the former regimes.

History also reveals that democracy can be very fragile. If the security of the country appears to be threatened by an external or an internal force, government leaders have been known to compromise the rights of the people in order to maintain power and order. This has occurred in the United States, a country with a long tradition of democracy. Police have often had to decide if their mandate was to assure that the laws were enforced under the rule of law embodied in the U.S. constitution, or if the "real law" was determined by the "power elite" in the form of political, business, or military leaders.

POLICING IN AN ERA OF CHANGE

Another theme manifested at the Symposium on Challenges of Policing Democracies was that societies are not static. The representatives of the various countries reported a number of law and law enforcement challenges facing their nations. For several countries, the problems may have resulted from the recent change in the form of government. These countries experienced an increase in poverty and rapid urbanization and immigration. When the citizens realized that the supposed benefits of the new government were not going to occur immediately, a state of disillusionment sometimes resulted. Other problems experienced were internal disorganization, increases in the rates of unemployment and the amount of poverty, immigration of poor, unskilled people to the country, and emigration of some of the better-educated professionals. These problems and uncertainties taxed the resources of the countries. Rising crime rates, particularly if crime had not traditionally been a major problem, and the emergence of organized crime, money laundering, violent crime, and various types of white-collar crime required that the police quickly obtain more training and skills. However, for the new democracies, police resources, both in terms of equipment and personnel, had declined immediately after they gained independence.

Countries with long traditions of democracy also face problems that require adaptation, professional responses, and constant diligence to assure that the challenges will be met in a manner that corresponds to the rule of law and maintains the personal dignity of citizens.

NOTE

† This section was prepared by Wojciech Cebulak, Pd.D.

Up to the point describing the most recent changes in the 1990s, the section is based on the analysis by Piotr Majer (1994) *Milicja Obywatelska I Policja w latach 1944–1994. Geneza I podstawowe przeobrazenia* (Citizens' Militia and police in the 1944–1994 period. Origins and fundamental changes). *Przeglad Policyjny, 4,* 36, 59–78. Majer teaches at the Higher Police School in Szczytno. Without this reference work, the description in this section would not have been possible.

REFERENCES

Arthur, J. A. & Marenin, O. (1996). British colonization and the political development of the police in Ghana, West Africa. In C. B. Fields & R. H. Moore, Jr. (Eds.), *Comparative criminal justice* (pp. 163–180). Prospect Heights, IL: Waveland Press.

Berman, H. J. (1992). Christianity and democracy in the Soviet Union. *Emory University Law Review, 6,* 23–34.

Black, H. C. (1991). *Black's law dictionary*. St. Paul, Mn: West Publishing Company.

Bracey, D. H. (1996). Civil liberties and the mass line: Police and administrative punishment in the People's Republic of China. In C. B. Fields and R. H. Moore, Jr. (Eds.). *Comparative criminal justice* (pp. 225–233). Prospect Heights, IL: Waveland Press.

Cebulak, W. (1996). Rising crime rates amidst transformation in Eastern Europe: Sociopolitical transition and societal response. *International Journal of Comparative and Applied Criminal Justice, 20* (1, Spring), 77–82.

Clinard, M. B. & Abbott, D. J. (1973). *Crime in developing countries*. New York: John Wiley & Sons.

Dai, Y. S., & Huang, Z. Y. (1993). Organization and function of police security agencies of the People's Republic of China. *Eurocriminology, 5–6*, 137–43.

Das, D. (1995, May). Challenges of policing democracies: Executive summary. *Proceedings of the Second International Police Executive Symposium*. Oñati, Spain.

Forrester, P. (1995, May). Policing and democracy. *Proceedings of the Second International Police Executive Symposium*. Oñati, Spain.

Iakovlev, A. (1990). Constitutional socialist democracy: Dream or reality? *Columbia Journal of Transnational Law, 28*, 117–132.

Johnson, H. A. (1988). *The history of criminal justice*. Cincinnati: Anderson Publishing Co.

Juviler, P. (1990). Guaranteeing human rights in the Soviet Context. *Columbia Journal of Transnational Law, 28*, 135–155.

Knoll, P. W. (1972). *The rise of the Polish monarchy*. Chicago: University of Chicago Press.

Laitinen, A. (1994). Police in Finland. In D. K. Das (Ed.). *Police practices: An international review* (pp. 123–180). Metuchen, N. J.: The Scarecrow Press.

Law of July 21, 1995, on amending the laws on the office of the Minister of Internal Affairs, on the Police, on the Office of State Protection, on the Border Guard, and of some other laws, in *Dziennik Ustaw Rzeczpospolitej Polskiej*. Warsaw: September 11, 1995, No. 104, item No. 515.

Ludwikowski, R. R. & Fox, W. F. (1993). *The beginnings of the constitutional era: A bicentennial comparative analysis of the first modern constitutions*. Washington, DC: Catholic University of America Press.

Majer, P. (1994). Milicja obywatelska i policja w latach 1944–1994. Geneza i podstawowe przeobrazenia [Citizens' Militia and Police in the 1944–1994 period. The origins and fundamental changes]. *Przeglad Policyjny, 4, 36*, 59–78.

Richards, M. (1996). *Policing the community: An English perspective*. Paper presented at the Academy of Criminal Justice Sciences Conference, Las Vegas, NV, March 12–16.

Richardson, J. J. (1995). Police history. In W. G. Bailey (Ed.), *The encyclopedia of police science* (pp. 553–558). (2nd ed.). New York: Garland Publishing, Inc.

Sigler, R. T., & Springer, S. (1996). Social change and adaptation in the Polish judicial system. In C. B. Fields & R. H. Moore, Jr. (Eds.), *Comparative criminal justice* (pp. 292–306). Prospect Heights, IL: Waveland Press.

Stead, P. (1985). *The police of Britain*. New York: MacMillan Publishing Company.

Terrill, R. J. (1995). Organizational structure: Three models with international comparisons. In W. G. Bailey (Ed.), *The encyclopedia of police science* (pp. 485–492). (2nd ed.). New York: Garland Publishing Inc.

Terrill, R. J. (1997). *World criminal justice systems* (3rd Ed.). Cincinnati: Anderson.

Walker, D. B., & Richards, M. (1995). British policing. In W. G. Bailey (Ed.), *The encyclopedia of police science* (pp. 41–48). (2nd ed.). New York: Garland Publishing, Inc.

Yang, C. (1994). Public security offenses and their impact on crime rates in China. *British Journal of Criminology, 34* (1), 54–58.

Section II
Country Perspectives

3

Challenges of Policing Democracies: the Croatian Experience

SANJA KUTNJAK IVKOVICH

INTRODUCTION: SOCIAL, POLITICAL, AND ECONOMIC CONDITIONS

The Republic of Croatia (*Republika Hrvatska*) is both a central European and a Mediterranean country. Its history has been strongly influenced by Mediterranean culture and politics and by its ties to central European peoples. Despite its modest size of 21,829 square miles (56,610 square kilometers), Croatia is a country of considerable geographic diversity. Its 1,260 miles-long land borders touch Bosnia and Herzegovina, Slovenia, Hungary, and Yugoslavia, and it faces Italy across the Adriatic sea (*1997 Croatian Almanac*, 1998).

Approximately one quarter of the Croatian population of 4.7 million (in 1996) lives in the capital Zagreb. The majority of Croatian citizens are ethnic Croats (79%) and belong to the Catholic church (77%). The median age in 1995 was 37.2 years. Only 4 percent of its population is illiterate. There are 146 cars and 202 telephones per 1,000 inhabitants, 2 national and 8 regional daily newspapers, 108 licensed radio and 10 television stations (*1997 Croatian Almanac*, 1998; *Croatia: Facts and Figures*, 1997).

Croatia's economy is lightly integrated into the world market. A substantial portion of its labor force works as guest workers in European Union countries. As a result of the recent war, per capita GDP decreased sharply from $5,106 in 1990 to $2,079 in 1992, but increased to $3,786 in 1995 (*1997 Croatian Almanac*, 1998).

Croatians have lived in southeastern Europe for more than thirteen centuries. Croatia was part of the Austro-Hungarian Empire and parts of its coast have been ruled by Italian city-states. Following the collapse of the Austro-Hungarian Empire in 1918, Croatia became part of the kingdom of Serbs, Croats, and Slovenes (subsequently renamed the Kingdom of Yugoslavia). After World War II, the Socialist Republic of Croatia existed for over four decades as one of the republics that constituted the former Socialist Federal Republic of Yugoslavia. After the collapse of Yugoslavia, Croatians voted for independence in 1991. Croatia was recognized as a free and independent country by the countries of the European Union on January 15, 1992, and became a member of the United Nations on May 22, 1992.

Croatia is a parliamentary democracy. The legislative function is performed by the Parliament (consisting of the House of Representatives and the House of Counties); the executive function lies with the President and administration; the judicial function is carried out by over 1,000 judges at various levels of the court system. The country is divided into twenty-one counties which constitute the country's principal units of local government.

In it efforts to achieve democracy, the Croatian Parliament introduced amendments to the Croatian Constitution[1] in early 1990. These amendments allowed for the existence of political parties other than the ruling Communist Party, and have thereby abolished the Communist Party's four-decade long monopoly over political life in Croatia. The first eight political parties were registered in February 1990 (Peric, 1995, p. 37). The first multi-party elections for the Croatian Parliament and local governments were held in the spring of 1990. The old Communist Party with a new name—Party of Democratic Changes (*Stranka Demokratskih Promjena*)—won only 20 percent of the seats in the Parliament, while the Croatian Democratic Union (*Hrvatska Demokratska Zajednica*) won the majority of the seats (59%); the remaining 21 percent of the seats were won by other parties (Peric, 1995, p. 47). The first session of the newly elected multi-party Parliament was held on May 30, 1990.

Unlike most other European countries in transition, Croatia's road to democracy is inextricably linked to the recent war, which has had a significant impact not only on policing but also on all aspects of life in Croatia. Therefore, any meaningful discussion of the efforts exerted by the Croatian police to adapt to the transition toward democracy must be related to the effects of the war on policing in Croatia.

In 1990, the presidency of Croatia embarked on numerous changes, some of which inevitably affected the police. It was proclaimed that all the political and ideological characteristics of the old regime had to be removed from the police and that the police should be established as a professional service in charge of protecting the constitutional order and human rights. The presidency ordered that citizens' requests for passports could no longer be rejected (as had been the case under the socialist/communist regime), that mail could not be opened illegally, and that there could be no illegal wiretapping (Peric, 1995, pp. 51–52). Furthermore, the presidency ordered that all political prisoners be released. In June of the same year (1990), the Croatian Parliament decided that all the characteristics of socialist/communist ideology were to be removed from the Constitution and from the law. Consequently, in July 1990, Croatia added new amendments to the Constitution which introduced changes in the organization of the state and government. However, Kregar, Smerdel, and Simonovic (1991) argued that these changes were rather symbolic; it was clear that social transformation could not be achieved on the basis of a reconstruction of the old Constitution and that a new Constitution was necessary.

On May 23, 1990, the Federal Ministry of Defense decided to withdraw all weapons from the territorial defense units (which were supposed to be under the control of individual republics and not the Yugoslav Federation) in Slovenia and Croatia. This swift and surprising move left Croatia virtually unarmed overnight. The president of the Knin District in Croatia, a stronghold of opposition to Croatian rule during the subsequent aggression against Croatia, declared, in August of 1990, that a state of war existed. As Johann Georg Reissmeller, one of the editors of *Frankfurter Allgemeine Zeitung*, wrote in October of 1990, "it smells of war" (cited in Peric, 1995, p. 61).

Croatia, at that time still one of the republics of the Federation, did not have an army nor the legal basis for creating one. The Yugoslav People's Army, the federal army of the former Socialist Federal Republic of Yugoslavia, had been established by federal law, which forbade each of the six republics to establish its own army.

The Croatian Ministry of the Interior was the only legitimate organized force at the republic level at that time and was, therefore, the only force that could carry out defense activities (Moric, 1994, p. 118). In consequence, beginning in August 1990 and lasting for a period of almost two years, the police performed two roles: the defense role and the "regular" police role.

The new Croatian Constitution, passed by the Croatian Parliament in December, 1990, declared Croatia a parliamentary multi-party republic. The basic principles of the new Constitution included political democracy, the rule of law and the principle of a market economy. Unlike the government's organization in the socialist/communist regime, in which the same body was in charge of both the legislative and the executive authority, the new Croatian Constitution was based on the principle of the division of authority into legislative, executive, and judicial branches. Kregar, Smerdel, and Simonovic (1991, p. 150) argue that "the separation of power (although emphasized in the Constitution as one of its fundamental principles) has not been consistently carried out because of a shift in favor of the executive power and the institution of the President of the Republic, which should primarily be ascribed to the specific conditions under which the Constitution was enacted." The Parliament also moved toward the transformation of public property into private property.

Because of the obvious need to establish an army and the legal impossibility of doing so, the Croatian Parliament passed changes to the *Law on Interior Affairs* (which regulated the police force) in April of 1991 to allow the establishment of the National Guard Corps (NGC) (*Zbor Narodne Garde*) as a police service *within* the Ministry of the Interior. Police officers constituted a large proportion of the manpower for the NGC, especially in the first few months after its formation.

The presidents of the six Yugoslav republics proposed that the governments of each of the republics organize a referendum and that citizens determine the future of the Yugoslav Federation. Consequently, a referendum was organized in Croatia in May 1991. An extremely high percentage of the registered voters in Croatia, about 84 percent, participated in the referendum. Voters had to decide on the following issues: (1) whether they wanted Croatia to become an independent state which could form a union with other republics, and (2) whether they wanted Croatia to remain in Yugoslavia as a part of a unified federal state. Over 90 percent of voters opted for an independent state by answering "for" to the first question and "against" to the second question (Peric, 1995, pp. 91–95). On May 23, 1991, the President of the Republic of Croatia declared that citizens had made the following decision: the Republic of Croatia should become an independent state and should not remain in Yugoslavia as part of a unified federal state. About a month later, on June 25, 1991, Croatia declared its independence from Yugoslavia. Within days, the war in Croatia began. The war resulted in numerous casualties, a severe refugee and humanitarian crisis, material destruction, and a (temporary) loss of approximately 30 percent of Croatian territory.

In September 1991, Croatia passed the *Law on Defense* which stipulated that the NGC become part of the Croatian Army within the Ministry of Defense. By the end of the year, the *Law on Interior Affairs* was changed again and all of the rules related to the NGC were erased from it. The majority of police officers were relieved from their duties in the defensive war by mid-1992 (Moric, 1994, p. 119). Thirty-three thousand police officers had participated in military actions, of whom 572 died, 2,741 were wounded, and 116 are missing (Moric, 1994, p. 10).

In addition to the war already raging in Croatia, the start of the war in Bosnia and Herzegovina in spring 1992 rapidly increased the number of refugees and displaced persons who remained in Croatia or were in transit to other parts of the world. In subsequent months, the number of refugees from Bosnia and Herzegovina drastically increased. In July 1992, the Croatian Government Office for Refugees and Displaced Persons announced that 265,786 Croatian Displaced Persons and 332,575 refugees from Bosnia and Herzegovina were registered with the Office and were receiving some form of support (*Halo 92*, 1992, #8, p. 5). The approximately 600,000 (registered) refugees and displaced persons residing in Croatia at the time equaled roughly 13 percent of the Croatian population, which, of course, added to the already existing difficult economic conditions and introduced yet another challenge for the police. The estimated expenses for refugees and displaced persons in 1992 were close to US$1.5 billion (Peric, 1995, p. 160).

The war had other serious economic consequences as well. The standard of living decreased rapidly: The average monthly salary in Croatia dropped from 663 German marks in 1990 to 175 German marks in 1991 and 1992 (Markovic, 1997, p. 5). The percentage of unemployed work force increased to 17 percent in 1993 (Peric, 1995, p. 166) and it remains high, even at the end of 1997. The country spent over 5 percent of its GNP (over one billion US dollars) for the army in 1997 (Rkman, 1998, p. 8).

By launching two combined military and police operations in 1995, Croatia regained control over the approximately one-quarter of its territory that had been lost earlier. Additional challenges were imposed on police officers who remained on that territory to perform "regular" police duties. Some of these challenges included the existence of a huge number of planted land mines that were left behind by the enemy (mostly without reliable documentation as to their location or type), which affected the mobility of police officers and citizens and led to new injuries and deaths. According to estimates, for example, there were over 60,000 land mines left in one particular sector alone (Peric, 1995, p. 156). Finally, the police played the major role in the peaceful reintegration of the remaining 5 percent of the Croatian territory. On January 15, 1998, the last UN forces, which had been deployed to stabilize the situation, left Eastern Slavonia.

THE CONCEPT OF DEMOCRATIC POLICING IN CROATIA

The Croatian police is a very young force. It has been in existence for only a few years. Historically, its predecessor was the communist militia. Relics of communist policing included a military-style organization, communist mentality, and its ways of understanding the mission of the police forces. As soon as circumstance permitted, the Croatian government undertook drastic changes in order to achieve democratic policing. The changes that followed were driven by the desire to depart from the image and practices of the communist militia—a military style force in charge of protecting the communist regime—and to embrace the image of a modern, professional force which serves citizens and protects public order and the democratic state.

The Ministry of the Interior views adherence to the rule of law and the protection of human rights as the cornerstones of democratic policing; the police are required to follow the letter of the law. For example, police officers are required to use minimal force and to use force only as the last resort. The Ministry enforces these rules by aggressively conducting investigations into every incident involving the use of force. Laws have been changed in order to better suit current political, social and economic conditions. Laws governing the conduct of police officers have been passed and further changes are still in progress.

Professionalism is also viewed as one of the basic elements of democratic policing. Emphasis is placed not only on professional training of police officers, but also on the development of their general attitudes and manners. Interactions with citizens should be polite and patient. Adherence to professional ethics has become an integral part of policing. In keeping with its commitment to place greater emphasis on crime prevention (rather than repression), the Ministry declared prevention to be one of its most important tasks.

A strict chain of command within the centralized force allows for accountability, an important practice which affects not only supervisors but also each police officer. The Ministry has increased control over the work and behavior of its employees by forming the Internal Affairs Office and by charging each supervisor with the responsibility for his or her subordinates and their professional conduct. The Ministry itself is controlled and held accountable to the Croatian Parliament and to citizens, and it interacts with the public and the media on a regular basis. The Ministry also established an Office of Public Relations.

ORGANIZATIONAL CHALLENGES

Differentiation from the Militia

The new Croatian police implemented changes to break away from the practices of its predecessor. These are visible even at the most superficial, yet symbolically important level—for example, the name of the organization. Kobali and Balokovic-Krklec (1995, p. 10) regarded the change of name a necessity for preparing the Croatian police for the new century.

> The first difference, seemingly unimportant, is obvious from the act of renaming the militia the police, which was supported by the examination of Klaic's *Dictionary of Foreign Words*. Namely, the militia, according to its Latin root (*miles, militis*—a soldier), is a citizens' or people's army, 'a special armed formation with the purpose of maintaining order in cities and villages' and, in post-World War II Yugoslav practice, 'armed people' as 'the keeper of the [socialist] revolution.' The Greek root of the word police (*politeia*—organized state life) defines a professional institution that is responsible for public order and the system of the state and eliminates the ideological overtone of the earlier name.

Unlike other countries in transition from a socialist/communist regime, and due to the war, the Croatian police were in a position to make a clean break from the militia of the old regime. By changing the name of the organization, the new Ministry of the Interior sent a clear signal of its differentiation from the old regime. The militia of the former Federal Socialist Republic of Yugoslavia had a reputation among the citizens as an uneducated, unprofessional force established and managed as a military organization. When militia officers interacted with citizens, they typically addressed a citizen by his or her first name,[2] indicating a lack of respect for the citizen. The militia's unfavorable reputation was so strong that citizens frequently shared jokes about the lack of ability and skills of militia officers.[3] Most of these jokes arose soon after World War II, when militia officers were recruited without a proper education or training. Over the years, the militia improved the technical skills of its officers. However, their interaction with citizens by and large remained marked by the mystification of police work, arrogance and lack of respect for the citizens, even in the 1980s.

In the late 1980s and early 1990s, citizens' perceptions of the militia were influenced by their views of the Socialist Federalist Republic of Yugoslavia, the (new) Socialist Republic of Yugoslavia, and the aggression against Croatia. Before

1990, over 70 percent of militia officers in the old Ministry of the Interior were ethnic Serbs (*Halo 92*, 1995, #44, p. 7); in one police district, 98 percent of the administrators/supervisors and 75 percent of the troops were of Serbian ethnicity (*Halo 92*, 1992, #3, p. 5). Serbs in Croatia constituted approximately 12 percent of the citizens, according to the 1991 census (Lajic, 1995, p. 60). This clearly disproportionate representation was especially strong among detectives in the criminal investigation units.

The new Croatian police wanted to break away from this negative image in order to secure at least some degree of support from the citizens. With war being waged, this matter was all the more pressing. It turned out that the disproportionately large percentage of Serbian militia officers in the old regime served as a catalyst for change on the eve and at the beginning of the war. The great majority of militia officers of Serbian origin left their posts when the socialist/communist militia became the Croatian police force.[4] Such a massive departure, which happened virtually overnight, introduced severe problems, especially in light of the complicated dual roles that the Croatian police performed at the time (defense *and* "regular" police work). This drastic decrease in numbers presented considerable challenges for the Croatian police; the necessity of rapidly hiring and training a large number of new police officers was a formidable task in itself. Highly qualified officers specializing in, for example criminal investigation, were very difficult to replace on such short notice. The problem was further aggravated by the fact that many of the police officers who had left their posts began performing terrorist acts and/or openly joined the enemy troops. Thus, new police officers were also necessary to fight the enemy.

Handling the Dual Roles in the War

While still part of Yugoslavia, Croatia did not have a legitimate basis to form an army. When the war broke out, the police forces were the only legitimate organized armed forces in the country which had at least some manpower and some weapons to defend Croatia, although police officers were neither trained to perform military duties nor were their weapons equipped or suited for military combat. Ivo Stipicic, the chief of the Split Police District, vividly described the problem (Mirosavljev, 1992b, p. 4).

> Due to the state of war, the police had to superimpose the fight against the aggressor over its basic role—public safety and the protection of the Constitutional liberties and rights of citizens. It is known that the police are not intended for military activity. And it has created quite a large number of problems for the police to adjust and to perform the defense function. The behavior of the army and the police is incompatible. The army has a [front] line behind which is the enemy with its manpower and weapons; the more people you kill and the more destruction you impose the greater the success. The police have the duty of protecting every life and everybody's property. It means do not kill, do not destroy.

Deputy Minister Josko Moric (1994, p. 120) described the effect these differences in training and philosophy had on the behavior of police officers.

> In their education and practice they [police officers] had been taught to use force only as the last resort, in situations in which other ways of preventing

attacks against people, against their rights and freedom, and against their property proved ineffective. They learned and performed their tasks by observing the rule that, in the use of force, they had to use all other means before they may use firearms. In instances in which they used force, especially firearms, the justification and legality of its usage was subsequently investigated and assessed. Terrorist attacks and the war put the police officers in situations in which they had to use firearms without first depleting all the other procedures. Many police officers were killed or wounded because they tried to solve the problem in another way, by using other means, or just as they were helping the wounded attackers.

Just as the use of police philosophy and practice might have cost police officers their lives in the war, the use of military philosophy and practice they had to learn in order to survive and perform well in the war caused problems when they returned to their regular police duties. Indeed, Moric (1994, p. 120) emphasized that, "[t]errorism and the war pushed them [police officers] into solving conflict situations by force. When they returned to uniformed police duties and tasks, this became a very serious problem for the police officers and for our organization as a whole. ... [D]uring 1993 [when most of the police officers returned were released from their military duties and started performing their regular police work] we made great efforts to solve that problem."

The rapid depletion of the police force over several months in late 1990 and early 1991 on the one hand, and the increased need to enlarge the force on the eve of open military conflict on the other hand, prompted the recruitment of thousands of people over a very short period of time. They were recruited as police officers, but their main role at the time was to defend the country.

Because the newly admitted people were officially police officers under the jurisdiction of the Ministry of the Interior, the Ministry was in charge of their training. Under regular circumstances, a person may become a police officer either by attending a two-year police high-school or, if the person is a high-school graduate, by attending a six-month course followed by six months of practical training (Jurina, 1994, p. 75). Obviously, during this emergency period, the time required for regular basic training for high-school graduates was not available. New police officers were required immediately.

The Ministry developed a suitably tailored basic course for newcomers. The Director of the Police high school (which is an integral part of the Police Academy), Josip Strmotic, has said that approximately 6,000 people attended these courses at the Academy within the last four months of 1991. These new police officers passed through a course that focused mostly on military education and training (*Halo 92*, 1992, #7, p. 13). Basic elements of police work were covered too, but were shortened in comparison with the regular police course. In this version of the course, for example, a significant portion was devoted to terrorism and the ways of dealing with it. By contrast, systematic training in the ethics of policing was limited. Preparations for the more systematic and thorough training of ethics were underway even at that time, and were further accelerated in the light of the obligations that Croatia has accepted by signing international conventions against torture and other inhuman and cruel punishments and procedures (Veic, 1997, pp. 278–280).

Furthermore, as was pointed out by Josip Hajdinjak, the chief of one of the police districts on the military frontline, at locations close to the front many times "there was no time for training" at all (Brmbota-Devcic, 1992, p. 13). According to Stjepan Krpicak, a highly ranked Croatian police official, since "there was not enough time [for training], the training was conducted on the run, on the field" (Balokovic-

Krklec, 1992a, pp. 6–7). Former Minister Ivan Jarnjak analyzed this situation in the following way (Rako, 1995, p. 7).

> In the years 1990–1991 we accepted a large number of people into the police and, instead of having peaceful and normal training, these people had to take guns into their hands—first to defend their homeland and to establish the country, and only after that to perform their police roles. It is certain that something has been skipped here. Not all [of the recruits] went through the basic training and we need time for all of these police officers to go through the things that they have skipped by a combination of circumstances, i.e., [they need] to be able to respond to the duties imposed in front of them completely: to know how to find a solution, how to solve the problems, and to be helpful to their homeland. ... However, even in the time of war we tried to train police officers to the extent possible. I would like to point out that we had 2,500 people trained at the Police Academy even in the midst of the most intensive military operations.

Short(ened) training, or no training at all, and no practical experience in policing clearly posed problems for a few years. During that period the police were not only involved in military operations, but were also in charge of "regular" police work for the whole territory, regardless of the extent of their military involvement in a particular part of the country. The consequences were also felt after the military operations subsided in intensity and when these police officers who performed only military functions returned to the police stations to perform "regular" police functions. Durica Franjo, the head of the Sector for Police at the Ministry described the problem of newcomers performing police duties (Balokovic-Krklec, 1992b, p. 6).

> We are trying to get the police to perform their regular duties and responsibilities in all parts of the territory regardless of the intensity of the military operations; in other words, by carrying out police activities [we want] to guarantee safety to the citizens and property, to maintain public order and peace, to regulate and control the traffic. In parallel with this, we have to work theoretically and practically on the education of the police in the spirit of democratic changes and to develop a new image—the police that help citizens but still consistently enforce the law and rules of this state. It is a very large and complex program because a large number of new people have been accepted. One segment of these people achieved their first experiences at the front lines, and there is a difference between gaining experience in such circumstances and [gaining experience] carrying out strictly police activities.

Policing Liberated Territories

In mid-1992, the Minister of the Interior at the time, Ivan Jarnjak, announced that the police were returning to their regular duties and responsibilities (*Halo 92*, 1992, #5, p. 12). The police were no longer directly involved in military combat, but were still in charge of some atypical police duties. For example, they preceded displaced persons returning home and cleared up or at least marked land mines. The police had to deal with individual terrorist attacks once the military activities stopped and the army left (*Halo 92*, 1994, #26, p. 7). A part of their regular duties included the maintenance of peace and order in the liberated territories.

In 1995 the Croatian Army and the police conducted two short-lasting but large-scale operations to free the approximately 25 percent of Croatian territory which had been lost. The second set of operations ("Storm" was the nickname for the military action and "Return" was the nickname for the accompanying police action) involved 3,600 police officers (*Halo 92*, 1995, #46, p. 14). The police closely followed the Army's footsteps and were charged with the role of keeping the peace and maintaining order. The Croatian Ministry of the Interior did not have police stations in these parts of the territory, neither did it have access to the existing buildings and equipment inherited from the earlier regime. However, Deputy Minister Moric emphasized that, as part of organizational changes in mid-1992, the Ministry had already organized "stations in exile" and treated these as a constituent element of the Ministry (*Halo 92*, 1995, #42, p. 10). To make the transition from Serbian occupation an easier and a more successful one, the police officers who would staff the police stations once the territory was liberated had been selected and the necessary equipment had been prepared for each of the police stations in exile.

Once the Croatian Army had liberated a territory, the police followed and established new police stations. Because the actors and the roles were determined in advance, the actual establishment of the police stations was made easier. The first duties were to secure the area, especially the important intersections; to provide the guarantee of the security to all the people and property in the area; to clear major roads of land mines and to start the search for hidden weapons. However, due to their relatively small numbers and the intensity of the operations, police officers from these newly established police stations were not able to perform all these functions. The Ministry had planned and organized that each police district in the country send a group of one hundred police officers to help their colleagues, once the planned military operation started (*Halo 92*, 1995, #42, p. 10).

Dealing with Terrorism and War Crimes

Historically, the first signs of the oncoming war were terrorist acts against the territory of Croatia. To cope with this problem, the Ministry of the Interior established the Department for the Fight against Terrorism and War Crimes, which was staffed with the officers who were specially trained to fight terrorism. The Department collected various documents and evidence of war crimes and sent the collected materials to the district attorneys upon the completion of each investigation. The Department also helped detectives who were in charge of criminal investigations related to the "Storm/Return" operation by providing background information. These detectives were also prepared in advance for their work in the newly freed territories. Their tasks required extensive preparations and training because they had to interview, over a relatively short period of time, a large number of persons and make inquiries into those persons' possible involvement in war crimes and crimes against humanity. That meant restricting the rights of those persons by detaining them for questioning. Consequently, in accordance with the democratic principles that guided the work of the Ministry of the Interior, this period had to be made as short as possible. Indeed, over a period of only several days 1,494 individuals were interviewed (*Halo 92*, 1995, #42, p. 17) and 180 of them were subsequently arrested.

Dealing with Inexperience

In the summer 1996 I interviewed several detectives in the course of carrying out a study about attitudes on police corruption. They told me, from their own perspective, about the impact of the departure of numerous detectives of Serbian origin in late 1990 and early 1991. They perceived as a great problem the fact that very few knowledgeable colleagues had remained on the force—they simply did not have older, more experienced colleagues to teach them how to investigate crimes (Kutnjak Ivkovich, 1996). Former Minister Jarnjak stated that, prior to that massive departure, approximately 80 percent of the detectives on the force were of Serbian origin and most left at the beginning of the war (*Halo 92*, 1993, #16, p. 3).

In addition to the other problems that this departure had caused, these events made the Croatian police force a very young and inexperienced one. In fact, a nationally representative sample of police officers reported in 1995 that seven out of ten police officers had been police officers for less than five years (Kutnjak Ivkovich & Klockars, 1997). This implies that, even under the assumption that no turnover occurred in the meantime, the majority of police officers in Croatia have less than ten years of experience in 2000. Toward the end of the war, the police force was comprised predominantly of enthusiastic, young officers who had extensive military experience but very limited knowledge of "regular" police work. For example, the average age of police officers in one of the police districts was only 27 (*Halo 92*, 1996, #56, p. 13) and in another only 31 (*Halo 92*, 1997, #67, p. 15).

Shaping the Police Culture

The fact that very few police officers from the old regime remained in the new police made learning police work very difficult. As an additional interesting consequence, since very few who were a part of the old police subculture remained, chances were small that that subculture, the old socialist/communist ideology and the old ways of policing, would be transmitted to the new police officers. It seems that even those few experienced police officers who remained on the force purposely tried to make a systematic effort to distinguish between "their old system of militia" and "our new police," and tried to behave appropriately for the new, changed times. It was, therefore, easier for supervisors and the Ministry in general to shape this newly formed police culture in the desired direction. On the other hand, the solidarity that had developed as a result of common war experiences had a serious impact on the code of silence; police officers whom I have interviewed (Kutnjak Ivkovich, 1996) told me that they would tolerate a considerable degree of misconduct by their colleagues. They remarked, for example, "How can I report him when he has saved my life ten times?"

The transition toward democracy, which had already started in society at large with the constitutional changes in 1990, caused high-ranking administrators to emphasize the democratic approach toward policing (while adhering to a strict chain of command in order to control the troops). The democratic way of policing includes professionalism and adherence to the rule of law, especially with respect to the "regular" police duties. Top administrators in the Ministry, such as Stjepan Krpicak, stressed that this was not an easy task because of the emotions involved. Krpicak describes an incident characteristic of the disagreeable aspects of policing during the war. As a consequence of a terrorist act, twelve police officers were killed and twenty-one officers were wounded. Recollecting the impressions and emotions he experienced when the team sent by the Ministry to investigate the incident arrived on the scene, Krpicak said that, "although we are professionals, it was very difficult to remain indifferent to the incident that was the direct reason for [our] trip. It was necessary to get the level of humanity and professionalism in us to the level of neutrality, so that we [could] conduct the investigation of the incident." After dealing with the overwhelming feeling of shock, the officers managed to carry out the investigation and "were successful in imposing [the] two mandates of the service: professionalism and correctness in work" (Balokovic-Krklec, 1992a, pp. 6–7).

The Ministry and the chiefs of police stations encouraged police officers to attend various specialized, in-service courses and/or to obtain a bachelor's degree in criminalistics from the Police Academy.[5] For example, Drago Matic, head of one of the police districts, said that "there is no good police officer without a developed love for the homeland and a professional qualification. All of our police officers have proven their love for the homeland in the war, now is the time for professional training" (Rako, 1997, p. 15).

Figure 3.1 shows that the Police Academy offers specialized courses which last between two and nine months. These courses are attended by large numbers of police officers. For example, former Minister Jarnjak reported that in 1993 approximately 2,200 police officers attended seminars and courses or were enrolled as students at the Academy in 1993 (*Halo 92*, 1993, #16, p. 3), while in 1995 that number had increased to approximately 4,000 police officers (*Halo 92*, 1995, #46, p. 7). The Ministry tries to stimulate education of police officers in various other ways. For example, in *Halo 92*, the journal published by the Ministry of the Interior, Cengic (1993, p. 12) wrote an article clearly intended to stimulate employees to attend various seminars and to encourage their superiors to support employees in doing so.

> The police force is composed of young people with very little experience, but with a great eagerness to learn. Our forces have lost many experienced police officers; some retired, others went to the other side [i.e., became the enemy]. But, those who stayed have to give more attention and provide greater care to the education of the young colleagues. Many of them fought at most of the fronts in Croatia, achieved military experience, but when they found themselves in a different front, in a fight against crime, difficulties occurred. It is not enough that someone wants to perform the police job. Education, as well as experience, is necessary. And it is gained by work and learning. Constantly!

Peaceful Reintegration and Joint Policing

A special organizational challenge was organizing police stations in the remaining 5 percent of the Croatian territory, located in Eastern Slavonia, which recently underwent a process of peaceful reintegration under the supervision of international forces. As was the case with other occupied parts of Croatia in the past, the Ministry of the Interior had organized police stations in exile, provided them with the necessary equipment, and selected the police officers and supervisors who would be sent there. As an integral part of peaceful reintegration, policing in this region has been performed jointly by Croatian police officers (approximately 600) and police officers of Serbian ethnicity from the (until recently) occupied region. The Ministry required that each of these police officers hold a bachelor's degree and have extensive professional training and experience. In order to prevent "decisions from the heart" and ensure "decisions from the mind," the Ministry requested that Croatian police officers whose family members had been killed in the war be excluded from the list of eligible candidates. The Ministry further established selection criteria that disqualified Croatian police officers who had any family members living "on the other side." The Ministry also demanded that the Serbian police officers selected (its future employees) not be accused war criminals. Each side had to approve the list of the candidates proposed by the other side.

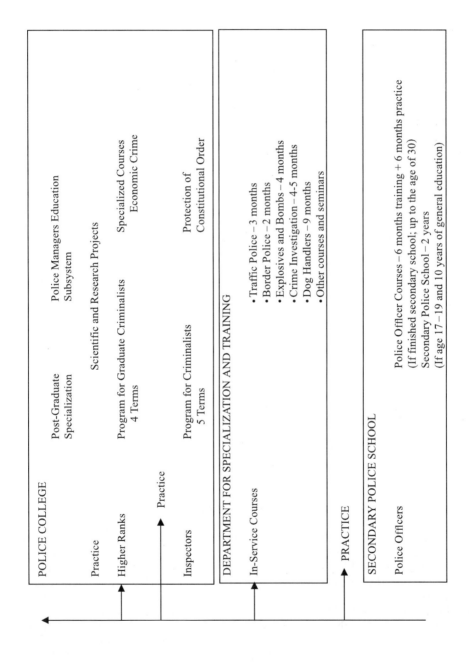

Figure 3.1. Police Education and Training in the Republic of Croatia

Source: Kuretic, 1997, p. 65

The first group of twenty Croatian police officers and twenty police officers of Serbian ethnicity completed a two-week course at the International Law Enforcement Academy in Budapest, Hungary, in the spring of 1996. The police aspect of their training included activities related to patrol. Other aspects of training included topics such as the role of the police in democratic societies, protection of human rights, the role of the police in the areas under peaceful reintegration and police ethics (*Halo 92*, 1996, #53, p. 5). In addition, a center for training police officers who were to be engaged in joint patrols was established in Erdut, Croatia.

Joint policing was carried out until the complete reintegration of Eastern Slavonia in January 1998. It was evaluated as being very successful. As of January 1998, all police officers of Serbian origin who participated in joint patrols and who fulfilled the employment requirements that apply to all employment candidates at the Ministry of the Interior became regular Croatian police officers (i.e., they became employees of the Croatian Ministry of the Interior).

Abandonment of the "General Practitioner" Model

In general terms, organizational changes in the Ministry of the Interior started after the first democratic elections in June 1990, but the process was brought to a halt in August 1990 with the emergence of the war. The Ministry "inherited the communist organization, mentality, and its ways of understanding the police forces" (Moric, 1994, p. 121). Therefore, the Ministry was aware that drastic changes were necessary to adjust to the new times, and it wanted to start with the implementation of numerous changes as soon as circumstances allowed (i.e., as soon as the war decreased in intensity).

The inherited educational system produced "a general practitioner" who was able to perform all police tasks and duties; the training system covered the whole range of police tasks and applied to all levels of responsibility. It was of utmost importance that the content of the basic police course be changed to allow for the adjustment to the new tasks (i.e. military, fight against terrorism) and to provide an opportunity for the ideological symbols of the old regime to be removed. Changes in mid-1992 (when most of the police officers returned to their regular police duties) focused on improving the educational system and resulted in the abolition of the "general practitioner" model. The model for the operational use of police officers had to be changed too.

> This educational model was matched to a model of operational use of police personnel. Since they were "general practitioners," the same police officers could deal with the duties related to public order and peace today, with the tasks related to border-crossing tomorrow, or with the tasks related to traffic safety, with the investigation of street crimes, or with the tasks related to drugs and smuggling (Moric, 1994, p. 121).

Subsequent changes in the system of police education resulted in a more specialized education and in more specialized tasks being assigned to police officers. A part of the change included a revision in professional responsibility; police officers are now held responsible for the decisions they make. As trivial as this sounds, this practice is in sharp contrast to that of the old regime in which the final decisions in cases were not made by the officers, who were required to await the decision from the "higher levels." Such practices introduced considerable friction and delays into the system and, not surprisingly, a large number of unsolved crimes was routinely recorded (Moric, 1994, p. 121).

The Open Door Policy

A dramatic change occurred in terms of openness to the public. Moric (1994, p. 121) describes the "old ways."

> Another dangerous relic from the past was the mystification of the police and its work. According to such practice, police officers and their duties and tasks were considered to be something extraordinary, hard to understand and difficult to grasp. The police were accountable to "someone higher up," not to the community. The public was not even supposed to know what the police was doing, as this could diminish police effectiveness.

Democratic societies regard the police as one of the public agencies which serves citizens, while socialist/communist regimes regard the police primarily as one of the mechanisms intended for the protection of socialist/communist regimes and values. In mid-1992, the Ministry of the Interior conducted an extensive reorganization. Some of the changes were introduced because the territorial organization of the whole country had changed; others because of the new tasks placed under the jurisdiction of the Ministry; and still others because of the increase in rates of particular crimes. In the spirit of striving toward democracy, the rule of law, openness to the public and responsibility were some of the buzz-phrases frequently used in the process of reorganization.

The Ministry had the very clear goal of changing the ways in which it interacted with citizens. For example, Deputy Minister Josko Moric (*Halo 92*, 1996, #60, p. 10–11) listed openness to the public as one of the primary changes that were carried out in mid-1992 (in addition to the termination of the "general practitioner" model of police education and the end of mystification of the police). The Ministry stated that one of the roles it needed to perform was to inform the public—both citizens and organizations—about problems and proposed solutions within the Ministry's domain. This information "can be denied only in the cases of state, military, business, or official secrets" (Kuretic, 1997, p. 16).

In fact, even in the midst of the war, and only a few months after Croatia had declared its independence, the Ministry established a new office, the Public Relations Office, and, soon afterwards, the International Relations Office. The Public Relations Office has a clipping service (and informs the heads of sectors and police administrations about articles published about the police) and publishes *Halo 92*, the monthly police magazine. *Halo 92* started as an internal magazine in the fall of 1991 (8,000 copies) and soon (in the spring of 1992) became a magazine open to the public (25,000 copies) and distributed to newsstands across the country. The magazine frequently features surveys inquiring about opinions held by police officers, citizens and the media. *Halo 92* is a police magazine designed primarily, as Bojadziski (1995, p. 211) argues, for the public. The results of several surveys of the opinions of police officers about, for example, the Ministry, morale within the Ministry, their supervisors, satisfaction with the new uniform, living conditions or health insurance were published in *Halo 92* and were thus made readily available to citizens.[6]

The Ministry made numerous changes in order to "open up" to the public. It seems that the public feels differently about the new Croatian police than it felt about the militia from the old regime. For example, in 1992 a deputy chief in one of the police stations said that "[t]oday citizens perceive the police as their own. There is no distance, like it used to be. They help us and cooperate with us" (*Halo 92*, 1992, #8, p. 10). This change in attitudes is in no small part due to the role the police played during the war, but is also due to some degree to the deliberate changes made by the police. A head of one of the police districts reported in 1992 that

> it is without any doubt that there has been a dramatic twist in the way the citizens relate to the police. Citizens see us as their friend, protector, the one who is always ready to help. In times of the most intense war activities the help from the citizens was immense, especially from the people who live in the close vicinity of the military facilities [which were held by the Serbs at the time] and who frequently, even at their own risk, did everything to enable us to conduct actions (*Halo 92*, 1992, #3, p. 4).

Organizational Changes

The country went through extensive changes in territorial administration; it has been reorganized into twenty regions. One of the first ministries that followed with its own reorganization plan was the Ministry of the Interior. Changes were made at the beginning of 1993 and twenty new police administrations were established. The legal basis for these changes, as well as for any other changes to the organization of the Ministry, were the relevant statutes and decrees passed by the Croatian Parliament. The implementation of organizational changes in the new police administrations did not proceed without difficulties because, as was emphasized by the former Minister Jarnjak (*Halo 92*, 1993, #21, p. 3), the quality of the physical plant, including the vehicles and even buildings, varied greatly across the administrations. The Ministry had developed a program to equalize quality, while adjusting for the territorial peculiarities of a particular administration (e.g., a police administration that controls sea traffic obviously has different needs from the police administration that is in charge of the largest metropolitan area). The new reforms also abolished the socialist/communist idea of small districts and strong local governments and their police stations. Because police stations were no longer established on the basis of the organizational principle of serving the district as the smallest territorial unit, police stations could now be established so that, if the organization of the territory so required, the same police station could perform tasks for several districts (*Halo 92*, 1994, #35, p. 3). As a consequence of the emphasis that the Ministry has placed on openness to the public, the organization of the Ministry is now publicly available (Gledec, 1994, p. 50–55; Kuretic, 1997), as are the names of the commanders and their deputies in each of the police administrations (*Halo 92*, 1993, #20, p. 3). The organizational chart of the Ministry is shown in Figure 3. 2.

In an effort to secure a successful adjustment to democratic changes and to the role of the police in democratic societies, the Ministry has decided that the reorganization of the police needed to be a continuous task which "should depend on the level of security and on the appropriate ways of adjusting to the arising circumstances" (Moric, 1994, p. 122). The new organization also allows the formation of patrol services. Former Minister Jarnjak argued for the formation of patrols at the district level (Lovric, 1993, p. 3).

> By assigning the same police officers to a particular district, we would like them to grow close to the citizens, which would be for their mutual satisfaction and benefit. That would, for example, mean that a police officer would ring the

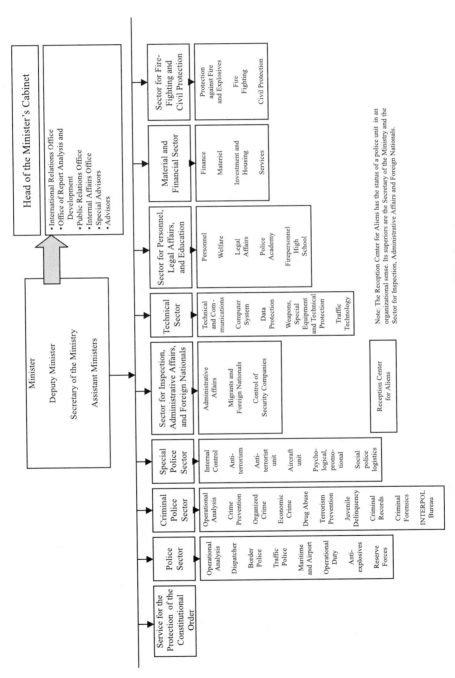

Figure 3.2. Organization of the Ministry of the Interior of the Republic of Croatia, at National Headquarters

Source: Kuretic, 1997, p. 20

bell for citizens living on his beat and wish them all the best in New Year and Christmas, but also for their birthday, and that the citizens will have the same approach toward their police officers. Therefore, we want to allow them to know each other, to develop a relationship of trust, and for citizens to cooperate with the police officer in the prevention of thefts, crimes, so that they can keep the public order and peace together.

Internal Organization of the Ministry of the Interior

There are several basic principles underlying the internal organization of the Ministry: the territorial principle, centralized organization, functional efficiency, management line structure, availability and free flow of information and organizational authority. The territorial principle embodies the Ministry's responsibility for the state's security, in a hierarchical order from the lowest unit (police station), through police administrations (the middle level), to the highest unit—the Ministry. Centralized organization requires absolute subordination of the lower organizational units to the higher units (Kuretic, 1997, p. 17). There are 20 police administrations, 195 police stations, and 113 police substations (outposts) in the current system (Gledec, 1994, p. 49; Kuretic, 1997, p. 18).

Police administrations "monitor the situation in the field, organize and direct police station activities, and take part directly in complex activities" (Kuretic, 1997, p. 35). Their internal organization depends on their size. Police stations differ in terms of the tasks they perform. They can be classified as specialized police stations (e.g., airport, maritime, traffic, or border police stations), or the so-called basic police stations which perform a variety of functions (uniformed and criminal police) or as mixed stations (uniform police, criminal police, and a possible combination of other tasks, such as traffic or border police). The organizational structure is centralized and the managerial structure is a hierarchical vertical structure (Kuretic, 1997, p. 35).

Unlike the previous police regime, police officers in this regime have to make decisions (and do not rely on the decisions made by someone "up there") and are held responsible for their decisions. Higher units control the work of the lower units, but the public and the Croatian Parliament have control over all the organizational elements of the Ministry. Zeljko Sacic, Deputy Minister for the Criminal Police, illustrates the lines of responsibility by the example of his sector (Kobali, 1997a, p. 11).

> Generally speaking, the first level of the fight against crime, and the most exposed one in the fight against it, takes place at the level of a police station. Logistic, professional, and personnel help, as well as securing the technical conditions and coordination with the other executive branches of the Croatian government, is provided to these police stations by the criminal police sectors at the level of police administrations and the Criminal Police Sector at the Ministry's headquarters. The head of the criminal police sector at the level of the police administration is responsible for crime levels and crime trends in his region. Similarly, the deputy minister for the criminal police is accountable to the Minister.

Management line structure determines that the headquarters of the Ministry give orders, coordinate, control and direct, but rarely perform regular police tasks. Finally, functional efficiency provides for the organization of the units in a way that enables them to merge tasks and facilitates their cooperation (Kuretic, 1997, p. 17).

The Ministry is organized into four basic sectors: Police Sector, Criminal Police Sector, Special Police Sector, and Service for the Protection of the Constitutional Order (see Figure 3.2). Routine police work, such as patrol, was separated from criminal investigation work at the level of the Ministry into two sectors: the Police Sector and the Criminal Police Sector. Moric (1994, pp. 122–123) argues that this organizational change was done to clearly follow the distinction between prevention (the Police Sector) and suppression (the Criminal Police Sector); the emphasis on prevention is now more pronounced than it used to be. However, because interaction between patrol and criminal investigations officers is necessary, their coordination was, according to the new police organization, achieved at the lower level, "at the level of police administrations and police stations directly in charge of the operative level." Changes in the way in which suppression is discussed (and, it is hoped, carried out) also follow basic democratic principles. For example, Deputy Minister Moric argues that there are changes with respect to the suppression of criminal interests.

[I]t is clear that the fundamental rights and freedoms of individuals cannot be at the service of the rights and freedoms of groups or the community, or so-called "higher interests." Therefore, the police cannot use suppression [of crime] to protect "higher interests," because there are no interests more important than those of individual rights and freedoms of each citizen.

The Ministry has declared prevention to be one of its most important tasks and conducts actions to support this goal (see, e.g., Bosiocic, 1996). For example, when changes in traffic laws were passed, large advertisements were posted by streets and roads reminding citizens about speed limits in the city and on the open road. The Ministry also provided fliers to citizens which contained traffic accident statistics—incident rates and numbers of dead and injured. In early December, 1997, the Ministry displayed large posters containing only one sentence: "If you drink, don't drive!" The letters in the "if you drink" part were purposely blurred, illustrating the way in which these letters would be perceived by an intoxicated person! This action was also accompanied by frequent jingles that were broadcast on TV during commercial breaks. As the holiday season approached, the frequency of these broadcasts increased.

Another prevention issue are "celebratory" explosions and shootings. Although their rates over Christmas holidays have declined since the beginning of the war—from 298 explosions in 1991 to 82 in 1994 and from 290 shootings in 1991 to 86 in 1994 (*Halo 92*, 1995, #49, p. 29)—the weapons are still readily available and tolerance for violence is still high. To decrease the number of injuries caused by "celebratory" explosions and shootings, as well as by firecrackers, fireworks and other similar items that are frequently used around Christmas time, the Ministry conducted an extensive information campaign. Posters reading "Peace and Good: The Silent Night … Celebrate without firing!" were accompanied by fliers which listed statistics about the previous year's victims and materials confiscated by the police. These fliers also contained a brief summary of the legal norms regulating the sale and use of such objects, accompanied by photographs of legal items. The same messages were broadcast on TV and radio stations and printed in newspapers. As an integral part of this action, the Ministry set up a stand (for which they won an award!) at a very popular Christmas Fair held every year in the capital, Zagreb. From the stand they distributed "Peace and Good" fliers and also showed samples of legitimate and illegitimate varieties of firecrackers and other pyrotechnic objects.

Police officers and the head of the Trauma Clinic spoke to visitors about the items and their dangers.

Organizational Changes Due to Independence

Certain organizational changes occurred by the simple fact that Croatia became an independent country. New tasks have emerged which pertain to the need to create a legitimate basis for crime-related communication and cooperation with countries around the globe. In order to manage such tasks, in 1992 the Ministry established a new office, *The Central National Interpol Bureau*. This bureau operates as an integral part of the Criminal Police Sector and is regarded to be of particular importance for a coordinated international approach toward organized crime (*Halo 92*, 1993, #18, p. 5).

Another addition to the Ministry is the *Service for the Protection of the Constitutional Order*.[7] According to its head, Deputy Minister Smiljan Reljic, this service is charged with the responsibility of providing safe conditions for peaceful living and working conditions for every citizen and its focus is on the prevention of terrorist attacks and espionage. Unlike the Federal Service for the State Security, which was responsible to a "higher authority" and was charged with the task of protecting the Yugoslav Federation and its socialist/communist regime, the new service functions in the spirit of democratic change, and is thus more open to the public. Reljic argued that the service is subject to dual control: on the one hand it tries to be as open to the public as are its counterparts in any democratic society, and, on the other hand, the service is responsible to the Croatian Parliament. Citizens are given the right to review their files, if such files had been kept by the earlier Federal Service for the State Safety. Reljic stated that this service, unlike the previous Yugoslav service, and in accordance with the democratic principle which states that citizens have the right to form and be members of political parties (Article 43, *The Constitution of the Republic of Croatia*, 1990), will not target individuals regardless of their political affiliation as long as their actions remain law abiding (*Halo 92*, 1996, #51, pp. 9–10).

One of the new departments within the Police Sector is the *Border Police Department*. As a consequence of its independence, Croatia now has approximately 3,000 kilometers of international borders. In the former Socialist Federal Republic of Yugoslavia the Army was in charge of the border and military border patrols, trained dogs, bunkers and barbed wire were used to protect the country (*Halo 92*, 1992, #2, p. 2; *Halo 92*, 1993, #18, p. 5). Democratic values have affected the way border control is carried out. Rather than charging the military with patrolling the borders with dogs and heavy weapons and shooting at anyone who tries to cross illegally, police officers must check documents at border crossings and monitor roads in the vicinity as a part of their routine police work. Naturally, in the aftermath of the recent war, the reality of the present situation mandates that the intensity of border control be commensurate with the degree of tension and hostility displayed by various neighboring countries. In other words, the practice of border control at the borders with Italy, Slovenia and Hungary presently differs from the practices at the borders with Bosnia, Herzegovina and Yugoslavia. To illustrate long-term orientation of border control practices, Deputy Minister Moric stated that

> in the control of borders we would like to establish the idea that we will not violate rights and freedoms of our citizens when they enter or leave the

country. With this procedure we would like to guarantee their rights and freedoms (Simic, 1992, p. 2).

However, former Minister Jarnjak emphasized that border control is not an easy task because some of the neighboring countries do not currently exhibit friendly leanings toward Croatia, because the terrain is difficult, and because there is a wave of immigrants to Europe as well as numerous tourists visiting Croatia (fifty million border crossings were expected in Croatia in 1994). Consequently, Jarnjak pointed out that police officers have to be psychologically and physically well prepared and well equipped for border control in order to perform their role successfully (*Halo 92*, 1994, #30, p. 4).

Another new department is the *Maritime and Airport Police Department*. It is in charge of security "in the air and on the sea", tasks which were previously under federal jurisdiction. The head of the department, Milan Pemper, noted that some of the organizational elements utilized by the department were adopted or adapted from those of democratic countries. Among other characteristics, considerable emphasis is placed on polite, professional conduct and a significant part of in-service training includes learning foreign languages (*Halo 92*, 1992, #12, p. 7).

Organizational Responses to Changing Crime Patterns

The second set of changes in organization relates to the emergence of new forms of crime and changing patterns in the quantity and quality of crimes committed. New departments within the Criminal Police Sector include the *Organized Crime Department* and the *Department for War Crimes and Terrorism*. The latter was established first as a section within the Criminal Police Sector in 1991 and then as a department in 1992. It was created in response to terrorist acts and the war waged against Croatia. Information about war crimes was collected in various police stations, but there was a need for the systematic collection of information and evidence related to those crimes at a centralized location. The primary role of the newly established department has been to investigate crimes such as terrorist acts, war crimes and crimes against humanity, as well as to collect evidence that will be used in subsequent prosecutions and trials. This department cooperates with and provides assistance to international institutions, such as Helsinki Watch and the International Tribunal in the Hague (Gataric, 1994, pp. 68–70). Deputy Minister Gledec (1994, p. 51) argues that the Croatian state approached the issue of war crimes as a state that respects the rule of law because it allowed international institutions and the public full access to its case files.

As a result of an increase in corporate, occupational and organizational crime in Croatia and other European countries in transition, the *Organized Crime Department* and the *Economic Crime Department*[8] were established as departments separate from the General Crime Department.

Responding to the Dynamics of the Society

The Ministry also established some additional departments in response to the dynamically changing social conditions at the time. For example, as the flow of displaced persons and refugees from Croatia and Bosnia and Herzegovina began, the Ministry established the *Department for Migrations and Foreign Nationals* within the Sector for Inspection, Administrative Affairs and Foreign Nationals. The department's basic duties involve the control of "the lawfulness of the police operational procedures with foreign nationals who are in any way illegally residing in the Republic of Croatia, as well as of organizing and implementing expulsions of aliens who have been ordered to leave Croatia by a court of law" (Kuretic, 1997, p. 31).

Private security companies were not allowed to exist or operate in the socialist/communist regime. Upon the initiation of democratic changes in the legal system (at the core of which was the privatization of various sectors of the economy), citizens were allowed to establish such companies to provide detective and security services to the public. The Ministry of the Interior formed a separate department, *The Department for Monitoring Security Companies*, which conducts inspections of security companies and determines whether a particular private security company fulfills the legal requirements.

In the old regime, firefighters operated under the auspices of the local district government. Because socialist/communist values are no longer accepted, small districts (which had local governments as the basic units of self-management), now have neither the same authority nor the same role in the current system. Consequently, in 1993, fire protection was appended to the Ministry of the Interior. Furthermore, civil protection (which was regulated as a part of the system of defense and self-management at the local level and was under the command of the Federal Ministry of Defense in accordance with socialist/communist principles) was also placed under the control of the Ministry of the Interior. The Ministry formed the *Sector for Fire Protection and Civil Protection*. It performs the following functions: "preventive protection of human lives, health, and property from fire;" "control over the manufacture, trade, and use of explosive materials;" "control of the transportation of dangerous substances in road traffic;" "repressive measures for the protection from fire and other disasters;" and "protection from immediate danger and prevention of disasters and accidents" (Kuretic, 1997, pp. 33–34).

Facilitating the Flow of Information

A greater openness to the public, a sense of responsibility to provide various state institutions with data, as well as the wish to provide an analysis of crime trends for the whole country (including violations of norms committed by police officers!), were some of the factors which led to the establishment of the *Office of Report Analysis and Development* within the Minister's Cabinet. An analysis of crime trends is conducted at the Ministry level and is published regularly in the "Yearly Report about the Work of the Ministry of the Interior of the Republic of Croatia." Police administrations collect crime statistics for their territories and hold press conferences, and newspapers periodically publish crime statistics (including statistics about crimes committed by police officers). The *Office of Report Analysis and Development* is vertically connected to the analytical services of the Police Sector, the analytical services of the Criminal Police Sector and the operational services of all police administrations. Interestingly, analytical services are not a structural part of the organization of any police station. However, as was emphasized by the head of the department, Miroslav Granic (*Halo 92*, 1996, #57, p. 7), analytical tasks are expected from the police stations partly because of the role they have with respect to the public. That is, police stations must conduct press conferences to inform the public of crime trends in the territory under their jurisdiction and must send reports to the police administrations. Police administrations are in a somewhat more difficult position because they must not only record and analyze crime trends, but also propose measures to be taken by police stations and follow up on measures actually taken.

In pursuit of yet another method of establishing better contact with citizens, granting them an opportunity to exercise control over the work of the police and of providing better control over the discipline of police officers, the Ministry has established another new office—the *Internal Affairs Office* within the Minister's Cabinet. A more detailed discussion about the role of this office will be provided in the section that deals with professional challenges.

OPERATIONAL CHALLENGES

Adherence to democratic values, intertwined with the influence of the war, created a set of operational challenges for Croatian police officers. Unlike police forces in some other countries in transition, Croatia did not experience a long-term shortage of police officers. Its problems were of a different sort, namely, how could the police be involved in the defense of the country and simultaneously perform their regular police functions? Administrators from one police administration particularly affected by the war remarked that "748 police officers left this police administration; some of them are still on the front. Because of the lack of police officers, the police are not able to 'cover' the whole city. That facilitates crime rates" (Mirosavljev, 1992a, p. 5). In fact, the war caused specific forms of criminality to increase— property crimes, crimes of violence and weapons smuggling. Furthermore, very few people were left to deal with prevention while the majority had to investigate already-reported crimes (*Halo 92*, 1992, #2, p. 7). The situation became even more challenging at the beginning of the war in Bosnia and Herzegovina in early 1992 when a few million people fled Bosnia and Herzegovina and arrived in Croatia. This, in turn, resulted in an increased number of people in need of police services. Furthermore, "at the beginning of the war all inmates from several large prisons in Bosnia and Herzegovina were released. Many of them arrived in Croatia as refugees and were caught in criminal activities" (Jarnjak, 1993, p. 3).

It seems that the fear of "ordinary" crime did not increase in most of the unoccupied territory, but citizens were fearful of terrorist attacks, bombings and war crimes in general. In fact, an analysis of safety issues in Croatia performed by the Kroll Consulting Company suggests that violence is relatively infrequent; Croatian cities, such as Zagreb, are very safe, and public evaluations of police responses to crimes and violence were very satisfactory (Butkovic, 1997b, p. 14).

The city of Split was strongly affected by the war due to its proximity to the border with Bosnia and Herzegovina. Split was a major port for humanitarian aid to Bosnia and Herzegovina, and often the first large city in Croatia that refugees from Bosnia and Herzegovina reached after they left their homeland. It was, therefore, not surprising that the citizens of Split experienced an increase in the fear of crime. A survey conducted in 1992 revealed that the fear of not only terrorist attacks but also of street crime was stronger in Split than it was in the capital, Zagreb, or in two other large cities (one of which had a front line passing through its suburbs). "[S]ubjective feeling does not have to correspond to the objective conditions, but Split had a reputation as a city where crime flourished, such as explosions, thefts, violence, drug use, smuggling of alcohol, cigarettes, cars" (Mirosavljev, 1993, p. 8). Objective measures suggest that, in addition to the fear of crime, real crime rates in the city probably increased; Split experienced the fastest growth in the number of crimes reported to the police in 1992 compared to 1991—a 500 percent increase in the number of murders, an 80 percent increase in the number of aggravated thefts, and a 200 percent increase in the number of motor-vehicle thefts reported to the police (Singer & Cajner, 1993). The Ministry reacted by sending additional police officers to help deal with the rising crime rates.

However, it is interesting that Durica Franjo, the head of a group of police officers sent to help the Split Police Administration deal with these problems, emphasized the necessity of re-organizing things and focusing on prevention as a new, more democratic orientation of the Croatian police.

> [T]he times when the police officers were firefighters have passed. If someone's house explodes [as a result of the bomb—a terrorist act] and we bring him [the victim] the offender tomorrow, it will mean nothing to him. We need to prevent the explosions and only then we've done the right thing! (Mirosavljev, 1993, p. 8)

Patrols were increased, preventive measures intensified, analyses conducted and controls (which targeted carefully selected places at certain times) were carried out. This combination of measures helped reduced crime rates (*Halo 92*, 1993, #18, p. 8).

As discussed earlier, even the police officers left to perform regular patrol duties frequently had very limited knowledge and experience in police work. This is not surprising, because the majority of police officers of Serbian ethnicity had left the police. When they left at the beginning of the war, very few people were knowledgeable about police work.

Street Crimes

Even physical conditions during the war were helpful to street criminals; one administrator noted that "black-outs [the whole city was blacked-out as a preventive measure against air attacks] favor crime" (Buble, 1992, p. 29). Furthermore, weapons were readily available and the level of stress was high during the war; the number of crimes committed with the use of firearms increased by five times in 1991 (*Halo 92*, 1992, #6, p. 8). Ivan Tounec, the head of the Operational Sector within the Police Administration Sisak, connects these facts in the following way (Balokovic-Krklec, 1993, p. 7):

> Crime statistics warn us that violence ... takes more brutal forms. The crime rates are increasing .... Unfortunately, people are more likely to settle accounts, even for the most minor things, through violence. We are witnessing almost every day that even minor traffic situations are resolved with an attack on the police officer or on the participant in the traffic accident. Public places have become media for the public expression of emotions, frustrations, tensions, failures, or successes.

When actual crimes reported to the police are examined, the data show some peculiarities relative to other countries in transition. Dujmovic (1996) compared Interpol crime rates for Austria, Bulgaria, the Czech Republic, Hungary, Germany, Poland, the Russian Federation, the Slovak Republic, Switzerland, Slovenia, Ukraine and Croatia for 1993 and 1994 (see Table 3.1). In regard to the crimes reported to the police, Dujmovic concluded that, "Croatia had a crime rate of 1,609.9[/100,000], in 1993. Out of all of these states only Ukraine had a lower rate of crime than Croatia .... Croatia had a very low rate of crimes (1994=1,334[/100,000]), and the largest decrease in crime known to the police in 1994 compared to 1993" (Dujmovic, 1996, p. 387). Furthermore, among these countries (except for Bulgaria), Croatia also had the highest percent of solved crimes.

Table 3.1. Crime rates in some Central and Eastern European countries 1993–1994.

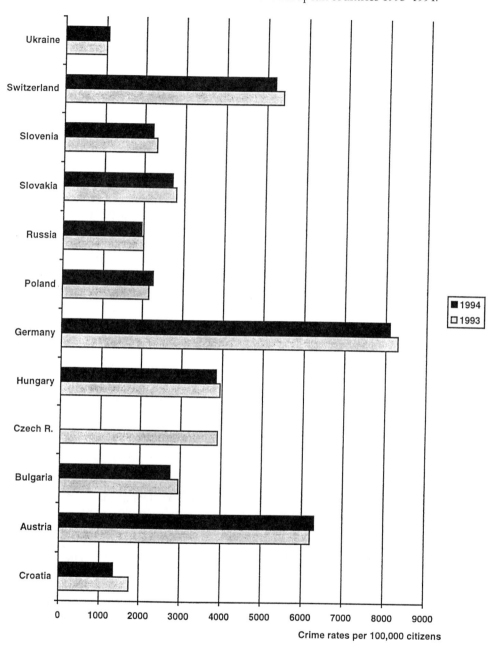

Source: Dujmovic, 1996, p. 387.

Singer (1996) reported that there was no general upward trend in the period 1990–1995 for crimes known to the police and to district attorneys. However, unlike most other countries in transition, statistics of crimes known to the police in Croatia showed a dramatic increase in total number of crimes against people and property in 1992 (see Table 3.2). Most of this increase in crimes was directly related to the

Table 3.2. Indicators of crimes known to the police and indicators of other important security issues 1991–1996

	1991	1992	1993	1994	1995	1996
Crimes Known to the Police	64,294	91,712	77,275	64,051	63,015	58,364
% of Solved Crimes	50.1	56.8	62.6	69.1	72.5	75.2
Violent Crimes	1,776	1,750	1,71	1,629	1,642	1,530
– Murder	437	387	218	143	166	122
Property Crimes	43,218	57,679	50,829	36,412	33,466	28,609
– Robbery	864	743	607	383	426	490
– Aggravated Theft	26,986	41,514	34,611	26,978	25,256	19,128
Crimes against Public Security of People and Property	2,785	6,047	2,475	1,633	1,503	1,352
Economic Crimes	3,399	4,705	4,715	5,470	8,302	8,581
Organized Crimes	2,565	3,614	3,556	4,127	6,221	6,397
Drug Abuse	484	895	911	857	944	2,350
Seizure of Drugs	450	965	1,290	1,669	2,483	3,052
Crimes against National Security	1,328	6,143	1,475	790	1,291	285
Disorderly Conduct	18,235	22,726	30,363	34,264	30,582	30,153
Traffic Accidents	53,297	56,815	58,188	62,120	61,656	59,420
– no. of killed persons	1,020	975	855	804	800	721
– no. of injured persons	15,845	17,516	15,596	17,679	17,665	16,182
Persons Crossing Croatian Border	4,880,811	46,126,381	62,248,774	8,368,541	74,503,733	85,593,047

Source: Kuretic, 1997, p. 75

war. For example, the Ministry reports (*Halo 92*, 1992, #13, p. 11) that there were over 4,500 such acts committed by the enemy and over 400 crimes against humanity and war crimes in 1992. Unfortunately, it was impossible to know exactly which of these crimes of violence were a direct result of the war and which were not (e.g., a man killed by his neighbor after an argument), except for crimes of violence related to the war by definition. Kovco (1996) compared crimes of violence, the definition of which requires terrorist acts or the state of war (e.g., genocide, war crimes), to "regular" crimes of violence (e.g., murder, assault). The data show a drastic increase of the number of persons accused of war-related violence in 1992 compared to the number of persons accused in 1991 (an increase of almost 2,000 percent), followed by a similarly drastic decrease in 1993. On the other hand, the number of persons accused of non-war-related violence did not change from 1991 to 1992 and it increased by 20 percent from 1991 to 1993 (Table 1 in Kovco, 1996, p. 397).

When crimes known to the police during the war years were compared, the data showed that crimes reported to the police increased drastically in 1992 relative to 1991 (Table 3.2). Although war activities were not over in 1992, this was, at the same time, the year in which most police officers returned to their regular duties. It is quite possible that the increase in crime reported to the police may be explained, in part, by the increased presence of the police officers in police stations or by a change in criminal policies. There was, as well, a gradual increase in the rate of crimes solved—from 50 percent in 1991 to 75 percent in 1996 (Table 3.2). This increase may be explained by the fact that a larger number of police officers were working on these cases after 1992, by a decrease in the number of crimes reported, and/or by a greater emphasis on prevention.

When robbery is selected as a typical war-related crime, the rates for the period 1990–1995 clearly suggest that robbery rates were directly affected by the war. For example, there is a drastic increase from ten robberies per 100,000 inhabitants in 1990 (no war yet, but terrorist acts already underway) to eighteen robberies per 100,000 inhabitants in 1991 (Dujmovic & Miksaj-Todorovic, 1996, p. 361), and a gradual decrease from 1991 to 1995 thereafter.

Aggravated theft numbers also show a similar pattern: first a drastic increase from 1991 to 1992 and then a gradual decrease from 1992 to 1996 (Table 3.2). The percentage of persons (both juveniles and adults) whom the police charged with aggravated theft doubled from 1991 to 1992 (Singer & Cajner, 1993, p. 12). Singer and Cajner (1993, p. 12) explain that "the shortage of material goods, inflation, the rationalization of distribution and expenditure characteristic of war and post-war times, very probably influence attitudes toward property and ownership which facilitate the increase in property crimes." Indeed, according to the Croatian Prime Minister (*Halo 92*, 1993, #16, p. 17), the inflation rates were extremely high; in 1992, the inflation rate was about 30 percent per month or 2,600 percent per year. Furthermore, crime rates for property crimes reported to the police appear to be inversely related to salaries in the same period. Property crimes reported to the police decreased gradually from 1992 to 1996, while the average salary grew from approximately 175 German marks in 1991 and 1992 to 674 German marks in Fall 1997 (Markovic, 1997, p. 5).

Crimes of violence, especially murders, also experienced a slow decrease from 1991 (when 437 murders were reported to the police) to 1996 (122 murders; Table 3.2). Kovco (1996) reports that the crimes of war-related violence greatly increased between 1991 and 1992. It is not surprising, then, that the increase in the numbers of persons charged with murder from 1991 to 1992 was higher in cities close to the front lines; for example, the increase in Split was over 500 percent (Singer & Cajner, 1993, p. 10). Singer and Cajner (1993, p. 10) studied crime rates in Croatia in 1991 and 1992 and concluded that, it is known that war decreases the level of respect toward the physical integrity and lives of people. People are more likely in such times to use violence for the resolution of conflict situations or the fulfillment of certain needs .... Therefore, it is very likely that the registered state of violent crimes is directly related to the war in our country and that only well organized and various preventive measures can prevent a further increase in violence.

Interestingly, Singer (1996) reports that the percentage of those who have criminal records, especially for insults, property crimes, and violent crimes, are unusually high among displaced persons returning to their homes. Displaced persons are an especially challenging population to interact with because they typically have waited for several years to return to their homes, most of which have been completely destroyed. They are more likely to be frustrated and angry with the current regime and the police. Therefore, police officers who undertook policing jobs in the regions freed by military/police actions in 1995 encountered additional problems. However, frustration and anger will diminish over time, and it is likely that the displaced population's contribution to crime rates will tend to disappear.

The increased level of violence, a change in attitudes toward violence and the availability of weapons affect police officers as well. For example, Aretto Koscica, a young police officer, describes problems he experienced during the performance of his everyday duties (Horvat, 1995, p. 16).

When I catch someone violating the law, many will today use democracy and their participation in the war as an excuse .... I stop a motorist who ran a red light and he tells me "You are stopping me because of my nationality." It is a dangerous job. Many today own weapons and bombs and I cannot play an American police officer from the movies. Yet he will tell me, "I'll activate a hand grenade ... and you think about it!"

The dark numbers of crime tend to be relatively high in countries in transition. For example, Siemazko (quoted in Hebenton & Spencer, 1997) estimates that approximately 70 percent of all crimes in Poland are not reported to the police, and that one reason for such non-reporting is that people are skeptical the police will be able to do anything. The International Victimization Survey has been conducted and the results suggest that, for example, non-reporting rates for burglary are still high, especially in Albania, Poland and the countries of the former USSR (Zvekić, 1996, p. 49). Croatia was included in the survey in the spring of 1997, but, unfortunately, results were not yet available.

Economic/Organizational Crime and Organized Crime

Economic/organizational crime and organized crime known to the police exhibited a trend quite different from the trend recorded for violent and "regular" property crimes in the period from 1991 to 1996—they are on the increase (Table 3.2). It seems that the structure of crimes known to the police is changing, and that economic crime and organized crime are taking a more dominant position (*Halo 92*, 1997, #68, p. 5). However, a part of this increase may be due to the increased attention given to these types of crimes from 1996 onward and to an increased level of skills in dealing with such crimes. Indeed, *Halo 92* reports (1996, #52, p. 50), for example, that in 1996 special attention was paid to economic and organized crime.

Organized Crime

According to Ivan Penic, the then Croatian Minister of the Interior, organized crime in 1997 plays an important role in the overall structure of crime known to the police. He further points out, though, "I cannot see the danger now that we will become similar to Belgrade, Moscow, Warsaw, or the cities of some other countries in transition in terms of the extent of organized crime" (Kobali, 1997b, p. 5). Hebenton and Spencer (1997, p. 5) argued that the lack of police resources (in terms of personnel and policy making) was a precondition for the flourishing of organized crime in Russia. The situation, with respect to this condition, is quite different in Croatia; but organized crime is present in Croatia nevertheless. Deputy Minister Zeljko Sacic reports that there are three or four criminal organizations, all of which are in relatively early organizational stages. However, unlike a typical "Mafia" organization with a vertical chain of command, these organizations have no such hierarchical order and prefer to "borrow" professional criminals specializing in particular crimes (*Halo 92*, 1997, #68, p. 5). The police are aware that these crimes may be quite serious and are concerned about "the better training and increased professionalism of these offenders, their connections with people abroad, and their adaptability to other crimes, especially drug trafficking and money counterfeiting" (Nadj, 1996, p. 437).

Typical crimes committed by organized groups in Croatia include stealing vehicles and smuggling them to other countries or selling them for parts within Croatia. Sacic reports (1996, p. 661) a case investigated in 1996 in which a group of ten people was under investigation for motor-vehicle thefts, falsification of documents, illegal possession of weapons and explosives, and drug use. Sacic notes that the investigation was extended to corruption charges against a number of police officers working on the borders and against a number of customs agents. Former Minister Ivan Jarnjak reported that eight police officers and eighteen customs agents were investigated and reported to the district attorneys because of their role in organized crime in 1995 (*Halo 92*, 1996, #52, p. 17).

Another form of organized crime is prostitution. The process of opening the international borders of East European countries resulted, over the last few years, in an increased number of women from other countries willing to be involved in prostitution. Other types of organized crime recorded in Croatia include smuggling (alcohol, drugs, coffee, cultural inheritance, etc.), organized illegal border crossings related to the falsification of Croatian documents, falsification of phone cards (Nadj, 1996), counterfeiting, protection services and drug dealing (Sacic, 1996). As a consequence of the war and less intense controls in that area, smuggling cultural inheritance objects has increased, and artifacts can be found on the illegal international market (Nadj, 1996, p. 438).

Serious challenges faced by the police in 1996 with respect to organized crime are related to either a complete lack of legal regulation or inappropriate solutions in legal documents. Consequently, as Sacic argues (1996, pp. 624–626), the police had a much better legal basis for the investigations of lower level perpetrators of crimes, but rarely for the heads of these organizations (and even less frequently for their financial sponsors). Interestingly, Sacic argues (p. 626) that

> [s]ome of these difficulties still can be found in inappropriate legal regulation, but also partially in the level of the professional and technical training of the police officers, as well as the district attorneys and the courts, to fight against these forms of crime.

The Ministry tried to provide assistance by organizing a special computer database dedicated to organized crime (which connects all the police stations, administrations and headquarters into the same computer network), by providing help in exchanging information with foreign governments about methods used and applicable statutes and by providing specialized training for police officers (Sacic, 1996, p. 628). A list setting forth crimes indicative of organized crime was also designed. One of the consequences of inadequate regulation was the inability to provide crime statistics about the roles of offenders in organized crimes known to the police.

Criminal statutes lacked a definition of organized crime. This definition was finally provided in the most recent set of legal changes that became effective on January 1, 1998. "A group" is defined as an organization of at least three people connected because of a permanent or temporary intent to commit crimes in which each of these individuals provides his or her own contribution (Article 89, *1997 Criminal Law Statute*). The new law also provides a definition of "criminal organization." Crimes committed by an organized group or a criminal organization are considered to be more serious than the same crimes committed by two individuals (e.g., kidnaping, Article 125; force, Article 128; aggravated theft, Article 217; robbery, Article 218). New crimes were added to the list by reforms in 1996 and 1997. For example, the new *Criminal Law Statute* (1997) criminalizes illegal transport of individuals across borders (Article 177); international prostitution (Article 178); money laundering (Article 279);[9] illegal drug use including the production and distribution of drugs (Article 173); the production, possession, and sale of objects intended for the falsification of documents (Article 314); and the avoidance of customs (Article 298). Veic (1996a) argues that, because of the important role government officials play in organized crime, it is important that changes in legal regulation include changes/expansions of definitions to include crimes committed by public officials, including police officers. Misuse of official position and privileges (Article 337) and misuse of carrying out a state function (Article 338) are now included in the law. New crimes that may be of potential interest to organized crime have also been added to the list, such as illegal organ transplantation (Article 242) and copyright violations (Articles 229–232).

As of January 1, 1998, additional changes were introduced into the criminal procedure statutes. These changes regulate some methods used in police work which seriously infringe upon citizens' rights, but can be useful in the investigation and prevention of organized crime. The law states that control of these measures rests with the district attorney and the investigative judge (Article 180, *1997 Criminal Procedure Law*). The goal was to protect the rights of all accused persons (and not just the persons involved in organized crime!) in the criminal process as much as possible by providing control over the police. The possibility that the police could decide, on their own, to jail a suspect or search a place of residence, which existed in the earlier socialist/communist legal solutions, was abolished even before these most recent changes. The new regulation provides a significantly greater degree of judicial control over the work of the police—warrants are issued by investigative judges. If the investigative judge does not approve the district attorney's request for a warrant, the issue is resolved by a tribunal composed of professional judges.

The *1997 Criminal Procedure Law* lists additional measures especially designed for a limited number of crimes, including organized crime and, interestingly, acceptance of a bribe by a government official. These measures provide the legal basis for a "temporary limitation of Constitutional rights and liberties" for the purpose of gathering information and evidence for a criminal trial. These include wiretapping, surveillance, secret following and recording, and use of undercover agents (Article 180). There are, however, problems related to these measures, which may seriously affect their effectiveness. The decision-making process in complex cases typically does not end in one day and involves a large number of persons. These two factors may cause information about a pending warrant to leak, which puts the efficiency of such measures in question.

In 1993 approximately 3,000 motor vehicles were stolen in Croatia and 987 vehicles were recovered (*Halo 92*, 1994, #32, p. 19). The problem is much more difficult then before because these crimes are now committed by organized criminals and not by individuals. This problem has been attacked by providing appropriate training for uniformed police officers, detectives and police officers at the borders, by road checks in search of suspicious vehicles and forged documents, and by the establishment of a separate department at the Ministry. The Ministry also expects help from the public and citizens are asked to report "suspicious" vehicles (*Halo 92*, 1992, #13, p. 3).

The roles played by other states in the fight against crime and in international cooperation are particularly important in the domain of organized crime. The Croatian police report that they cooperate closely with the Slovenian and Austrian police, especially with respect to stolen vehicles and smuggling. For example, the Croatian police, in a joint effort with the Slovenian police, disrupted one of the smuggling channels in which a group stole cars in Italy and Switzerland and smuggled them to Bosnia and Herzegovina through Slovenia and Croatia (*Halo 92*, 1997, #62, p. 14). On the other hand, the majority of vehicles stolen in Croatia, according to the Croatian police, typically disappear somewhere in other East European countries, yet cooperation with the police from these countries is not productive (*Halo 92*, 1995, #36, p. 4). It seems that international cooperation is particularly important because "Croatia is starting to appear as a transit country toward Eastern Europe" (Horvat, 1994, p. 19). The Croatian police also participate in various international conferences, such as one about new forms of crime in Europe organized by the Council of Europe held at the Croatian Police Academy in 1996 (*Halo 92*, 1996, #59, p. 14–16).

Similarly, the role of international cooperation is very important for the process of curtailing drug smuggling and trafficking. For example, Croatia is on "the Balkan Heroin Route" which, according to some estimates (*Halo 92*, 1995, #49, p. 46), supplies between 80–90 percent of all heroin used in Europe. This route was temporarily changed during the war (*Halo 92*, 1995, #40, p. 13). Furthermore, some of the Croatian ports (Split, Rijeka) serve as entrance points for drug smuggling into Europe. There was a dramatic increase in the number of drug-related crimes in which drugs were confiscated from 496 in 1991 to 1,873 in the first ten months of 1995 (*Halo 92*, 1995, #49, p. 47). This increase is probably due to the increased attention and skills devoted to drug-related crimes and to more numerous attempts to transport drugs through Croatia.

The Ministry also organizes specialized courses (e.g., *Halo 92*, 1995, #49, p. 46). In 1995 alone, the Ministry organized six seminars on this topic which were scheduled in addition to the regular specialized training offered at the Academy. One of the courses held in 1996 provided Croatian police officers and Slovenian police officers with an opportunity to learn from agents from the US Drug Enforcement Administration about the methods the DEA uses to deal with drug-related crimes (*Halo 92*, 1996, #57, p. 20).

Economic Crime

Economic crime is another category which has shown an upward trend in the statistics of crimes known to the police (Table 3.2), especially in the large cities. Nadj argues (1996, p. 439) that the increase was due to a number of factors, including conditions in the (market) economy, the appearance of new entities on the market, the change in the type of ownership (from state/public to private), the increase in the number of privately owned companies, the lack of financial responsibility and the lack of (or limitations on) existing legal regulations. One of the characteristics of economic crime in Croatia is its swift adjustment to new social and economic conditions (which, of course, creates problems for police officers). For example, in 1992 Croatia received humanitarian aid for displaced persons and refugees from Croatia and Bosnia and Herzegovina and new forms of property crimes developed—the theft of the humanitarian aid in large numbers (*Halo 92*, 1992, #12, p. 7). Companies bought smuggled property, did not pay customs duties and fixed taxbooks or other company records. The majority of economic crimes known to the police may be classified under fraud and falsification of company documents, which constitute approximately 60 percent of all economic crimes (*Halo 92*, 1994, #27, p. 5).

Although "Croatia is still considered to be *virgo intacta* with respect to sophisticated forms of economic crimes" (Horvat, 1995, *Halo 92*, #48, p. 17), new forms of such crimes have been investigated by the police: computer crime, violation of copyright laws, violation of trademark laws, money laundering and ecological crime. The Ministry conducted an intensive police action in 1993–1994. Damage to the state uncovered in that action alone was several hundred thousand German marks (*Halo 92*, 1994, #27, p. 5).

Most economic crimes were registered as "fraud" in the past. The new *1997 Criminal Law* expanded this definition by introducing a new set of organizational and occupational crimes, e.g., fraud in commerce (Article 293), fraud in public service (Article 344), cornering the market (Article 288), conducting negligent business transactions (Article 291), misuse in business transactions (Article 292) and entering into harmful contracts (Article 294).

An additional problem relates to violations of copyright laws and laws regulating the protection of authorship. In comparison with western democracies (e.g., US, Canada, Germany), Croatia and other former socialist countries have an extremely high rate of copyright violations. For example, some estimates suggest that over 70 percent of the computer software in Hungary and over 90 percent in Slovenia and Croatia is illegal, compared to less than 50 percent of the computer software in western democracies (*Halo 92*, 1997, #72, p. 43). This is not surprising in light of the low standard of living in the former socialist/communist countries, especially in comparison with the standard of living enjoyed by western democracies. A lack of technical skills and a lack of legal regulation in the area of copyright law are some of the problems Croatian police officers have experienced in dealing with copyright violations. The new set of laws that became effective on January 1 prohibits the use and distribution of illegal software, as well as the illegal distribution of licensed software. Furthermore, these laws also provided a basis for the criminal justice system to prosecute individuals who commit organizational, occupational, or economic crime using a computer.

Equipment

Equipment presents another challenge to policing in Croatia. As a consequence of the war, sixty-five police stations were destroyed completely, as was 60 percent of the equipment (*Halo 92*, 1992, #7, p. 3). Because some of the equipment ended up in the hands of the enemy, the Ministry decided to change both its technology and the organization itself (*Halo 92*, 1996, #61, p. 6). Unlike most of the other former socialist countries, which still have old equipment or only gradually changed their equipment, Croatia is in a unique situation—most of its equipment was destroyed and, since the police had to perform military functions, funds were made available to the police. However, the selection was limited due to the international arms embargo imposed upon the countries of the former Yugoslavia. Because of the war, of course, the shopping list favored weapons over forensic lab items, for example. One of the problems is that as a result of this embargo, weapons were not purchased on a systematic basis. A present task is to strive toward standardization of available weapons (*Halo 92*, 1996, #61, p. 8). A new IBM computer system which can provide access to 1,600 terminals and serve as a database for records about persons, objects and events was purchased in 1993 (*Halo 92*, 1994, #26, p. 5). A large percentage of the budget is spent on the maintenance of existing equipment and vehicles, so the acquisition of new vehicles has decreased (*Halo 92*, 1996, #26, p. 5). This is how Davorin Pavlovic, the head of the Technical Sector, compares conditions in the old socialist/communist regime with conditions in the 1990s (Rako, 1996, p. 6):

> At the time when the Croatian police were formed, the level of equipment was very low, except for telecommunications .... [T]he cars in the motor pool could be counted using fingers .... Today we have satisfactory equipment at all levels of the Ministry [i.e., the headquarters, police administrations, police stations]; the number of motor vehicles has increased three to four times. Furthermore, the special police equipment was rather limited; the computer network was not well developed, while today we have approximately 3,000 personal computers. All the police stations, including police outposts [substations] are connected to the mainframe computer.

Interestingly, the police logo placed on all police vehicles, including new ones, won first prize at an international competition organized by the *Law and Order* magazine for best design on a police vehicle outside the USA (*Halo 92*, 1994, #32, p. 3). However, it will take a long period of time before buildings and equipment are completely repaired; for example, to repair 10 percent of its most important buildings, the Ministry spent 3.5 million German marks (*Halo 92*, 1994, #15, p. 12).

As a part of the process for eliminating the mystery associated with the police and promoting greater openness to the public, the annual police budget is now publicly available information (about US$500 million to US$700 million annually), as are the major budget items. Approximately one-half of the budget is spent on salaries (*Halo 92*, 1996, #53, p. 18) and only 6 percent in 1997 was used for "capital investments"—i.e., new equipment and maintenance of buildings. The headquarters of the Ministry moved to a new building in the fall of 1997. This is how the Ministry describes the general level of the quality of the Ministry's equipment (Pavlovic in Rako, 1996, p. 7):

> The level of the equipment at the disposal of the Croatian police is high compared to other police across the world. The quality of the equipment, the selection of the best world-known organizations as manufacturers is at the same level as in other police forces in [Western] Europe. We use an IBM computer system that is present in almost all the police forces across the world; we also use Motorola for communication, just like other police forces in the world do; finally, we also own a Multanov radar control system and digital phone exchanges. During the equipment acquisition we followed the organizational changes in the police and always made decisions to buy quality products. It is better to have fewer items of equipment, but that all of these items are technologically the most advanced ones on the market.

PROFESSIONAL CHALLENGES

Changing Attitudes and Perceptions

A key factor in the transition toward democracy is a change in the attitudes, perceptions and performance of the police. Former Minister Jarnjak and other highly ranking police officials initiated a discussion of the professional behavior of police officers, with particular emphasis on a polite and professional approach toward citizens and on the need to develop healthy, critical, but respectful attitudes toward coworkers. Former Minister Jarnjak said in an interview (Piskor, 1993, p. 5),

> We are trying to achieve that every Croatian police officer be polite, professional, and follow the rule of the law. I am not saying that all of the police officers at this moment are following these directions. But I have to tell you that we are involved in the selection every day and that we decide to fire those who do not follow these guidelines via a summary process. Every morning we have a report that tells us in which of the police administrations disciplinary measures have been conducted because the police officer did not behave as he/she should have .... [I]t is very important to the Croatian police that, on the one hand, police officers be correct and polite toward the citizens, and, on the other hand, that they are respected and accepted by the Croatian people.

Police officers confirmed these tendencies by telling me that the Ministry sends directives about professional police behavior, that these are discussed during courses at the Police Academy and that police officers who were rude toward citizens have been disciplined by their supervisors (Kutnjak Ivkovich, 1996).

One principle the Ministry is trying to establish to provide for a modernized and democratic police force is the principle of independence. The concept is that professionalism and the independence of the police would be jeopardized if political parties were allowed to influence the work of the police. The police aim to follow the Constitution and other laws in their work, rather than take directions from political parties (*Halo 92*, 1994, #30, p. 4); "police officers will try to perform their job in a polite, professional way following the laws" (Jarnjak, 1995, p. 3). Consequently, in order to achieve greater professionalism in their work, changes in the *Law on Interior Affairs* in 1991 prohibit police officers from becoming members of any political party.

A rule that governed the behavior of police officers under the old socialist/communist regime prohibited police officers from holding a second job. The current legal rules contain the same restriction. Veic argues (1996b, p. 131) that such regulation discriminates against police officers because it may seriously worsen their living standard. Indeed, one of the challenges facing police officers is their relatively low standard of living. The salary range in 1993 was from 217 German marks (about US$130) per month for younger police officers to 465 German marks (about US$280) per month for a higher inspector (*Halo 92*, 1993, #25, p. 8). The average salary of police officers in 1996 ranged from US$200 to US$300 monthly (*Halo 92*, 1996, #53, p. 18). This places police officers' salaries below the average industrial salary, which is about US$350–US$400 monthly (Markovic, 1997, p. 5). Croatia may be on the verge of experiencing what Poland experienced already—"the poor pay of the police force not only resulted in policing having little in terms of status but also in many good police officers leaving" (Hebenton & Spencer, 1997, p. 8). The Croatian police are well aware of this problem. "We'll try to do something with respect to this problem because, unfortunately, there are already cases in which police officers leave their job due to low salaries and difficult conditions of the police work" (Osrecki, 1993, p. 8). A new set of laws regulating the police will have been drafted by the end of 1998, and one of the proposed rules abolishes the prohibition of outside employment. Veic (1996b) suggests that any outside employment should be approved by the Ministry.

Education

One of the early challenges was providing education and training for the large number of new employees in 1991 and 1992 and adjusting education and training to incorporate democratic values. The main educational institution is the police academy. Unlike police academies in the US and the United Kingdom, which are only training institutions, police academies in Europe tend to be institutions that also offer academic degrees (Pagon, Virjent-Novak, Djuric & Lobnikar, 1996, p. 6). The Croatian Police Academy consists of the Police High School and the Police College (Chart 1). It also provides specialized courses.

In the days of the socialist/communist regime, the Police High School was a four-year institution in which students studied in a semi-military atmosphere with a strict daily routine characteristic of a military academy (*Halo 92*, 1995, #46, p. 11). Students were mostly separated from the outside world and were limited in their movements. As a result, students were often frustrated, and their negative feelings had a harmful influence on their work and self-image. The system has been changed based on the German Hessen Model (e.g., students live in dormitory-type buildings, but determine their own schedule and can move freely). The school started admitting students into the third year of education at the high-school level after their completion of the first two years of general education in a different educational institution. Approximately 400 students are admitted each year. Interest in the school has increased, so more qualified students are admitted since the percentage of applicants accepted is about one-third (*Halo 92*, 1995, #43, p. 16).

Comprehensive changes in the content of materials taught have also taken place. About 90 percent of the professors and instructors have been replaced and "about 50% of the classes in this school are focused on the appropriate approach of police officers toward citizens—manners, speech, writing, learning foreign languages" (Kobali and Balokovic-Krklec, 1995, p. 10). Students take courses such as sociology, philosophy and forensic psychology (*Halo 92*, 1992, #13, p. 21). Ethical issues are taught in various courses, and there is a separate course on ethics. The rest of the curriculum is devoted to professional courses that combine theoretical issues with practical ones. For example, conducting an arrest is taught in two courses—a course about the rules of police work and a physical education course (*Halo 92*, 1995, #46, p. 11).

The Police High School also provides a six-month-long training course for individuals who want to become police officers and have already obtained a high-school diploma. Both police-related courses and general courses (e.g., foreign language, telecommunications) are taught (*Halo 92*, 1994, #27, p. 19). This is the training that the majority of the police officers hired in 1990, 1991 and 1992 went through. The course is to be followed by six months of practical training. Moreover, a recent idea was to provide specialized training to police officers beyond the basic course. There is a range of specialized courses offered in the Department of Specialization and Training, lasting from two to nine months.

The Police College offers associate's and bachelor's degrees in criminalistics and is open to the public. The program emphasizes criminalistics and criminology. The

College has also started a graduate program in criminalistics. One of the courses includes a segment on ethics and human rights in which behavioral standards are examined from ethical and legal perspectives.

The Ministry tries to provide quality education for police officers in order to give them the necessary skills to perform their job. Especially affected by the massive Serbian departure from the force were detectives, and the need for professional training was especially pressing for the criminal police. Consequently, 1,200 detectives from the criminal police took a four-day course dealing with specific crimes (*Halo 92*, 1997, #66, p. 12); forensic technicians took several months-long courses in their specialty; and a number of police officers took courses abroad (at the FBI, DEA, or Interpol).

How do the supervisors learn their job? The problem of providing quality training to supervisors was particularly relevant because there were a large number of promotions in a relatively short period of time. The Ministry organizes courses for supervisors, the heads of sectors and their assistants, the chiefs of police stations and their assistants, and the heads of police administrations. The first of these courses was held in 1994 (*Halo 92*, 1994, #29, p. 9). Another specialized course was offered for the heads of special police units (*Halo 92*, 1993, #22, p. 11). Former Minister Jarnjak (1994, p. 4) spoke to a group of supervisors about the need for careful personnel planning.

> You have to take care that the employees around you, regardless of whether they are your subordinates or supervisors, are professional and smart people .... If they are not correct, strong, and good, you'll soon be surrounded by poltroons who will try to use circumstances for their own good. You should not allow such people to advance in their career, because they are dishonorable people who will not be accepted in another workplace, still less in the police where everything is based on team effort.

Disciplinary Proceedings

The war necessitated the hiring of a large number of people without applying the usual strict requirements. Some of these people were not able to respond to the challenges of their job successfully and professionally, so the Ministry started the process of clearing up its ranks as soon as the majority of police officers had been released from their military duties. "[F]or many mistakes, for laziness, the golden excuse was the following, 'Let it go, it is the war!' That excuse is no longer valid." For example, in mid-1992, the Ministry fired over 600 people (Jarnjak, 1992, pp. 4–5). As a consequence of greater openness the public has learned about the size of the force for the first time; there are presently about 34,000 people employed by the Ministry. The Ministry intends to gradually reduce the number of employees to 27,000 (that would correspond to the standard of approximately 150 citizens per police officer). This is to be accomplished by providing retirement packages to approximately 2,000 people per year and by hiring approximately 600 people from the Academy each year (*Halo 92*, 1997, #72, p. 9).

One of the challenges was to carefully improve the organization by firing people who could not perform their duties successfully even after taking additional specialized courses; people who are corrupt; people who used excessive force; and people who violated other legal norms. Deputy Minister Moric argues (1994, p. 7) that, despite all the economic, organizational, operational and professional challenges, "our police officers, because of the defensive war in which they were the first organized formation to fight the enemy, have today, despite all the difficulties, a high sense of responsibility and a great deal of patriotism. And this is contrary to corrupt attitudes."

Control over adherence to the rule of law and professional conduct was initially in the hands of supervisors who could use various types of disciplinary measures to inflict punishment for misconduct. However, since the tendency in the entire Ministry was toward centralization, a separate office was formed in 1994 at the Ministry level, *The Office of Internal Affairs*. The functions the Office performs include "identifying every type of unacceptable behavior of police officers, supervision of adherence to the legal exercise of authority by police officers, and supervision of the legality of performance and financial discipline" (Kuretic, 1997, p. 47). Because there is frequent contact between the police and citizens, it is important that citizens be able to complain about the misconduct of police officers directly to the Internal Affairs Office (which is located in the Ministry headquarters) or to the officers' immediate supervisors. In fact, citizens embraced this opportunity and, although it took them some time to learn about the Office, 249 cases were investigated in 1994 and 353 a year later. Twenty-two percent of the cases in 1995 were found to be completely or partially founded, while 64 percent were unfounded (*Halo 92*, 1996, #53, p. 9).

Difficult conditions in Croatia have produced numerous opportunities for corruption over the last five years. With the wave of the refugees from Bosnia and Herzegovina and the war in the region, many opportunities arose for the smuggling of merchandise and persons across the border. A case described in the newspaper *Slobodna Dalmacija* illustrates the point: A person with falsified documents offered 50 German marks to the border police officers (*Halo 92*, 1993, #19, p. 5). Similarly, one of the smugglers offered a bribe of 5,000 German marks to a police officer (*Halo 92*, 1993, #25, p. 7). Another newspaper, *Vecernji list*, ran a story in 1997 about a motorist who, in order to "solve" a misdemeanor (underage driving), offered a bribe to a police officer and therefore committed a felony. However, the dark numbers with respect to corruption are especially high in any country, particularly for the "garden-variety corruption," such as acceptance of free meals.

The old socialist regime had a reputation for being quite corrupt. Some outside agencies have evaluated the current regime as "one of the most corrupt countries in the world" (Butkovic cites a study by Daimler-Benz, 1997a, p. 15). Similarly, "corruption exists at the highest levels in the government; the Croatian program of privatization is infected by corruption" (Butkovic cites a study by Kroll Associates, 1997b, p. 15). Furthermore, the President suspended thirty-three highly ranked members of his party, the Croatian Democratic Union, because they were under investigation for financial embezzlement. A study by the magazine *Globus* suggested that approximately 60 percent of the respondents thought that the state did not have the resolve to deal with corruption at the highest levels of the government and that approximately one-half of the respondents thought that Croatia is one of the most corrupt countries in the world (Butkovic, 1997b, p. 15).

Disciplinary procedures were rather rare in the old socialist/communist regime. In the annual report for 1988, it is written that, for the whole national police force (of an unknown size) over a period of one year, 17 disciplinary procedures and one criminal procedure were initiated (Mulac, 1989, p. 120). On the other hand, the disciplinary process is used much more frequently today. The data for police administrations (there are 20 administrations in the country) show that, for example, in 1992 there were 175 disciplinary procedures initiated resulting in 31 people being fired and 14 criminally investigated in only one of the police administrations (*Halo 92*, 1993, #18, p. 9). In another police administration there were 138 disciplinary procedures—105 completed by the time of the interview, 18 people fired and 10 investigated criminally (*Halo 92*, 1994, #27, p. 8) in 1993. In 1994, in yet another police administration, there were 115 disciplinary procedures in six months resulting in 9 people being fired and 4 under criminal investigation. However, as is true with all the police in the world, the dark numbers and the code of silence are present and the official data cannot present the true state of affairs. What is important is the fact that, compared to the old police, the new Croatian police have a different attitude and at least try to investigate and punish police officers for violating the law.

CONCLUSION

For the Croatian police, the road toward democracy has been a unique one. In addition to the usual set of problems faced by the police in societies in transition and in developing democracies, the situation in Croatia was burdened with another factor—the war. Due to the war, the police were required to perform military functions in addition to their regular functions. The situation was further exacerbated by the fact that a large number of police officers from the old socialist/communist regime left the force virtually overnight. A high number of refugees and displaced persons, the war economy, increased unemployment numbers, the introduction of the multi-party system, the transformation of public property into private property—these are some of the major political, social and economic circumstances still facing the police.

The Croatian police tried to parallel democratic changes in the larger society by adhering to the rule of law, by selecting adherence to the rule of law as their primary goal, by implementing "the open door policy" and thus exposing themselves to public criticism, by being exposed to control by both the Parliament and the public, and by striving toward professionalism and the integrity of police officers. Typical methods used included increasing emphasis on education and specialized training (both in Croatia and abroad), proposing changes in existing laws, implementing organizational changes in the police force structure, providing stricter control over the work of police officers and disciplining rule violators. Compared to the Polish police which, according to Haberfeld (1997, p. 653), create the impression that they are "not presently capable of protecting the rights, serving the needs or earning the trust of the population they police," the Croatian police project an entirely different image—the image of a professional police. It seems that the Croatian police succeeded in their primary mission—fighting against crime, at least with respect to street crime; Croatian streets seem to be safe (Butkovic citing the results of the Kroll study, 1997b).

Heymann (1997) proposes a dichotomy of "weak" and "strong" criminal justice systems on the basis of their ability to handle crimes committed by the powerful. Very "weak" police forces would not be able to control street crime: "when a democracy is unable to provide protection against predatory activities of other citizens, the call for authoritarian alternatives grows" (p. 12). If we assume that there is a continuum instead of a dichotomy, because of their ability to control street crime, the Croatian police would definitely be positioned away from the "weak" toward the "strong" police end. However, exactly how far they would proceed along this continuum depends on their ability to control the crimes of the powerful. One of the serious challenges to further progression in the direction of "strong" police is their potentially limited ability to control corruption and embezzlement by powerful people in Croatia (Butkovic, 1997a, 1997b). In addition, other parts of the criminal justice system may either not operate efficiently (for example, in order to deal with an unusually large number of outstanding cases, the Ministry of Justice introduced a temporary measure in January 1998—the busiest courts were ordered to work in shifts) or may be influenced by politics. For instance, Krapac (1996, p. 344) argues that the nomination of judges is seriously affected by political factors.

As a part of a society that attempts to develop democracy and has experienced additional challenges because of the war, the Croatian police certainly deserve a positive class-report so far. However, the state of anomie (Merton, 1938) caused by the war has passed. With a set of new substantive and procedural laws which became effective on January 1, 1998, the Croatian Parliament and the Croatian police have set the stage for a new step on the road toward democracy. Exactly how successful they will be remains to be seen, but many indicators give rise to moderate optimism.

NOTES

1. In the former Socialist Federal Republic of Yugoslavia there existed the Federal Constitution as well as constitutions for all the constituent republics.
2. The Croatian language, as does German, distinguishes between the *Sie* form and the *Du* form in addressing another person. The *Sie* form is used when interacting with strangers and persons of different social standing; it indicates a great degree of respect and is a formal way of addressing others. Professors and students, physicians and patients, judges and attorneys, for example, will typically use the *Sie* form in their interactions. The *Du* form is used in the interactions of persons who know each other well, such as good friends, co-workers of the same rank or family members. Using the *Du* form in other instances, especially when the party using it is doing so in the official capacity (e.g., a uniformed police officer on duty), represents a clear signal of disrespect and abuse of power.
3. A typical joke about the (lack of) militia officers' abilities would go like this: Question: How many militia officers does it take to replace a light bulb in the ceiling? Answer: Five. Question: Why? Answer: One militia officer to stand still on the table holding the bulb in his hand while the remaining four militia officers rotate the table.
4. At the beginning of 1994, only 4.7 percent of the police officers in Croatia were ethnic Serbs (*Halo 92*, 1995, #26, p. 4). As will be discussed shortly, the number of police officers of Serbian origin will probably increase in 1998, i.e., upon the conclusion of peaceful reintegration of the remaining 5 percent of Croatian territory.
5. The Police Academy meets the educational needs related to the police. It offers, for instance, a variety of courses of study, from short-term courses that focus on specific aspects of policing, a regular two-year high school education specifically designed to provide both general education and comprehensive training in policing, to undergraduate programs leading to associate's and bachelor's degrees in criminalistics. These degrees require respectively two and four years of full-time study. The role of the police academy in Croatia is considerably broader than the role of a typical police academy in the United States. Further description of the Police Academy will be furnished in subsequent sections of this paper.
6. When Carl Klockars and I conducted our study of police officers' attitudes about corruption (see Kutnjak Ivkovich and Klockars, 1997), we received very professional and enthusiastic support from the Public Relations Office.
7. The counterparts to the *Service for the Protection of the Constitutional Order* in other countries are typically the secret police.
8. The notion of "economic crime" is somewhat similar to the more established terminology of "corporate crime." In many ways, the notion of economic crime better reflects reality in Croatia than does the notion of corporate crime. Most corporate entities are actually small businesses with very small capitalization, but a few employees (some or most of whom are typically the owner and his or her immediate family), and a culture that is vastly different than the corporate culture of a typical western company. Thus, the opportunities for criminal activities are fewer and are of a different variety. In an effort to reflect this difference, I shall adopt the "economic crime" terminology throughout the remainder of this paper.
9. Before the new statute on money laundering, the police did not have a legal basis to confiscate the profits achieved through money laundering (*Halo 92*, 1996, #59, p. 7).

REFERENCES

Balokovic-Krklec, K. (1992a). Supervizor hrvatske policije [Supervisor of the Croatian police]. *Halo 92, 2,* 6–7.
——. (1992b). Ratom do mira [Toward peace through war]. *Halo 92, 3,* 6.
——. (1993). Ukostac ca skripom smrskane svagdasnjice [Dealing with bleak reality]. *Halo 92, 15,* 7.
Bojadziski, O. (1995). Istrazivacko novinarstvo, pomoc policiji u tranziciji k demokratskom drustvu [Investigative journalism—Assistance to the police in transition toward a democratic society]. *Policija I sigurnost, 4,* 207–216.
Bosiocic, G. (1996). Pravni I prakticki aspekti ustrojstva tijela unutarnjih poslova u Republici Hrvatskoj. [Legal and practical aspects of the organization of the police in the Republic of Croatia]. *Hrvatski ljetopis za kazneno pravo I praksu, 3,* 585–606.
Brmbota-Devcic, T. (1992). Pakrac—Nepokoreni grad [Pakrac—Undefeated city]. *Halo 92, 2,* 13.
Buble, N. (1992). Kad svima smrkne [When lights are out for everybody]. *Halo 92, 10,* 29.
Butkovic, D. (1997a, August 1). Vijece za strateske odluke: Tudjmanovo orudje apsolutne centralizacije vlasti ili pokusaj da se bolje kontroliraju strana ulaganja? [The Committee for Strategic Decisions: Tudjman's apparatus of absolute centralization or an attempt to better control foreign investments?] *Globus,* pp. 14–15.
——. (1997b, September 29). Tajni izvjestaj o Hrvatskoj [The secret report on Croatia]. *Globus,* pp. 14–15.
Cengic, S. (1993). Na policajce se I puca! [Police officers are even being fired at!]. *Halo 92, 20,* 12.
Croatia: Facts and Figures (1997). Http://www.lexis-nexis.com.
Dujmovic, Z. (1996). Usporedni pokazatelji kriminaliteta u srednjoj I istocnoj Europi u 1993. I 1994. Godini [Comparative indices of crimes in Central and Eastern Europe in 1993 and 1994]. *Policija I sigurnost, 3,* 384–395.
Dujmovic, Z. & Miksaj-Todorovic, L J. (1996). Criminological characteristics of offenders guilty of robbery. In M. Pagon. (Ed.). *Policing in Central and Eastern Europe: Comparing firsthand knowledge with experience from the West.* Ljubljana: College of Police and Security Studies.
Gataric, I. (1994). Izvjesce o radu Odjela ratnih zlocina I kaznenih djela terorizma [The report on the work of the Department for War Crimes and Terrorism]. In J. Sintic (Ed.). *Uloga policije u demokratskom drustvu* [The functions of the police in a democratic society]. Zagreb: Ministarstvo unutarnjih poslova Republike Hrvatske.
Gledec, Z. (1994). Policijski sustav u Republici Hrvatskoj [The police system in the Republic of Croatia]. In J. Sintic (Ed.). *Uloga policije u demokratskom drustvu* [The functions of the police in a democratic society]. Zagreb: Ministarstvo unutarnjih poslova Republike Hrvatske.
Haberfeld, M. (1997). Poland: The police are not the public and the public are not the police. *Policing: An International Journal of Police Strategies & Management, 20,* 641–654.
Halo 92 (Issues 1 through 72). Monthly publication of the Croatian Ministry of the Interior.
Hebenton, B., & Spencer, J. (1997, November). *Law enforcement in societies in transition.* Paper presented at the 49th Annual Meeting of the American Society of Criminology. San Diego, CA.
Heymann, P. (1997). Principles of democratic policing. In *Policing in emerging democracies: Workshop papers and highlights.* Washington, DC: National Institute of Justice, U.S. Department of Justice.
Horvat, K. (1994). Na opasnim kotacima [On dangerous wheels]. *Halo 92, 32,* 19.
——. (1995). Prljave igre bijelih ovratnika [Dirty games by white-collars]. *Halo 92, 48,* 15–17.
Jarnjak, I. (1992). Izvaci s tiskovne konferencije [Excerpts from the press conference]. *Halo 92, 11,* 4–5.
——. (1993). Neprestano budni I mobilni [Always awake and mobile]. *Halo 92, 21,* 3.
——. (1994). Izvaci s tiskovne konferencije [Excerpts from the press conference]. *Halo 92, 30,* 4–5.
——. (1995). Izvaci s tiskovne konferencije [Excerpts for the press conference]. *Halo 92, 38,* 3.

Jurina, M. (1994). Prikaz sustava policijskog obrazovanja Republike Hrvatske [The review of the system of police education of the Republic of Croatia]. In J. Sintic (Ed.). *Uloga policije u demokratskom drustvu* [The functions of the police in a democratic society]. Zagreb: Ministarstvo unutarnjih poslova Republike Hrvatske.

Kazneni zakon [1997 Criminal Law Statute] (1997). Zagreb: Narodne novine, 110/97.

Kobali, D. (1997a). Ustroj prilagodjen potrebama [The organization adjusted to needs]. *Halo 92, 66,* 11–13.

———. (1997b). Izvaci s tiskovne konferencije [Excerpts from the press conference]. *Halo 92, 68,* 5.

Kobali, D. & Balokovic-Krklec, K. (1995). Policajac XXI stoljeca [A police officer of the 21st century]. *Halo 92, 46,* 9–12.

Kovco, I. (1996). Delikti nasilja [Violent crimes]. *Policija I sigurnost,* 3, 396–414.

Krapac, D. (1996). Hrvatska kao zemlja u tranziciji: ucinci politickih, gospodarskih I socijalnih promjena na kriminalitet I kaznenopravni sustav [The Republic of Croatia as a country in transition: Effects of political, social, and economic change on crime and the criminal justice system]. *Hrvatski ljetopis za kazneno pravo I praksu,* 3, 336–346.

Kregar, J., Smerdel, B., & Simonovic, I. (1991). Novi Ustav Hrvatske: nastanak, osnovne ideje I institucije [The new Croatian constitution: Its genesis, basic ideas, and institutions]. *Zbornik Pravnog fakulteta u Zagrebu,* 41, 129–149.

Kuretic, Z. (Ed.) (1997). *Ministarstvo unutarnjih poslova* [Ministry of the Interior]. Zagreb: Ministarstvo unutarnjih poslova Republike Hrvatske.

Kutnjak Ivkovich, S. (1996). Field notes.

Kutnjak Ivkovich, S., & Klockars, C. B. (1997, March). *Attitudes on police corruption: Does the length of service mMake a difference?* Paper presented at the 1997 Annual Meeting of the Academy of Criminal Justice Sciences. Louisville, KV.

Lajic, I. (1995). Demografski razvitak Hrvatske u razdoblju od 1991. do 1994 [Demographic development of Croatia 1991–1994]. *Revija za sociologiju,* 26, 55–64.

Lovric, J. (1993). Policija znaci zivot [The police mean life]. *Halo 92, 16,* 3.

Markovic, L. J. (1997, December 21). Panika medju poslodavcima [Panic among the employers]. *Vjesnik,* p. 5.

Merton, R. K. (1938). Social structure and anomie. *American Sociological Review,* 3, 672–682.

Mirosavljev, B. (1992a). Policajci izasli iz rova [Police officers left the trenches]. *Halo 92, 3,* 5.

———. (1992b). Rat ne pita za zrtve [The war doesn't care about the victims]. *Halo 92, 8,* 4–5.

———. (1993). Protiv ratom izazvanih sindroma [Against war-triggered syndromes]. *Halo 92, 18,* 8–9.

Moric, J. (1994). Uloga policije u demokratskom drustvu—dosezi hrvatske policije [The role of the police in a democratic society—The achievements of the Croatian Police]. In J. Sintic (Ed.), *Uloga policije u demokratskom drustvu* [The functions of the police in a democratic society]. Zagreb: Ministarstvo unutarnjih poslova Republike Hrvatske.

Mulac, V. (1989). Izvod iz Izvjestaja o radu sluzbi in organa unutrasnjih poslova S. R. Hrvatske u 1988 [Excerpts from the 1988 report on the work of the Ministry of the Interior in the Socialist Republic of Croatia] *Prirucnik za strucno usavrsavanje radnika unutrasnjih poslova,* 2, 115–135.

Nadj, I. (1996). Novi oblici kriminaliteta u Republici Hrvatskoj [New forms of crime in the Republic of Croatia]. *Policija I sigurnost,* 4–5, 434–441.

1997 Croatian Almanac (1998). Http://www.hina.hr/almanac97.

Osrecki, K. (1993). Izvaci iz tiskovne konferencije [Excerpts from the press conference]. *Halo 92, 18,* 8–9.

Pagon, M., Virjent-Novak, B., Djuric, M., & Lobnikar, B. (1996). European systems of police education and training. In M. Pagon (Ed.). *Policing in Central and Eastern Europe: Comparing firsthand knowledge with experience from the West.* Ljubljana: College of Police and Security Studies.

Peric, I. (1995). *Godine koje ce se pamtiti* [Years to be remembered]. Zagreb: Skolska knjiga.

Piskor, M. (1993). Izvaci iz interviewa [Excerpts from the interview]. *Halo 92, 18,* 5.

Rako, S. (1995). Policija koja odgovara zahtjevima vremena [The police which corresponds to the current challenges]. *Halo 92, 46,* 5–8.

———. (1996). U dosluhu s najnovijim tehnickim dostignucima [In sync with the latest technological developments.]. *Halo 92, 61,* 6–8.

———. (1997). Izmedju jucer, danas I sutra [Between yesterday, today and tomorrow]. *Halo 92, 67,* 15.

Rkman, I. (1998, January 4). Hrvatska ce jos neko vrijeme morati natprosjecno izdvajati za obranu [Croatia will continue to set aside unusually high amounts for defense]. *Vjesnik,* p. 8.

Sacic, Z. (1996). Organizirani kriminalitet u Hrvatskoj I sredstva borbe protiv njega [Organized crime in Croatia and measures to fight it]. *Hrvatski ljetopis za kazneno pravo I praksu, 3,* 607–668.

Simic, J. (1992). Hrvatsku nece cuvati dresirani psi [Croatia will not be guarded by trained dogs]. *Halo 92, 2,* 2.

Singer, M. (1996). Globalne tendencije kretanja I strukture kriminaliteta u Hrvatskoj [Global trends and patterns of crimes in Croatia]. *Policija I sigurnost, 3,* 349–383.

Singer, M. & Cajner, I. (1993). Kriminalitet u Hrvatskoj u 1992. Godini [Crime in Croatia in 1992]. *Policija I sigurnost, 2,* 1–16.

Ustav Republike Hrvatske [The Constitution of the Republic of Croatia] (1990). Zagreb: Narodne novine, 56/90.

Veic, P. (1996a). Kaznenopravne mogucnosti suzbijanja organiziranog kriminaliteta [Prevention of organized crime through criminal law in the Republic of Croatia]. *Hrvatski ljetopis za kazneno pravo I praksu, 3,* 669–685.

———. (1996b). Neka otvorena pitanja "Policijskog zakonodavstva" Republike Hrvatske—de lege ferenda [Some open questions about the Police Laws in the Republic of Croatia—De Lege Ferenda]. *Policija I sigurnost, 2,* 117–136.

———. (1997). Eticka obuka u Republici Hrvatskoj [Education in ethics in the Republic of Croatia]. *Policija I sigurnost, 6,* 277–280.

Zakon o kaznenom postupku [1997 Criminal Procedure Law] (1997). Zagreb: Narodne novine, 110/97.

Zvekic, U. (1996). Policing and attitudes towards police in countries in transition. In M. Pagon (Ed.), *Policing in Central and Eastern Europe: Comparing firsthand knowledge with experience from the West.* Ljubljana: College of Police and Security Studies.

4

Estonian Police

ANDO LEPS

INTRODUCTION

In the west and the north, Estonian shores are washed by the Baltic Sea; in the east, the Republic of Estonia borders on the Russian Federation and in the south on the Republic of Latvia. Estonia's territory is 45,200 square kilometers. In addition, about 2,000 sq. km. (some of it around the town of Petseri in the south-east and some on the east bank of the Narva river), which had been assigned to Estonia under the terms of the Tartu Peace Treaty of 1920, was transferred to the Russian SFSR in 1945.

Administratively, Estonia is divided into 15 *maakonds* (counties) of which two are located on islands (Saaremaa and Hiiumaa). The 1993 administrative reform defined two types of administrative units within *maakonds*, towns and *valds* (communes). Estonia has 46 towns (their population ranging from 1,300 to 443,000 inhabitants) and 209 *valds*. The biggest towns are Tallinn, which is also the capital city (443,000 inhabitants), Tartu (106,000), Narva (79,000), Kohtla-Järve (56,000), and Pärnu (52,000 inhabitants). Twenty-nine percent of the population of Estonia lives in the capital city (Leps, 1992).

During the Soviet period, and to a certain extent also before, a number of major industrial enterprises were set up in northern Estonia (Tallinn and the towns of the North-East). The best conditions for agriculture exist in central Estonia. Estonia's natural resources include oil shale, peat, limestone, clay, sand, gravel, curative sea mud, mineral water, timber and fresh water. Estonia has a dense roads network (14,800 kilometers of common carrier motor roads, 5 percent of which is surfaced, and 1,000 kilometers of railways).

Estonia reestablished its independence in 1991, after having been occupied and consequently annexed by the Soviet Union in 1940. According to the constitution which took effect on July 3, 1992, Estonia is an independent democratic republic with supreme power vested in the people. Legislative power belongs to the 101-member *Riigikogu* (parliament) which is elected for a term of four years. The President of the Republic is the head of state and is elected by the *Riigikogu* for a five-year term of office. Executive power belongs to the republic's government, which consists of the Prime Minister and fourteen ministers. In *maakonds*, executive power and state assignments are carried out by county governments directed by County Governors.

The local self-governments are: (1) local councils, representative bodies elected for three years and (2) local governments, executive bodies set up by the local councils, headed by the commune elder or mayor elected by the council for a term of up to three years.

On January 1, 1997, Estonia had 1,462,000 inhabitants. Different historical periods have had a different effect on the growth of population in Estonia, as well as its composition and distribution (*Eesti Statistika*, 1997). Significant changes in Estonia's demographic development were caused by World War II and the social upheavals it brought about. The Baltic Germans, Estonia's second largest minority, left for their historic homeland. This was followed in June 1940 by the Soviet deportation to outlying regions of Russia of over 10,000 people, large numbers of Estonians seeking refuge in the West using different ways to leave the country, the resettling of Estonian Swedes to Sweden in 1944, immediate war losses, massive arrests and another wave of mass deportation in 1949.

The total population loss, as a result of World War II and the initial years of Soviet occupation, was about 200,000 people or 20 percent of the population. For a short time after the amalgamation of the mainly Russian-populated districts of Estonia to the Russian SFSR in 1945, Estonia became almost 100 percent monolingual.

After occupation by the Soviet Union, Estonia became an area with a high rate of immigration as well as out-migration. In the period 1945–1990, Estonia's population increased by 718,000 people or by over 80 percent, of which immigration accounted for 70 percent and natural increase for only 30 percent. As the immigrants settled mainly in towns and there was a movement into towns also from Estonia's rural areas, the share of urban population increased from 56.4 percent in 1959 to 71.5 percent in 1989 (see Table 4.1). As a result of the rapid increase of the non-Estonian population, the proportion of the Estonians among the overall population declined to 61.5 percent by 1989, and there were towns in Estonia where the percentage of ethnic Estonians had decreased to 20 percent or less (Paldiski 2.4 percent, Narva 4.0 percent; Maardu 15.4 percent and Kohtla-Järve 20.9 percent).

Since 1990 there have been significant shifts in demographic processes. The birth rate began declining and, in connection with the reestablishment of Estonian independence, non-Estonians began to emigrate from Estonia. In the six years from

Table 4.1. Number of population (thousands)

Year	Total	Percentage of urban population
1922	1,107	27.6
1934	1,126	34.6
1940	1,018	....
1945	854	....
1959	1,191	56.4
1970	1,355	64.7
1979	1,464	69.4
1989	1,566	71.5
1990	1,572	71.5
1991	1,570	71.4
1992	1,562	71.3
1993	1,526	70.6
1994	1,506	70.3
1995	1,492	70.0
1996	1,475	70.7
1997	1,462	69.5

1991–1997, the population of Estonia had decreased by a total of 108,000 people, or 7.4 percent, mainly as a consequence of emigration. As a result, the percentage of ethnic Estonians has gone up to 65 percent. In recent years there has been no increase in the average life expectancy. As regards average life expectancy Estonia lags six to eight years behind the countries of the European Union. There is a difference of thirteen years between life expectancy for men and for women, which is a normal demographic occurrence (*Eesti Statistika*, 1997).

The official language in Estonia is Estonian. The main religious communities are Lutherans (Protestants), Russian Orthodox and Baptists.

There have been rather substantial changes in Estonia's economy since the reestablishment of independence in August 1991. So far the best economic achievement was recorded in 1989. From then on the volume of production began to drop rapidly. The year with the sharpest drop was 1992, and in 1993 the trend continued. In 1994 the drop was considerably less significant. Since 1995, for the first time since the reestablishment of independence, volume of production has begun to rise.

It is apparent from Table 4.2 that the drop in the real domestic product is directly correlated with the curves of crime and of crime detection. This process can also be observed in other former socialist countries. But as compared with other centralized countries in transition from a planned to a market economy, the drop in Estonian economy has perhaps not been quite as drastic (see Table 4.3).

It must be pointed out, however, that a significant proportion of the money is concentrated in the hands of a very small proportion of the population (about 10 percent) who can afford to buy durable goods, eat healthy food and have savings. The incomes of the richest 10 percent of families account for 29 percent of the income of all families, while the incomes of the poorest 10 percent of families

Table 4.2. Gross domestic product, crime and crime detection in 1989–1994

Year	Change of real GDP as compared to previous year, in percent	Number of crimes recorded by police	Crime detection as percent of recorded crime
1989	—	19,141	35.7
1990	–6.5	23,807	26.9
1991	–13.6	31,748	20.0
1992	–14.2	41,254	18.2
1993	–8.6	37,163	24.6
1994	–3.2	35,739	26.7
1995	+1.5	39,570	28.5
1996	+6.0	35,411	35.1

Table 4.3. Gross national product per head of population in 1991 (in US dollars)

Estonia	3,830
Latvia	3,410
Russia	3,220
Hungary	2,720
Lithuania	2,710
Sweden	25,110
Finland	23,980

account for only 2.5 percent of the same figure. In addition, the income of poorer families is mainly from child benefits, alimony, subsidies and pensions, and it generally increases more slowly. In the fourth quarter of 1994, the respective figures for the bottom and top 10 percent of families were 330 and 3,909 kroons a month per family (Leps & Remmel, 1997). The situation of families with many children, as well as of families where only one parent has a job, the families of pensioners, not to mention those who only subsist on subsidies and grants, is the most difficult. The difference in the level of welfare between families and the trends of its change is one of the main reasons for the existence of crime. In Estonia there is no balancing middle class at present (Leps, 1994a).

THE CONCEPT OF DEMOCRATIC POLICING

Crime cannot be eliminated by police methods alone. Democratic policing requires that the impoverishment of large strata of the population be stopped, that international relations in the sphere of legal assistance become normalized and that guarantees be given that the state will regulate the economy. It is important to develop better knowledge about legal matters among the population as well as the attitudes and styles necessary for coping with day-to-day living in accordance with legal norms. Democratic policing encourages the protection of cultural values, ensuring that violence, discrimination and bad taste are kept under control. The police organization of a democratic society has a place and a role in the work for better knowledge about legal matters among the population. Knowledge that the police are where they are needed secures trust in their support and assistance. It makes one feel respected and leads one to behave as a dignified citizen. In a civil society, one expects high professional and cultural standards and faithfulness to their mission from the police. They must create the conditions for achieving them and, in part, they can do this through an appreciation of the nature of the process of development, which, in turn, helps the police control criminogenic situations in society.

It must be borne in mind that the police organization needs suitable people who can be brought in through a system of proper personnel selection, management and training corresponding to modern requirements, which provides the foundation for the professional standards needed for policing democracies. According to the Estonian Police Department Information Service, a preventive model of police organization has proved to be efficient in modern democratic states. The main elements of the functioning and development of such a model requires raising the intellectual level of the police. Raising the intellectual standard of the police also entails the adoption of appropriate information technology and other high-technology equipment, the application of scientific and technological achievements, a transition to special photographic technologies and the development of a police strategy that agrees with the tactics of operative work. Organizationally, raising the intellectual level of the police can be supported by the establishment of a state investigation unit. In crime fighting, modern methods and means of investigation must gradually begin to dominate, helping to guarantee the efficiency of criminal investigation.

The decentralization of the police will guarantee a transition to an intensive model for the maintenance of public order and the application of the principles of prevention in the fight against crime. Decentralization is a precondition for bringing the work of the police in the area of public order maintenance into harmony with the local needs and of raising the professional level of the fight against crime.

Decentralization of the police must be accompanied by the creation of a new type of police headquarters. This will help in democratizing the organization of the police and will improve its quality. The former police management practice based on authority and command must gradually be replaced by a new system based on normative regulation. The effective representation of police views and interests in the parliament and the government, the creation of economic and technical conditions for the professional work of local police forces, and the organization of police personnel selection, management and training are some of the main tasks of the police headquarters, the managing body of the police. In order to ensure the qualitative development of the police organization on the basis of the experience of modern democratic countries (such as the Netherlands, Austria and Switzerland), a police (interior defense) development center should be set up. Its main tasks should be to work out and implement novel proposals concerning the methods, tactics and technology of the police (interior defense) and the organization of information exchange and scientific and technical contacts with scientific and technological institutes of the police in other countries.

The development of the police on the principles of democratic organization of society and on the basis of a preventive police strategy requires qualitative improvement and renewal of legislation concerning the police, as well as its financial and economic re-organization. It is necessary to achieve cooperation and compulsory mutual assistance among all legal protection bodies (police, defense police, border guards, customs, the executive department, defense forces headquarters and legislative bodies, etc.). The drafting of an integrated code of laws concerning the police (fundamentals of police organization, police organization law, etc.) should be considered.

Policing democracies calls for the ability by the police to work with the democratic mass media. It is necessary to raise the quality of the police magazine and police news; to organize 'open door' days in police institutions; to involve veteran policemen more widely for police educational purposes at schools and institutions; and to conduct regular surveys of the public's information about legal issues, their respect of the law, and their expectations and attitudes concerning security matters. The design and administration of such surveys can be a form of research work at the police college and police schools. The results can be used to draft proposals for changes in the police development plan as well as in the schools.

Estonia's gradual amalgamation with the European Union is bound to bring changes in various spheres of life, including the organization of the police. The Estonian police will work with EUROPOL (an agency created to harmonize policing in member countries of the European Union), whose standards will be compulsory for all member countries of the European Union. Little by little, we shall have to get accustomed to the idea that the Estonian police can only work effectively on a basis compatible with the police standards that exist, in all spheres of work and responsibility, within and outside Europe. Adoption of the ideology of European Union member countries as regards organization of the Estonian police will contribute to:

(1) the development of the police under the principles of democracy, social justice and human rights;
(2) a deeper awareness of our European identity and future, without disregarding the perils concurrent with the information age and the price of successfully coping with them, as well as our global tasks and our ethnic and local roots; and
(3) the strengthening of the ethical foundations of the Estonian police based on professionalism and the ability to respond to challenges of policing democracies (State Police Department, Information Bureau, 1997).

ORGANIZATIONAL CHALLENGES

The History of the Police

Under orders from the Provisional Government of Estonia, the first steps to take the police over from German Occupation authorities were taken on November 12, 1918. An Estonian militia was hurriedly set up, and their organization was entrusted to local self-governments. In setting up an Estonian police force, it seemed best to call it a militia, as after a period of revolutions the police no longer commanded the trust and respect of the people (*Eesti Politsei 75*, 1993). The forerunners of the Estonian militia are many and varied: the police of czarist Russia, the militia of the period of the Russian provisional government, the Bolshevik militia and the *Bürgerpolizei* formed during the period of German Occupation (Lindmäe, 1994). Having been set up in a hurry, the Estonian militia suffered from the same drawbacks and vices inherent both in the militia of the period of revolutions and in the German Occupation police. The Estonian militia were composed of people lacking professional skills and equipment. The militia were unable to control crime which was rampant at the time. As history repeats itself, today we speak again about the weakness and inexperience of the Estonian police.

On December 17, 1919, the Constituent Assembly adopted a law on police organization, effective as of January 1, 1920. By that law, the militia in Estonia was abolished and a state-controlled police force was set up. In 1924, on the initiative of the Minister of the Interior, the Parliament passed a law on the unification of the various branches of the police. By that law, the criminal police was brought under the Ministry of the Interior and the criminal police headquarters was united with the Ministry of the Interior police headquarters. In this manner, both the field or general police as well as special police—criminal police and security police—were brought under the Ministry of the Interior and one headquarters. That step guaranteed unity of the police and ensured better cooperation among its different branches (Liit, 1920, 1927).

On April 8, 1925, the government issued a decree on the temporary organization of the police school by which the police school was to be set up at the Police Headquarters. The syllabi were to be worked out keeping in mind that criminal investigation requires a thorough knowledge of criminal and criminal procedure law. A policeman also had to be able to direct traffic, supervise the activity of firemen, and be a good sportsman and marksman. Driving, too, was taught at the police school. From 1937 on, the Central Board of Police Officials published a monthly professional journal, *Politseileht* (Police Paper). The principle in the police force was that the police must never be late! In the course of time the police won the respect and trust of the people. Also the defense police was quite active. The police closely cooperated with the Defense League. If the police did not have enough strength at any locality, a helping hand was given to it by the Defense League.

Immediately before the Soviet Occupation of 1940, the criminal police was staffed with 143 officials, overwhelmingly very experienced people. Directed by the prefecture, the field police consisted of constables, junior constables, senior patrolmen and patrolmen—altogether 1,486 men at the end of 1938. After their occupation of the Republic of Estonia, the Soviets immediately began to dismantle and destroy the police organization built up over a period of two decades. Some police officials were arrested even before the beginning of the war between the Soviet Union and Hitler's Germany. Many policemen went into hiding. When the war began, they took up arms to fight against the Soviets. The Estonian police was disbanded as of September 1, 1940.

During the Nazi Occupation, new people were recruited into the police force. Among them there were former military men, border guards and reserve officers. In 1943, the Estonian Police Chief issued orders to set up an Estonian Police Battalion composed of 700 Estonian field police by the first days of May. The battalion was engaged in defensive battles until February when the East front collapsed. The battalion suffered heavy casualties, with a total of only 200 surviving. When in the autumn of 1944 Estonia was once again occupied by the Soviets, some of the surviving policemen escaped to the West, some fell in with the ranks of resistance fighters, while others were deported to Siberia where most of them died. Only a few survived (Lindmäe, 1994).

The reestablishment of Soviet Occupation in the autumn of 1944 ushered in a period of the Estonian SSR Soviet-style militia. Such a situation lasted until November 1990. During that long period of time, the structure of the militia in Estonia was that adopted elsewhere in the Soviet Union. The militia was subordinated to the Ministry of the Interior, which was, in fact, the ministry of police (militia). Quite logically, the organization of the militia was typical of that in other countries. Particularly in the 1970s and 1980s, the Estonian SSR militia employed numerous jurists with a university education. The level of crime was relatively low, if we take the number of murder cases as the basis of comparison. At the same time crime was gradually increasing.

On November 16, 1988, the first blow was struck against the powerful imperial structure of the Soviet Union. The Estonian declaration of sovereignty showed all the peoples of that occupied country the road to freedom and independence. That date became a milestone on Estonia's road to the reestablishment of independence. The local councils set up as the result of local elections in December 1989 began actively to contribute to the strengthening of law and order. A cause for major concern was the fact that the militia, particularly on the rank-and-file level and especially in Tallinn and Northeast Estonia, was largely made up of Russian-speakers who either actively or passively supported Estonia remaining in the Soviet Union and under the undivided rule of the Communist Party. In 1990 there was a change of the Estonian SSR government. The situation became very grave. On May 15, 1990 gang leaders of the so-called Internationalist movement broke through the gates of Toompea Palace, the seat of both the parliament and the government. There was a real danger of a coup. The militia was unable to maintain public order and protect the state power in Estonia. It was necessary to speed up reforms connected with the institution of the police, and to make these irreversible. A police law was needed. This was adopted in August, 1990 by the Republic of Estonia Supreme Council. At the same time a breakthrough was achieved in negotiations in Moscow, as a result of which the Estonian Ministry of the Interior became excluded from the Soviet-wide centrally controlled union-republic system, and the Estonian Ministry of the Interior was subordinated to Estonian authorities alone. By a government decree of November 20, 1990, the State Police Department of the Estonian Ministry of the Interior was instituted, and three days later county and town police prefectures were set up. On November 12, 1990, the State Pre-Trial Investigation Department was created (*Meet Estonian Police*, 1992).

On May 31, 1993, the Republic of Estonia State Police Department was renamed the State Police Department within the jurisdiction of the Ministry of the Interior. An independent State Defense Police, also within the jurisdiction of the Ministry of the Interior, was set up on the basis of the former defense police. By the same decree the Pre-Trial Investigation Department was renamed the State Police Department Central Investigation Bureau. There were structural changes also in the Police Department itself.

Present Structure of the Police

The structure and duties of the police have been laid down in the Police Law adopted on September 20, 1990. Amendments to the law were made by a law adopted on April 21, 1991. By the nature of its duties, the police is divided into field police, traffic police, criminal police, security police and police involved in pre-trial investigation.

The central institution for the field police, traffic police, criminal police and police involved in pre-trial investigation is the Estonian Police Department which is within the administrative sphere of the Ministry of the Interior. For the security police the central institution is the Estonian Security Police Department, also within the administrative sphere of the Ministry of the Interior.

The territorial units of the Soviet militia, the so-called departments of interior affairs, were dissolved on 28 February 28, 1991 and replaced with police prefectures. The new structure of the Police Department had been endorsed somewhat earlier on February 1, 1991 (Leps, 1995a).

The Police Department has the following structural units and police institutions: Police Prefectures; Central Criminal Police; Police Central Investigation Bureau; Security Service; Field Police Bureau; and Police Information Bureau. In addition, there are police schools and educational units, as well as other institutions and units necessary to ensure the organizational stability of the police and their activities.

The Estonian Police Department directs, coordinates, guides and controls the activity of all police institutions and units within its sphere of administration. The Police Department is headed by the Police General Director.

Police prefectures are territorial police institutions. A police prefecture has a service area covering one or several counties or towns. It discharges duties mandated for the field police, traffic police and criminal police. A police prefecture is headed by a police prefect. Territorial police precincts are set up in the service area of each police prefecture. Precincts are further divided into constable patrol areas. At the moment the Estonian police has the following prefectures: Harju, Hiiumaa, East-Viru, Järva, Lääne, West-Viru, Narva, Põlva, Pärnu, Rapla, Saare, Tallinn, Tartu, Valga, Viljandi, and Võru.

The Central Criminal Police, the Field Police Bureau, the Traffic Police Bureau and the Police Information Bureau discharge duties of national or interdistrict importance or of a specific nature.

The Security Service is charged with guaranteeing the security of the president, the parliament's speaker, the prime minister and other officials and official guests of the country, as well as with maintaining safety and order at sites appointed by the government.

Police Duties

The Field Police carry out all duties mandated for the police, unless they fall into the area of other police branches.

The duties of the Traffic Police are to combat traffic violations, to supervise traffic, and to maintain traffic safety.

The duty of the Central Criminal Police is to fight crime. The police involved in pre-trial investigation examine all crimes throughout Estonia.

The duty of the Security Police is to defend Estonia's constitutional order, territorial integrity, and state secrets, to engage in counter-intelligence, and to fight against terrorism and corruption.

The structure or the Security Police was endorsed by the government decree of June 22, 1993. According to the decree, the Security Police Department has four divisions headed by directors: Tallinn division (for Tallinn, Harju, and Rapla counties); Viru division (East-Viru and West-Viru counties); South-Estonian division (Tartu, Jõgeva, Võru, Põlva, and Valga counties); and Pärnu division (Viljandi, Pärnu, Järva, Haapsalu, Saare, and Hiiu counties).

The creation of an independent government department must be considered a very important event in the history of the security police. It clearly defined their rights and duties, as well as areas of activity and functions.

Another important aspect in the development of the department is that, based on Article 11 of the Police Law, the general director of the Security Police Department is nominated and discharged by the government, on the recommendation of the Ministry of the Interior and with the advice of a commission set up by the parliament. The general director of the security police is appointed for a term of five years.

OPERATIONAL CHALLENGES

Crime

Estonian society has experienced several changes in the past few years, all of which have had an effect on crime: its state, its growth, its level and its structure. An increase in crime presents a fundamental challenge to the police, for the public holds them accountable and demands results, and expects that their personal safety will be assured. In turn, the police may respond aggressively to new patterns and levels of crime and become less sensitive to democratic values and legal norms in their work.

In 1986–87 the trend of diminishing crime figures, established years ago, continued. In 1986 the police (militia) in Estonia recorded 12,500 crimes and in 1987 the number of recorded crimes was 11,465 (down 8.3 percent compared to 1986). There are grounds to suspect that such a drop had been artificially achieved. Obeying the Communist Party instruction that crime must decline in the conditions of socialism, law-and-order bodies simply left some crimes unrecorded (for further references, see Leps, 1990, 1991a, 1991c, 1992, 1993, 1995d; Leps & Remmel, 1994).

But such deception and tampering with statistical data could not last forever. In 1988 crime figures began to mount and 12,167 crimes were recorded (6.1 percent more than in the previous year). The growth of crime continued also in the next five years and went up to 41,254 recorded crimes in 1992. In 1993 and 1994 there was a certain drop, but it was of a temporary nature. The growth continued in 1995 and crime increased by 10.7 percent. But from 1995 to 1996, crime decreased by 10.5 percent.

The level of crime calculated per 100,000 inhabitants increased steadily and was 2,422 in 1996, as compared to 815 in 1986. (See Table 4.4.) The rise in the level of crime can be attributed above all by the increasing number of offenses. At the same time one must not underestimate the fact that Estonia's population decreased by 108,000 people in the past seven years.

A number of changes have taken place in the structure of crime. The percentage of serious crimes against the individual is decreasing. In 1986 murder, assault with grave bodily injury, and rape constituted 3.4 percent of the total number of crimes, and in 1995 the figure was 2.1 percent. At the same time the percentage of common crimes against property (theft and robbery) went up from 50.3 percent in 1986 to 82.3 percent in 1996. (see Table 4.5.)

Table 4.4. Crime rates and trends

Year	Recorded crimes	+/− as compared to previous year	Change in percent	Level of crime per 100,000 of population
1986	12,500	−2,328	−15.7	815
1987	11,465	−1,035	−8.3	742
1988	12,167	+702	+6.1	781
1989	19,141	+6,974	+57.3	1,222
1990	23,807	+4,666	+24.4	1,514
1991	31,748	+7,941	+33.4	2,022
1992	41,254	+9,506	+29.9	2,641
1993	37,163	−4,091	−9.9	2,435
1994	35,739	−1,424	−3.8	2,395
1995	39,570	+3,831	+10.7	2,682
1996	35,411	−4,159	−10.5	2,422

Table 4.5. Distribution of crimes

Year	Percentage of crimes against the individual	Percentage of crimes against property
1986	3.4	50.3
1987	3.1	54.0
1988	2.9	64.3
1989	2.4	73.3
1990	2.0	79.2
1991	1.6	84.6
1992	1.7	87.2
1993	2.2	81.7
1994	2.5	77.5
1995	2.1	83.2
1996	2.1	82.3

Crimes are becoming more brutal and violent in nature. This is apparent also from the constant increase of murder cases.(See Table 4.6.) In 1986, 75 cases of murder were committed in Estonia (four cases or 5.1 percent less than in the previous year) and in 1987, 70 cases. But since 1987, when 78 cases of murder were recorded in Estonia, the number of this type of crimes has been constantly mounting. The number of murder cases went up particularly sharply in 1992 when there were 103 more murders than in the previous year. In the following years the increase in the number of murder cases has been somewhat slower. In 1996 the number of cases of murder and attempted murder went down by 36.

The biggest factor in the increase in crime is the explosive growth of thefts. Compared with general crime, the number of thefts has been increasing at a much higher rate (Table 4.7). In the past eleven years crime in Estonia has gone up by 283.3 percent, but the number of cases of theft has increased at nearly double that rate. The increase in theft was 421.1 percent. In Estonia, theft has consistently been

Table 4.6. Trends in the commission of murder

Year	Recorded cases of murder and attempted murder	+/− as compared to previous year	Change, in percent
1986	75	−4	−5.1
1987	70	−5	−6.7
1988	78	+−8	+11.4
1989	89	+11	+14.1
1990	137	+48	+53.9
1991	136	−1	−0.7
1992	239	+103	+75.7
1993	328	+89	+37.2
1994	365	+37	+11.3
1995	304	−61	−16.7
1996	268	−36	−11.8

Table 4.7. Trends in theft

Year	Recorded cases of theft	+/− as compared to previous year (in percent)	Theft as percentage of total crime
1986	5,881	−19.1	47.0
1987	5,821	−1.0	50.8
1988	7,428	+27.6	61.1
1989	12,989	+74.9	67.9
1990	17,518	+34.9	73.6
1991	25,082	+43.2	79.0
1992	33,309	+32.8	80.7
1993	27,339	−17.	73.6
1994	24,719	−9.6	69.2
1995	28,165	+13.9	71.2
1996	24,764	−12.1	69.9

the most widespread type of crime, accounting for 47.0 to 80.7 percent of total crime.

A distinct category is so-called economic crime. Because of changes in penal law made in 1991–1992, data on economic crime are not comparable since the transition to a market economy thoroughly changed the character of economic crime. We can only say that this type of offense as part of the general number of crimes has gone down considerably. In 1986 economic crimes made up 10.2 percent of total crime, but in 1995 they only accounted for 2.9 percent. Also the nature of economic crime has changed. During 1986–1990, the majority of economic offenses were categorized as expropriation—the squandering or embezzlement of property entrusted to one's care. In more recent years, fraud has taken the first place.

The number of crimes committed by or with the participation of juveniles (Table 4.8) is high and displays a constant tendency for growth. Until June 1, 1992, children aged 14–17 were regarded as juveniles. As of June 1, 1992, the lower age limit for

Table 4.8. Trends in juvenile crime

Year	Number of crimes committed by or with the participation of juveniles	+/- as compared to previous year (in percent)	Percentage of total detected crimes
1986	1,516	−13.5	15.5
1987	1,682	+10.9	20.0
1988	1,642	−2.4	24.5
1989	1,516	−7.7	30.5
1990	1,403	−7.5	21.9
1991	1,340	−4.5	23.3
1992	966	−27.9	13.6
1993	1,563	+61.8	17.1
1994	1,816	+16.2	19.0
1995	2,433	+34.0	21.6
1996	2,311	−5.0	20.1

criminal responsibility was raised to 15 (*Eesti Kriminaalkoodeks*). In 1986, 1,516 crimes committed by juveniles were recorded in Estonia, a total of 15.5 percent of detected crimes; in 1995, children committed 2,433 crimes (an increase of 60.5 percent over 1986) or 21.6 percent of all detected crimes (Urvaste, 1997).

The main problem to be solved in the anticipation and prevention of crime is juvenile crime. If Estonia wishes to reduce or at least stabilize crime, its society must deal with children. The sooner responsible adults can interfere in the life of a young person having taken the criminal path, the greater the hope that the youngster can be brought back to the right road, and the better the chance that in the future the number of crimes will diminish. Unfortunately, no progress in this respect has been achieved in Estonia.

Immigration

The pre-war Republic of Estonia was a nation-state with an 88 percent Estonian majority in 1939. However, if we only regard the territory of the former Estonian Soviet Socialist Republic which is controlled by the Republic of Estonia at present, we must leave out of our calculations a slice of territory with about 60,000 inhabitants in 1939 (including thousands of Russians). In that case the proportion of ethnic Estonians would go up to as high as 92 percent. At that time Estonia encompassed a number of long-established minority groups: Russians in the Setumaa area and along the shores of Lake Peipsi, Swedes on the western coast, and Baltic Germans. In addition, there were 4,000 Jews in the major towns and 3,000 Latvians near the southern border. The immigrant population consisted of one or two thousand Poles (mostly farm laborers on a temporary basis) and about one thousand Ingermanland Finns whose native land was adjacent to Estonia's border in the east. A large proportion of the ethnic minorities left Estonia during World War II: the Germans were repatriated in 1939 and the Swedes during the war. Some Jews left Estonia and others were murdered by the Nazis. Therefore the overwhelming majority of resident non-Estonians are post-war immigrants, not descendants of Estonia's genuine ethnic minorities. The population of ethnic Estonians was reduced because of those who departed from the country, those who were killed in the war and those who escaped to the West.

In the period 1946–1990 the Estonian population increased by 680,000 people. Natural increase accounted for 361,000, and 319,000 or 47 percent, were immigrants. In 1990, the population of the country was 1,572,000.

The first wave of immigrants came from areas bordering on Estonia: the Pskov, Leningrad and Novgorod oblasts, later from Central Russia, and finally from the Islamic regions of the Soviet Union.

It should be pointed out that a shift in demographic processes took place in Estonia in the second half of the 1980s. That period is characterized by the birth of the idea of the reestablishment of independence, the so-called singing revolution which came to its logical conclusion with the proclamation of new independence in August 1991 (Leps, 1991c).

Sensing the danger extensive immigration was posing to the nation, all national movements in Estonia shared the common demand of putting a check on immigration and limiting it by immigration quotas. Curbing immigration did bring results, at least as far as legal immigrants were concerned. Estonia's migration balance has stayed negative since 1990 (See Table 4.9).

There are grounds to believe, however, that in 1989–1992, when old structures no longer worked and new ones had not yet been created, much illegal immigration into Estonia took place. As a result, the number of illegal residents in Estonia is estimated at 40,000 to 50,000. Unfortunately, neither the police, the statistics department nor the migration department can detect the exact number of such people in the country. At the same time it is a fact that many of these people are connected with the underworld and their criminal activity is considerably higher than that of any other stratum of Estonia's population (Leps, 1991b).

The present situation regarding the ethnic composition of the Estonian population is illustrated by Table 4.10 listing population data in the present-day administrative borders of Estonia.

The population in Estonia has increased approximately by one-third in 1996 compared to the pre-war period to a total of 1,512,000. It appears that 35 percent of the Estonian population are non-Estonians by their ethnic background.

At the same time different areas in Estonia vary a great deal as to their ethnic composition. The proportion of non-Estonians is high in the industrial centers of Northeast Estonia bordering on Russia. As a rule, the level of crime is higher in those towns (Leps, 1991b), which does not naturally mean that one ethnic group has a more criminal nature than another, but simply refers to the fact that the criminal activity of immigrants is usually higher than that of the stationary population.

Unfortunately ethnic Estonians have been very near to becoming a minority even in their historical capital, Tallinn. The proportion of ethnic Estonians in rural areas and small towns is considerably higher.

Table 4.9. Net balance in migration

Year	Net balance
1990	–4,021
1991	–8,034
1992	–33,827
1993	–13,779
1994	–7,631
1995	–8,200
1996	–5,700
1997	–5,000

Table 4.10. Ethnic composition of population

Year	Population	No of ethnic Estonians	Percentage of ethnic Estonians
1922	1,044,000	951,084	91.1
1934	1,061,300	973,212	91.7
....			
1941	1,017,475	907,976	89.2
1945	809,000	787,000	97.3
....			
1959	1,196,791	892,653	74.6
1970	1,356,100	925,157	68.2
1979	1,465,800	947,812	64.7
1989	1,565,700	963,281	61.5
....			
1992	1,562,216	967,012	61.9
1993	1,528,114	965,768	63.2
1994	1,509,285	964,433	63.9
1995	1,491,583	957,900	64.2
1996	1,475,000	953,500	64.6
1997	1,462,000	950,124	65.0

In the immediate post-war period, the population balance was changed by the genocide of the Estonian people, particularly the elite. In later periods, harsher methods of Russification were abandoned for more sophisticated ones, such as smothering the Estonians under massive immigration (Leps, 1990). In general, Russification did not work nor did attempts to destroy the Estonian intelligentsia, the flag-bearer of resistance to Soviet imperialism.

The proportion of ethnic Estonians among criminals has constantly been 10 percent below their proportion in the population. The percentage was particularly low in the years 1984–1990 when hopes for the reestablishment of Estonian independence were the highest. From 1991 the number of ethnic Estonians among criminals has gone up abruptly and is now fluctuating between 55.8 and 59.8 percent. (Table 4.11) There many be several reasons for this. One of the most likely seems to be that the police pay less attention than before to the Russian-speaking population. Estonian-speaking criminals are caught in larger numbers. This situation has arisen as Russian-speaking policemen have left Estonia.

In recent years, Estonia has expelled over a thousand illegal immigrants. Most of them have been returned to Russia. In the future Estonia must cooperate with the Nordic countries, the destination of the highest number of refugees passing through Estonia. Unfortunately, the countries of Central Europe are not particularly interested in the problems of the Baltic countries since they have a lot of problems of their own. As Estonia has little money and the border with Russia is relatively open, the possibilities of preventing illegal immigrants from entering Estonia are rather modest. The situation is made even worse by the fact that Estonia has not entered into any agreements with any country about the return of illegal immigrants. Because of the absence of such a document, Russia, for example, is not compelled to receive refugees from Estonia. However, Estonia is planning to conclude such agreements with Russia, Finland and Sweden. Estonia does not have a law regulating the status of refugees, as the country has not even acceded to the Geneva Convention, and therefore the issue of providing asylum to refugees has not been

Table 4.11. Ethnic composition of criminals

Year	Percentage of ethnic Estonians among the population	Percentage of ethnic Estonians among criminals
1979	64.7	56.8
....		
1982	63.8	52.9
....		
1985	62.8	50.8
....		
1988	61.8	50.1
....		
1991	61.9	55.8
1992	63.2	59.8
1993	63.9	59.1
1994	64.2	57.5
1995	64.2	57.4
1996	65.0	57.4

regulated. At present, it is out of the question for Estonia to accede to the Geneva Convention, as this would involve major expenditure. According to international requirements, Estonia would, in that case, have to keep refugees at first in a camp of internment and then issue them work permits and provide them with jobs (Leps, 1991c).

PROFESSIONAL CHALLENGES

Training of the Estonian Police

Training is organized in this sequence:

(1) A primary three-month preparatory course at the training centers of the prefectures is given for those initially employed by the police. This is followed by a four-month probationary period. Those who have successfully completed the course and probation are entitled to continue their training at the police school and serve in the position of junior police officer.
(2) Basic training of police employees is conducted at the Paikuse and Tallinn Police Schools. The period of studies is ten months. Those having completed basic training are obliged to work in the force for at least three years. Figures, unfortunately, tell a different story: 31 percent of those who completed the course of the Paikuse Police School in the period 1992–94 and 28 percent of the graduates of the Tallinn Police School have left the force within three years.
(3) After three and a half years in the service, police employees are entitled to acquire applied higher education at the Estonian State Defense Academy Police College. Applicants must have completed basic police school training.
(4) Studies in the higher level of the State Defense Academy are open to those police employees who served in the police for at least three years after completing the Police College course.

By a decree of the Estonian government of June 22, 1993, the Defense Police was set up as an independent institution. This was doubtlessly an event of major significance which ended a two-year period of uncertainty. It gave the Defense Police a feeling of security and outlined its rights and obligations as well as its functions and areas of activity.

The work of the police has been seriously hampered by the fact that in the past three years the post of Minister of the Interior has changed hands three times and that of the General Director of the Police Department six times. Political circles are continuously attempting to politicize the police, which is impermissible. Frequent changes of police structures, one of which is again to be expected in the near future, do not allow police employees to concentrate on their everyday task of fighting crime despite the current high crime rate. Such turnover, however, is probably a characteristic of a country in which democracy is still young and brittle.

Personnel

Personnel selection, schooling and better use of existing staff constitutes the actual basis for the effective functioning of the police as an organization as well as for individual professionalism. Technical means alone do not count for much without educated professional people.

In the first years after the reestablishment of the Republic of Estonia, the strategy of the Estonian police was to increase its numbers. Before the formation of the police there were 5,000 people working in the militia. In the first years the police started off with a roll of 4,500. As of July 1, 1995 the police had 6,605 positions of which 21 percent or 1,391 were vacancies. In addition there are 1,492 positions for office and auxiliary staff, of which 265 or 18 percent were vacancies. (See Table 4.12.) It is not quite clear what criteria have been relied on in creating positions in the police (State Police Department, Information Bureau, 1994–95).

Internationally, the basis for defining the number of police employees is their ratio per 1,000 inhabitants. In Europe this figure is the highest in Greece—4 policemen. In most European countries the number varies from 1.8 to 2.2. In Estonia the figure is 3.5 if one considers the actual number of police employees and 4.4 if the ratio is based on the number of positions. In view of the criminogenic situation in the country and the individual training and low professionalism of the Estonian police, a certain quantitative stability must be maintained, at least in the next few years. When individual professionalism begins to rise, however, the number of police officers could gradually be reduced to a ratio of 3.0 to 2.5, i.e. to 4,500–3,750 police officers. The number of office and auxiliary staff should not be increased, however.

Table 4.12. Composition of police personnel

Number of policemen	As of Dec. 31, 1993		As of Dec. 31, 1994		As of July 1, 1995	
	Positions	Percent	Positions	Percent	Positions	Percent
All policemen	7,312	81	6,558	81	6,603	79
Criminal police	1,387	78	1,345	84	1,335	84
Field police	2,788	78	2,697	82	2,576	80
Traffic police	538	86	731	85	715	84
Investigators	346	78	346	66	346	64

Table 4.13. Background of personnel

Status of policemen	As of Dec. 31, 1993 Number	Percent	As of Dec. 31, 1994 Number	Percent	As of July 1, 1995 Number	Percent
Former militiamen	2,760	47.4	2,221	42.8	2,125	41.4
Women	815	14.0	913	17.6	967	18.8
Ethnic Estonians	3,716	63.8	3,734	71.9	3,706	72.2
Insufficient command of Estonian	1,448	24.9	756	14.6	811	15.8
No police (militia) training	745		651		265	

It appears from Table 4.13 that the number of former militiamen in the Estonian police is decreasing. At the same time the number of ethnic Estonians in the police is growing. As of July 1, 1995, there were 563 policemen who were not citizens of the Republic of Estonia (11%), and 811 policemen (16%) did not have sufficient command of the Estonian language to meet the requirements of their position. Although the number of people admitted to work in the police without any training for police work has been declining, the number of such people is still high. The number of policewomen has been steadily going up and is at present above the average European level of about 10–11 percent.

The formal education of Estonian police officers has been unsatisfactory in recent years. Quite a large number of policemen (ten percent) do not even have any initial schooling at all. Of the rest, 52 percent have secondary education or incomplete secondary education, 28 percent have specialized secondary education and only 20 percent have a higher education degree (of these 10 percent are trained in law). (See Table 4.14.)

We can notice a drop in the number of policemen aged under 25 and an increase in the 25–35 group. This development should be regarded as positive.

Table 4.14. Schooling and age of personnel

Education: Education and age of policemen	As of Dec. 31, 1993 Number	Percent	As of Dec. 31, 1994 Number	Percent	As of July 1, 1995 Number	Percent
Higher education,	1,031	18	1,001	19	1,008	20
of which in law	590	10	517	10	508	10
Specialized secondary education,	1,618	28	1,385	27	1,430	28
of which in law	682	12	605	12	588	11
Secondary	3,045	52	2,709	52	2,591	50
incomplete secondary	127	2	97	2	102	2
Age:						
Up to 25	2,032	35	1,724	33	1,610	31
25–34	1,916	33	1,699	33	1,710	33
35–49	1,568	27	1,439	28	1,491	29
50 and over	305	5	330	7	320	6

Disciplinary Punishments in 1994

According to the State Police Department Information Bureau, 938 police officers received disciplinary punishments in 1994: 79 in the form of a call to order, 325 a reprimand, 159 a severe reprimand and 72 a warning of insufficient compatibility for the position. Sixty-nine police officers were transferred to a lower position and 234 police officials were discharged from the force. Criminal proceedings were taken against 12 police officers; investigation was concluded in five cases and a verdict of guilty was passed on five officers. Fourteen officers were punished for violation of rules regulating the use of special means of the police, including firearms. Seventeen police officers had committed traffic accidents while driving a police vehicle while under the influence of alcohol, and 38 while sober.

Of the 938 police officials who had received disciplinary punishments, 360 (38.4%) had a long service record (they were former militiamen); 238 had completed prefecture courses; 198 had no training and 81 were graduates of the Paikuse and 61 of the Tallinn Police Schools.

Violations of service discipline and actions compromising the reputation of the police are committed by police employees mainly because of unsatisfactory training, negligent attitudes to duties, insufficient life experience (in the case of young police officers) and high turnover rates.

In the period 1992–94, 621 policemen completed the course of the Tallinn Police School; of these 447 (72%) were still working in the police on January 1, 1995. In the same period, 721 policemen completed the course of the Paikuse Police School and 494 (68.5%) of them still worked in the police on January 1, 1995.

In 1994, 1,062 new police officials were hired and employed; of these 651 (61.3%) did not have any special training. In the same year, 1,078 police employees left the police, 671 of them at their own request.

As a result of the high fluidity of its personnel the police has a large number of beginners, employees with poor professional knowledge and little experience, which accounts for their high percentage among those with disciplinary punishments: 436 (46.5%) of the 938 officers disciplined were beginners (State Police Department, Information Bureau, 1997).

RESPONSES TO CHALLENGES

International Cooperation

During the first period of independence, 1920–40, international cooperation of the Estonian police with that of the neighboring countries was mainly limited to contacts which had to do with concrete criminal cases. Almost no materials about it have survived in Estonian state archives, as they have passed through the purgatory of both the German and the Russian secret services. Therefore we know about international cooperation in the police sphere only to the extent to which it was reflected among other publications in the *Police Paper*.

The Estonian police took the first step toward international cooperation in the autumn of 1936 when a four-member delegation of the police headed by the police director took part in the International Police Congress in Berlin, where a a session of Interpol's immediate predecessor, the International Criminal Police Committee, also took place. In the period following the Berlin Congress, the Estonian police were involved in episodic cooperation with said committee but did not send a representative to the committee, a precondition to full member status. The reasons Estonia failed to follow the example of Lithuania and Finland and join the committee are unknown. However, the government found it possible to conclude bilateral agreements concerning the extradition of criminals and to accede to international conventions. Estonia signed the Hague opium convention in 1922. In 1925, an agreement was concluded for the mutual extradition of criminals with the United States. After Estonia's occupation by the Soviets in 1940 all issues of international cooperation were taken off the agenda by the Soviet militia, as pertaining to a sphere which behind the iron curtain had been restricted to the competence of the KGB.

By the spring of 1990, no alternative existed to the reestablishment of Estonian independence. Independence led to the restructuring of the police. Many ideas for this venture came from abroad. Having been a part of the former Russian empire, Estonia had to establish the new police on the foundation of those public order maintenance professionals of the previous regime who had not been involved in politics and had remained true to their calling. Their mental outlook, however, had been shaped by the totalitarian regime whose principles did not always fit those of a democratic society. The first major assistance came from the United States. At the invitation of the Harvard University Russian Research Center and, with support from the George Soros Foundation, a three-member delegation went to Cambridge in order to take a look at the work of the police, the prosecutor's office and the detention system in two states, New Hampshire and Massachusetts. During their stay, an agreement was reached that a small number of Estonian police officials could undergo training in American police academies. As the three persons were later active in writing the draft of the Estonian police law and have been in the police to this day, the benefit from that particular act of assistance was enormous. Active aid in the restoration of the Estonian police was given also by the neighboring country in the north, Finland. Later assistance programs were launched by Sweden and Germany. Assistance in schooling was offered by France and Britain.

The reborn Estonian police began functioning under conditions in which a legislative vacuum on the one hand and open borders on the other hand, side by side with legal nihilism, led to an unexpectedly sharp growth in crime. The borders failed to keep criminals out. They arrived in Estonia from all corners of the world. Criminals from the east were lured into Estonia by her relatively noticeable western ways and high living standard, and those from the west by the spontaneity of the budding business activity and inexperience of beginning businessmen. A contributing factor was the Soviet-created myth of the limited honesty of western businessmen, extreme faultiness of the laws and lack of international agreements in the police sphere even with the neighboring countries. Therefore, international cooperation was sporadic and often limited to personal contacts between police officials in concrete cases, mostly within the borders of the former Soviet Union. As to the economic crimes of western criminals, most of them were quite new to our police. The need to join the International Criminal Police Organization (Interpol) was apparent.

The Estonian police began to prepare documents for joining Interpol and in December 1991 the decision was made by the government. The application was made at the eleventh hour and an invitation was received to take part in the Interpol General Assembly. Contacts were established with the Finnish National Central Bureau of Interpol and a delegation went there on a fact-finding tour to see how work should be organized at home. It was also agreed that before Estonia's admission to Interpol, all contacts with other member countries would be mediated by the Finnish Interpol office.

During the preparatory period and in launching the work of the Estonian office, both Finnish and Lithuanian colleagues were of great assistance. According to the regulations in force, the Estonian police had to be ready to play their role immediately on joining the organization; an Estonian National Central Bureau of Interpol was set up by an order of the Police Department's General Director as of July 1, 1992. On the opening day of the 61st Interpol General Assembly in Dakar, Senegal, on November 4, 1992, Estonia was admitted to Interpol as a full member. Together with Estonia several other fragments of the former Soviet Union, namely, Latvia, Ukraine, Kazakhstan, Armenia and Azerbaijan, were given Interpol member status. Lithuania had reconfirmed the status of a founding member a year before.

International police cooperation is largely a response to new forms of transnational crime. A number of stable criminal associations have been formed in Estonia, each with wide international links. There are armed conflicts between groups over the distribution of areas of influence. The groups are involved in practically the whole repertory of international crime: car theft, the smuggling of alcohol, tobacco, weapons, drugs and radioactive materials, counterfeit banknotes, trade in humans, tax fraud and money laundering. Estonia has become notorious as a transit country of trade in humans, particularly in regard to smuggling Asian immigrants into other European countries. Cases of the smuggling of Kurds via Estonia into Finland and Sweden are well known (State Police Department, Information Bureau, 1994–95).

Recently, the Estonian police, together with some Dutch colleagues, identified an international criminal group which, operating from Estonia with the assistance of a local inhabitant, illegally shipped over 50 nationals of Sri Lanka via Estonia and Sweden to Holland last year. There is information about Estonian girls being lured into Western Europe and compelled to work like slaves in brothels (State Police Department, Information Bureau, 1994–95). There are more and more reports of Estonian nationals, companies and vehicles connected with the trade in drugs and radioactive materials. The illegal export of alcohol and tobacco has achieved proportions which affect the economy. The secret trade in weapons has earned official condemnation by Estonia's neighbor in the east, Russia. Estonian car thieves are known practically everywhere in Europe.

International crime has found itself a firm place among the vices of Estonian society. At the same time, legislative activity in Estonia has not yet been able to eliminate essential shortcomings in criminal proceedings, thus obstructing international cooperation. For example, no one can be sued for crimes committed outside the Republic of Estonia, and there are difficulties in returning stolen property to other countries. Valid legal assistance agreements have only been concluded with Russia, Ukraine, Latvia, Finland and Lithuania, though these function well. European conventions on legal assistance, exchange of detainees and the extradition of criminals were signed three years ago, but have not been ratified by the parliament, again complicating international cooperation. Estonia has not acceded to a number of other international conventions, among which the most important for the police are the European money laundering convention and the United Nations drug conventions. Work at those problems is continuing and positive results are expected soon.

Despite these difficulties, the Estonian police are well integrated into international cooperation. Many police officials have been trained or have taken part in advanced training courses in other countries. Several countries, above all Germany, Finland and Sweden, have given Estonia extensive material assistance, which has substantially improved technical supplies of the Estonian police. Personal contacts among police officials have widened considerably. A section of the International Police Association has been formed in Estonia. International police cooperation now occupies a firm place in everyday work. There have been practical contacts with all European countries, and to a lesser extent with countries of America and Asia. Contacts are most frequent with Finland, Russia, Sweden, Latvia and Lithuania, but efficient cooperation has been carried out also with colleagues in Germany, Austria,

Holland, Poland, Switzerland, the United States and Great Britain (State Police Department, Information Bureau, 1997).

Estonian Police Law

Background

The attempt to bring about democratic behavior by a police force through legislation is a necessary though not sufficient element for change. Laws without the culture of respect for the rule of law, as is evidenced by the police in established democracies, are insufficient by themselves in ushering in democratically-oriented police services. Estonia has enacted specific police laws to introduce democratic norms as regards the organization and work of the police. As this chapter demonstrates, new laws are an expression of political will to break away from the older, totalitarian heritage of policing. But laws are only the first step. The more difficult change is to have the police accept and act on the norms embodied in the laws. As noted earlier, that stage is still in progress.

A new Estonian Police Law was adopted on August 20, 1990, and the Estonian police was called into existence as of March 1, 1991. When the Estonian Police Law was adopted the Estonian parliament was still called the ESSR Supreme Soviet. Writing the police law began in 1989 when a committee to draw up Estonian legal protection concepts was established, headed by the then-deputy chairperson of the Estonian SSR Council of Ministers. There were representatives of all legal protection bodies among its members: the Estonian SSR Ministry of the Interior, the Estonian SSR Ministry of Justice, the ESSR Supreme Court, ESSR State Security Committee, the ESSR Council of Ministers and the ESSR Supreme Soviet, officials from the existing districts' and towns' executive committees, leading jurists both in theoretical and practical issues, and lawyers.

Writing the legal protection concept was ridden with difficulty and marked by conflicting interests and world outlooks. It should not be forgotten that at the time Estonia was still a socialist country. Despite all the differences and arguments, the concept was completed in November, 1989 and published in the *Rahva Hääl* daily together with an appeal for its public discussion. Meanwhile, developments in Estonia progressed very quickly and no public discussion ensued. The word *militia*, however, was gradually ousted by the word *police*. Yet it was clear that the ESSR Ministry of the Interior, with all its institutions and statutes, could not simply be renamed the Estonian Police. Besides, the then-ESSR Ministry of the Interior was still subordinated to the USSR Ministry of the Interior.

With free and democratic parliamentary elections in March, 1990, Estonia's aspirations for independence and disintegration from the Soviet Union took another step toward reality. Later, an agreement was reached with the USSR Ministry of the Interior, as a result of which the ESSR Ministry of the Interior gained independence. As Estonia was set on a course for full independence, it became more and more obvious that the Estonian police would have to be reinstituted in order to guarantee protection to the interests of the Estonian state and the security of its citizens. In the initial phases, major discussions were provoked by the word *police*, because of associations with the "Polizei" of Estonia's Nazi Occupation period of 1941–44.

A police force cannot function, however, if there is no legal basis for it. The need arose to work out an Estonian police law. As a number of subcommittees and work groups had already been established when the Estonian legal protection concept was written, it was not difficult to set up a group to work out the Estonian police law. Its first session could be called a historic one. The committee sat around a big table, each person with a clean sheet of paper before her or him. Materials on the initial years of the Estonian police were consulted and the fifty years of Soviet occupation had to be taken into account as well. Through difficulties and hot discussions the first version of the law was born. The law was adopted by the ESSR Supreme Council on August 20, 1990, even before the adoption of the Constitution. Possibly the version adopted was not the best one, but it did create a basis for the resurrection of the Estonian police. The law, amended on April 21, 1993, regulates the activity of the Estonian police.

In accordance with the police law, the ESSR Council of Ministers issued a decree to define the service areas of police prefectures. In November, 1990 the top police department officials were appointed, along with all police prefects. County-level interior affairs departments had not yet been abolished then, and up to March 1, 1991, these departments were headed by police prefects. It was a difficult period of transition and many specialists retired from service, unable to adapt to the period of uncertainty. A police school had been opened at Paikuse on September 1, 1990, to train the first police officers after the occupation period (*Eesti Politsei 75*, 1993).

The first version of the police law adopted on August 20, 1990 did not contain police service regulations and these had to be enforced later by a separate government decree. Another shortcoming of the police law was that it did overlook police employees' social guarantees in all their aspects. This has led to several amendments to the police law, and could only be corrected in the new version of the law, prepared to be submitted to the parliament in 1996. The new draft defines, in very concrete terms, police officials' protection by the state, as well as their responsibility and long-term security, until now defined in vague terms or not defined at all. Other reasons for the writing of a new police law are the need for a change in the police department structure, the establishment of the security police and a number of other circumstances.

When the police law was drawn up, both the ESSR Criminal Code and the ESSR Criminal Proceedings Code were still effective. These contained stipulations which made it impossible to reflect all the elements of a democratic state in the police law. Some of the most complicated problems in writing the police law concerned the definition of the structure and functions of the central body, the police department. As the situation had changed very much, the new structure of the police and its central body could not be built up after the example of the pre-war Republic of Estonia. There were heated discussions whether the central body, the Police Department, should be an administrative body only or if it should also have some functions as a working body. The problem has not yet been solved to everyone's satisfaction, and the police department continues to carry out a double role, that of an administrative and a working body at the same time.

The newly adopted Estonian police law became the basis on which it was possible to continue to draw up other documents required for the work of the police, such as the statutes of the police department, prefectures and territorial police stations, bureaus and services, the wearing of uniforms, and other regulations. The police law and its amendments have always gone together with more accurate definitions of development trends and the application of new principles characteristic of a democratic state. The most important trend in the writing of Estonia's new police law and of the development of the police force created on its basis was to guarantee that, contrary to the repressive Soviet militia, the Estonian police would become a body protecting the interests of the Estonian people. It was extremely important to provide that the rights given to police officials by the police law should not violate the constitutional rights of individuals and should not contradict universally recognized concepts for the protection of human rights. By 1995, the

Estonian police had come to the next stage of its development, when work at drawing up the police development plan for 1996–2000 was begun.

Long-term Security

Below is a brief overview of the long-term security provided to police officers by the state. The state must ensure long-term security to policemen. As Article 3 of the police law stipulates, police officials are civil servants who represent state power within their areas of competence while on duty. While the activity of most other civil servants is regulated in the civil service law, the activity of police officials is regulated by the police law.

The police law defines the place and role of the police in the system of state and municipal institutions, and defines the main tasks, obligations, rights, and responsibility of the police, as well as its organizational foundations and supervision over police activity. (The full text of Chapter 8 of the new draft of the Estonian Police Law is included in the Appendix.)

Media

Relations between the media and the police are often contentious. The media may exploit failures of police work for their own economic, political and journalistic instincts and interests, or incite the police to behave aggressively against perceived dangers to public well-being and order, such as stories about threats posed by immigrants or ethnic or marginal groups to the safety, identity and future of the country or locality. The police, in turn, may seek to use the media to polish their image, extract resources by playing on people's fears of crime, and cater to bias, prejudice and stereotypes by their description of criminality and dangerous classes.

Recently a tabloid published in the Estonian capital printed a story on how policemen used rubber batons to disperse a youth gang assembled in one of Tallinn's outlying residential areas. The police had been called by residents of the neighboring houses whose peace was continually breached by the boisterous youths. Most of the youths appeared to be drunk or under the influence of drugs. They refused to obey police orders and the police were compelled to use force. The reporter, however, described the event as a depressing act of violence against young people. To police were ascribed phrases and words which are not fit to print. Additionally, a police officer who wasn't even on the spot was named as the leader of the operation. The story abounded in fanciful details and smacked of slander.

Unfortunately, such articles are not exceptional in the Estonian press. In reporting the conflict between the offender and the police, young journalists, often with very little experience, frequently assign the police the role of the culprit. Why? To a large extent this must be due to the legal nihilism characteristic of the period of transition Estonia is currently experiencing—a legal vacuum, as it were, created by the replacement of old laws and morals with new ones. Democracy is interpreted as license to do anything and freedom is seen as anarchy. Many people also have their personal accounts to settle and often the press is used for that purpose.

In response to such attitudes, the magazine *Police* was launched. In 1992, means were made available for printing a sixteen page monthly publication. Since 1993 the journal has grown to twenty-four pages and is printed in two colors; beginning in 1996 the number of pages grew to thirty-two. The circulation of the journal has been stable at 6,000 copies.

What kind of materials does the journal publish? The stress is on articles reflecting police work in Estonia and elsewhere. In each issue there is at least one longer article on how some serious crimes were investigated and solved. There is a regular column in which successful police operations are briefly recounted. The column "From the History of Estonian Criminal Science" summarizes analytic writings published on the investigation and detection of grave crimes from the past. There is also a regular section called "In the Courts." Most issues open with a long interview with a leading police official. Light is also shed on the life and work of border guards. Some space is also reserved for crossword puzzles and jokes.

The journal regularly publishes translated articles on the work of the police in other countries—the US, Finland, Germany or Sweden—which are received through contacts with foreign police journalists as well as from foreign newspapers, radio and the TV. This line of work is considered to be particularly important because the Estonian police have a lot to learn from their colleagues abroad. Fact-finding trips to other countries are frequent, but not everything learned is adopted, in view of the fact that the police is a relatively inert and conservative organization.

CONCLUSION

On the eve of the turn of the century, with the world in a deep crisis, we have finally come to the question, what next? Although a degree of democracy has become established in the whole world and the influence of superpowers has diminished along with the opposition of the East and the West—as a result of the collapse of the Soviet Union—sources of tension and conflict which earlier seemed to remain in the background have moved to the fore.

Above all, there is the open opposition between the formal and the shadow economy. The questions of how and to what an extent is an open or official economy capable of controlling the shadow economy have assumed extreme importance. It is clear even without further discussion that organized crime, particularly international organized crime, finds its foundation and support in the shadow economy. The opposition of the formal and the shadow economy is *the* problem for today's world (Leps, 1994b). Consequently, if we look at the essence of things, the war against crime is in the first place a war against organized crime or against the shadow economy. The war against the so-called classical crime, which has been considered "an expression of inevitable evil" in the West, and "a hangover of capitalism" in the socialist East, is of only secondary importance (Leps, 1994a, 1995b, 1995c).

Society today has come to a stage where humanity is confronted with environmental pollution, lack of water, famine, illiteracy, unemployment and crime. The people of the modern world are characterized by loss of faith in former values. Quite recently society was reinforced (or at least maintained) by respect for state power, patriotism, faith in the political system of the country, religion, etc. Several of these factors are no longer of any consequence today. The notion of honesty has become devalued. In the former socialist countries, in particular, much of the population and especially high-ranking civil servants are attempting to become rich by any means possible. The building brick of society—the family—has changed as well. A disdainful attitude of children to their parents has become widespread, parents are ceasing to be an authority to their children, and on the whole, the authority of another person has become a thing of the past. The minority is ever less and less satisfied with decisions taken by the majority.

In many people, such a situation causes an emptiness of the soul. As a result, some turn to religion where they are often caught in the grip of fundamentalism or fanaticism. Quite a few people, however, find crime to be a way out of the situation.

The proportion of the latter in the world population is unfortunately increasing, and crime, particularly serious crime, is on the increase in the whole world. Law and order bodies, particularly the police, are ever less and less successful in coping with the tasks society has laid on them. All the above processes also occur in Estonia and its neighboring countries, a fact that has to be taken into account by both state institutions and the population in guaranteeing law and order.

Is it possible to fight organized crime? Organized crime comprises above all domestic and international economic crime. This crime obeys the dictates of the shadow economy and it is no secret that people connected with the shadow economy react much faster and more subtly to changing needs than the open economy is capable of doing (Leps, 1994b). In Estonia the proportion of shadow economy may be as high as 35 percent of the gross domestic product. Calculations as to the wealth of organized criminal gangs in worldwide terms differ a great deal, but probably their yearly turnover is up to 750 billion dollars, more than the gross national product of most of the world's countries. This was pointed out at the United Nations international conference against organized crime in November 1994 in Naples (Leps, 1995a, 1995b). Nor is it a secret that the shadow economy surrounds itself with its "own police" (from body guards to paid murderers), qualified lawyers and experts and pays them considerably more than open structures are able (or willing) to do. It is also a known fact of everyday life that criminals united into different organized gangs work with much greater dedication than the police who are supposed to protect open economy from criminal assaults. It is a logical question to ask: Why do criminals belonging to organized gangs work better and more successfully than the official police or law and order bodies in general?

In my opinion the answer is very simple. The people belonging to criminal gangs have an aim to become rich. The faster this happens the better. Consequently, criminals have a concrete motivation. And now I ask my second question: what is the motivation of the police who often have to work with low-quality vehicles and weapons, without the necessary training and for a low salary? As that motivation is nowhere directly in view, police officials will obviously never get rich, while there is some probability of their getting killed.

Rider (1993) is of the opinion that neither the police nor the prosecutor's office is capable of defeating organized crime today. Traditional methods are of no help here. Such work can only be done successfully by intelligence and counter-intelligence bodies. There may be a grain of truth in this.

This leaves us with the most important point. Talk about patriotism or love of one's native land has little if any significance at the present time. Maybe we simply have to accept that there exists a certain category of people who have, inexplicably, taken it into their heads to do this hard work called policing. As Hegel (1927) has noted, in classical Athens, the home of democracy, only slaves worked as policemen for their work was regarded as being too dirty for free citizens.

REFERENCES

Eesti Politsei 75 (75th anniversary of the Estonian Police) (1993). Tallinn: Koostanud K. Liiva. The Estonian State Police.

Eesti statistika aastaraamat 1997 [Statistical yearbook of Estonia 1997] (1997). Tallinn: Statistikaamet.

Hegel, G. W. F. (1927). *Filisofija prava* [Philosophy of law]. Moscow: Sotsekgiz.

Leps, A. (1990). Kuritegevuse seisund, dünaamika, tase ja struktuur Eestis aastatel 1944–1989 [The state, dynamics, level, and structure of crime in Estonia in 1944–1989]. *Tartu Ülikooli Toimetised, 916*, 11–27.

———. (1991a). Comparative treatment of crime. *Tartu Ülikooli Toimetised, 925*, 47–79.
———. (1991b). Comparative analysis of crime: Estonia, the other Baltic republics, and the Soviet Union. *International Criminal Justice Review, 1*, 69–92.
———. (1991c). *Kuritegevus Eestis* [Crime in Estonia]. Tartu: Tartu Ülikool ja Eesti Akadeemiline Õigusteaduse Selts.
———. (1992). *A brief survey of the criminogenic situation in the Republic of Estonia, 1945–1992.* Tallinn: The Estonian State Police.
———. (1993). Viron racialists kasvanut rajusti itsenäisyyden aikana [Crime in Estonia in the period of independence]. *Helsingin Sanomat, 3.*
———. (1994a). Przestepczosc w Estonii (Niektorze teoretyczne I praktyczne problemy) [Some theoretical and practical problems related to crime in Estonia]. *Preglad Prawa Karnego, 10*, 43–52.
———. (1994b, June 15). Avaliku majanduse ja varimajanduse vastasseis on tänapäeva maailma jaoks probleem nr. 1. [Opposition of the open and shadow economies is problem number one for today's world]. *Rahva Hääl.*
———. (1995a). *Estonian Police 1918–1995.* Tallinn: The Estonian State Police.
———. (1995b). *Ühiskond—kuhu lähed?* [Society—where are you going?]. *Virumaa Teataja 17–26,* January.
———. (1995c). Bodies of law and order, particularly in a changing society. Paper presented at the International School of the Sociology of Law in Onati, Spain.
———. (1995d). VIII Balti riikide rahvusvaheline kriminoloogia seminar St. Peterburgis [The 8th international criminology seminar of the Baltic countries in St. Petersburg]. *Juridica, 1.*
Leps, A., & Remmel, M. (1994). *A crime forecast for Estonia.* Tallinn: The Estonian State Police.
Leps, A., & Remmel, M. (1997). *Crime and income distribution in Estonia. Crime and criminology at the end of the century.* Paper presented at the IX Baltic Criminological Seminar, May 22–25, 1996. Tallinn: Estonian Defense and Public Service Academy.
Liit, A. (1920). *Politseiniku käsiraamat* [Policeman's handbook]. Tallinn: Oma Kirjastus.
Liit, A. (1927). *Kuritegude jälgimine* [Crime investigation]. Tallinn: Siseministeeriumi Administratiivala Kirjastus.
Lindmäe, H. (1994). Eesti politsei ajaloost [On the history of the Estonian police]. *Juridica, 3.*
Meet Estonian Police (1992). Tallinn: The Estonian State Police.
Rider, B. (1993). *Organised Crime in the United Kingdom.* Cambridge: Cambridge Unversity Press.
State Police Department, Information Bureau (1994–95; 1997). Press release. Tallinn: The Estonian State Police.
Urvaste, H. (1997). *Alaealiste kuritegevus aastail 1994–1996* [Trends in juvenile crime in 1994–1996]. Tallinn: The Estonian State Police.

Appendix

CHAPTER EIGHT
POLICE OFFICIALS' STATE PROTECTION, RESPONSIBILITY AND LONG-TERM SECURITY[1]

Article 34: Legal protection and security of police employees' activity

(1) No one shall have the right to compel a police official to carry out duties not established by law or other legal acts.
(2) No political party or any other association or movement or its representative shall have the right to interfere in a police official's work or to give him orders.
(3) If a police official has been given an order which is in conflict with the law he shall be guided by law.

---

[1] Translation by the author.

(4) Police officials shall have the right to unite into trade unions for the protection of their rights and interests.

Article 35: Police officials' responsibility

(1) A police official shall assume the responsibility provided by law for unlawful activity or inaction.
(2) Obeying an unlawful order or instruction shall not relieve a police official from responsibility.

Article 36: Compensation of costs of schooling

(1) A police official shall be obligated to compensate the expense incurred by the state for his schooling if he was discharged from police service:
a) because a court sentence pronouncing him guilty became effective;
b) for a disciplinary offense;
c) at his own request.
(2) No compensation for the expense shall be required if the police officer's service in the police before his discharge is:
a) at least three years after completing the full course of a police school;
b) at least five years after graduation from a higher police educational institution.
(3) Compensation for the cost of schooling shall be provided for in a schooling contract.
(4) The order for the calculation of the costs indicated in sections 1 and 2 of this article shall be established by the Minister of the Interior.

Article 37: Police officials' long-term security

(1) Persons supported by a police official who died while carrying out his duty or in connection with his service in the police shall be paid a single subsidy of up to ten years' wages by the state. The dead police official's funeral shall be organized at state cost.
(2) A police official having been disabled while carrying out his duty or in connection with his service in the police shall be paid a single subsidy as follows:
a) in the case of 1st or 2nd category disablement: up to five years' wages;
b) in the case of 3rd category disablement: up to one year's wages.
c) a police official to whom a bodily injury was caused in connection with carrying out his duty or in connection with service in the police shall be paid a single subsidy of up to three months' wages.
d) The order for the calculation and payment of the subsidies indicated in sections a, b and c of Section 2 shall be established by the Republic's Government. Depending on the number and age of persons supported by the police officer, the Republic's Government may increase the size of the single subsidy.
e) The cost of treatment of a police official having been injured or fallen ill while carrying out his duty or in connection with his service in the police, including the cost of medicines, shall be covered by the state.
f) If necessary and if possible, a police official shall be allocated employer's housing in his service area. A local self-government may allocate housing at the expense of the self-government's housing stock to a police official serving the commune or town on conditions and in the order established for the allocation of employer's housing. If a police official dies while carrying out his duty or in

connection with his service in the police, his spouse and under-age children shall retain title to the lodging.

g) The state shall guarantee police officials free regular health inspections in the order established by the Minister for Social Affairs in consultation with the Minister of the Interior.

h) Police officials shall be provided with compulsory insurance by the state on the basis of personal insurance policies.

Article 38: Police officials' pension for years of service.

(1) A police officer shall be guaranteed a state pension, with consideration of exceptions provided by this law.
(2) The size of the pension of a pensioned police official is 60 percent of his last basic wage.
(3) A pensioned police official's wages shall be revised if the wage rate in his last position is later changed. The order for the revision of pensions shall be established by the Republic's Government.
(4) Police officials shall be entitled to the pension indicated in section 2 of this article after 20 years of service in the police, including work in other positions counted into the police service record. Every additional year in excess of this term shall add 3 percent to the pension, up to 75 percent of the wage on the basis of which the pension was calculated.

Article 39: Police officials' disablement pension

(1) A police official's disablement pension shall be allocated to a police official disabled while carrying out his duty, irrespective of his service record in the police.
(2) A police official's disablement pension is: 100 percent of the basic wage for 1st category disabled persons; 80 percent of the basic wage for 2nd category disabled persons; and 60 percent of the basic wage for 3rd category disabled persons.

5

The Challenges of Policing Democracy in Hungary

ISTVAN SZIKINGER

INTRODUCTION

Hungary is a small country in the very heart of Europe, with an area of 93,000 square kilometers and a population of 10.5 million. It is situated in the Carpathian Basin, surrounded by Austria, the Slovak Republic, Ukraine, Romania, Yugoslavia, Croatia and Slovenia. The river Danube bisects the country, dividing it into two almost equal parts.

The process of changing the political system began at the end of the 1980s as a consequence of domestic and international developments. A peaceful process of negotiations between the Hungarian Socialist Workers' Party (HSWP) and the Opposition led to the constitutional reforms of 1989. The first post-communist, the right-centrist government, was formed by a coalition of the Hungarian Democratic Forum, the Independent Smallholders' Party, and the Christian Democrats following the 1990 multi-party elections. In 1994, however, a landslide victory of the former (reform) Communists clearly revealed the disappointment of most Hungarians with the government's political leadership. The Hungarian Socialist Party (the successor of the HSWP) acquired an absolute majority in Parliament but for political reasons decided to form a coalition with the liberal Free Democrats.

The Hungarian economy is growing rapidly, increasing by four percent of Gross National Product (GNP) in 1997. The Gross Domestic Product (GDP)/person ratio (income per capita) was $4,260 in 1996. The unemployment rate reached 9.2 percent that year (http, 1998).

THE CONCEPT OF DEMOCRATIC POLICING

No general agreement exists concerning the concept of democratic policing. It is a platitude, of course, that any definition of democratic policing presupposes first of all a democratic society to be served and to be relied on as a supportive working environment. In other words, democratic policing is inseparable from a corresponding societal environment. The real question is, however, whether the organization and functioning of law enforcement should itself incorporate the basic values of a democracy.

Many authors think that the phrase "democratic police force" is a contradiction in terms because of the inherent authoritative and paramilitary character of institutions dealing with the maintenance of public security (Berkley, 1969, p. 2; Westley, 1970, p. 22). Within this conception, limiting the scope of police powers and strengthening the accountability of law enforcement structures are the core steps and goals for rendering the police apparatus at least tolerable for a democratic society.

Another perspective perceives police as an integral part of democracy by embodying its features (Reith, 1943) or being capable of adopting them (Skolnick, 1975). Recently the idea of community policing has brought back the concept of the police as an organic, and sometimes even organizing, part of the democratic public (Alderson, 1979; Trojanowicz & Bucqueroux, 1990).

Hungary has had a long tradition of self-government including responsibilities for maintaining order and keeping the peace. Apart from periods without real independence (e.g., occupation by Turks in the sixteenth and seventeenth centuries, the direct dictatorial government of the Habsburgs between 1849 and 1867), policing used to be first of all part of the jurisdiction and competence of local authorities. This gave, no doubt, democratic features to institutions of public security since local bodies used to be generally, but not without exemption, close to the population in terms of satisfying people's needs. Following the 1867 political compromise between Hungary and Austria, however, central government took the initiative for modernization. Since then, the mostly nobility-based provincial self-governments lost their progressive character. Local governments had been relatively democratic within the system of the pre-modern state, being more sensitive to local problems than any central, far-away institution could be. However, progressive constitutional reforms were much better articulated by the central state, since reforms necessarily included the introduction of ideas and policies from more developed countries.

The nationalization of the police began with the Act on the Status of Budapest as the Capital City of Hungary enacted in 1872, and the 1881 Act (No. XXI) establishing the Budapest State Police. The first article of the 1881 Act determined that

> under the name of "Capital Police" a police force with an integrated structure shall be established in Budapest, which shall be controlled by the organs of the state. The jurisdiction of the said police force shall cover all the internal and external territories of the capital (*Fővárosi rendőrségi törvény*, 1881).

The regulation reflected the general pattern of the European continent of placing police activities into a relatively structured and elaborate legal framework. Thus, the democratization of the political system necessarily had an impact on policing the Hungarian capital. Democratization was a necessary consequence of having rejected the absolutist dictatorship of the Dual Monarchy. In establishing independent institutions, the constitutional patterns of advanced democracies were considered and partly taken over.

In the same year, a Hungarian National Gendarmerie was established for rural law enforcement. This organization bore explicit militaristic features but it did not have public administrative powers. The gendarmerie followed and enforced instructions issued by civilian organs of the state and by local authorities. The democratic character of its activities depended on the nature of its directing authorities. Since the gendarmerie followed orders, the government unit and personnel issuing the order naturally determined the outcome of a given intervention. Nevertheless, members of the gendarmerie had to act even in the absence of civilian authority and control and that led to the development of rules of conduct enshrined in internal regulations. Through this process, militaristic attributes inevitably influenced all functions of the gendarmerie. In addition, military characteristics appeared more and more frequently in other policing agencies, including the Budapest State Police (Fosdick, 1969, pp. 124–126). Thus, a French model of law enforcement characterized Hungary by 1945. Two state-controlled policing agencies were responsible for the maintenance of order: the police in the cities and the gendarmerie in the rural areas.

The nationalization of the remaining city police organizations took place following World War I. By that time, central government had, to a great degree, lost its progressive, democratic character. After World War I, the post-war revolutions and the fall of the Austro-Hungarian Monarchy, a right-wing political course was set under the leadership of Governor Miklós Horthy. The merger of city police forces (including the capital police) into a national structure was carried out on the basis of a government decree which embodied definitely lower constitutional values compared to the 1881 Act on the Budapest State Police.

The first post-war government declared that the gendarmerie bore collective responsibility for contributing to the deportation of Jews and other people in violation of the elementary norms of humanity. The institution was abolished and since then one centralized national police organization fulfills the general tasks of law enforcement and order maintenance.

The authoritarian nature of the Hungarian law enforcement structure was accompanied by a wide scope for discretion (Concha, 1988; Kmety, 1900). Being aware of these characteristics, progressive writers argued for judicial oversight of police activities as the minimum requirement for a responsible police (Tomcsányi, 1929). No significant academic discussions on the democratization of police have taken place during the communist era.

ORGANIZATIONAL CHALLENGES

Even before the collapse of the communist dictatorship in 1989, opposition political forces had called for the "blowing up" of the oppressive police apparatus. Anti-communist groups formed an Opposition Roundtable to negotiate, as an integrated body, with the outgoing regime about the rules of transition. The Roundtable was comprised of these groups, movements and organizations: *Bajcsy-Zsilinszky Társaság* (Bajcsy Zsilinszky Society); FIDESZ (*Fiatal Demokraták Szövetsége*—Union of Young Democrats); FKGP (*Független Kisgazdapárt*—Independent Smallholders' Party); KNDP (*Kereszténydemokrata Néppárt*—Christian Democratic People's Party); MDF (*Magyar Demokrata Fórum*—Hungarian Democratic Forum); MNP (*Magyar Néppárt*—Hungarian People's Party); MSZDP (*Magyar Szociáldemokrata Párt*—Hungarian Social Democratic Party); and SZDSZ (*Szabad Demokraták Szövetsége*—Alliance of Free Democrats).

There was a broad consensus among the opposition forces that the police should be democratized by rendering them accountable to elected local governments (Szikinger, 1989). However, the first democratically formed government following the free elections of 1990 revised the earlier proposals of the groups on which the electoral coalition was based. Transformations of the law enforcement structure were postponed into the indefinite future.

Innovation within police was in the beginning limited to emphasizing the new democratic environment for policing and to the promotion of the idea and image of policing as a service for the population. The issue of the democratization of the police was absorbed into the question of how to reform the general state structure. The then-Head of the Ministry unit responsible for policing, Mr. László Korinek, stated that "the police are only a part of this issue and if the state itself is democratic its police should be the same. Democratization does not depend on the organizational principles which will determine the police structure being created" (Korinek, 1992, p. 10).

Consequently, the police survived the transformation of the political system basically intact. After a general protest action by the police in October, 1990, which blocked major transportation junctions and the ports of entry of the country, even cautious endeavors toward structural democratization of the police have been dropped. The dominant items on the agenda since then have been the efficiency and strength of the police. The democratic reorganization of the still-overcentralized and militaristic national police force in Hungary is not part of the discussion.

OPERATIONAL CHALLENGES

The Crime Situation

Crime and Social Values

My thesis is that crime in itself does not represent a threat to democracy. Rather, the real challenge is the way the society copes with crime. Crime is *prima facie* a reflection of legal definitions declaring certain behavior which is considered deviant to be punishable according to the relevant sections of the Criminal Code. Because the term deviant inevitably points to acts and omissions which are exceptional or in opposition to the way of life of the majority, crime never can prevail in a real democracy which expresses the values and norms of the society as a whole.

Further, as there are additional, less serious legal forms of sanctioning undesirable behavior (e.g., torts, administrative offenses, moral condemnation, etc.) crime and its punishment should encompass only a very small part of social relationships. This also means that if some forms of behavior were to become the general pattern or at least widely acceptable in a society, such behavior necessarily would have to be legalized, independent of its moral values. Maintaining slavery or some forms of economic speculation (e.g., establishing a monopoly) can be mentioned as examples of activities defined as criminal acts in respect to the normal workings of society and the economy, depending on the ideological state of affairs in those societies. For example, democratic institutions had existed within the city-states of ancient Greece (despite an economy based largely on slavery in some city-states, such as Athens) and even in certain states of the USA before the abolition of slavery declared the human equality of all people. Even enlightened societies can declare clearly immoral acts to be legal. On the other hand, according to present Hungarian law, wearing a piece of clothing with a red star is a criminal offense, though in most cases no anti-social intent underlies the display of such totalitarian symbols.

Crime is, therefore, by virtue of its very definition, not a challenge but a natural attribute of all societies. Crime is not a social but a legal phenomenon. Crime cannot pose a menace to society but only to the legal system. Democratic societies are characterized by high levels of incorporating and reflecting the societal structure of interests in the institutions of politics and economics. If we accept that people act in correspondence with their interests, the actual state of criminal law and criminality in a given country reveals first of all not the moral level of a society but its capacity to organize itself in an appropriate manner. When criminality is presented as a major threat to society, there is always an effort being made behind the scenes to introduce or to reinforce non-democratic institutions.

In sum, I assume that crime is never a cause but rather a pretext for shaping government and judicial policies serving the various interests of those acting in the name of public power.

Patterns of Crime

Many scholars speak about an explosive increase in criminality in the post-communist countries during periods of transformation in the political system. No doubt, official statistics reveal a substantial growth of crime.

According to unified police-prosecution statistics, there were 157,036 reported criminal offenses in Hungary in the year 1984. As the country had about 10.5 million inhabitants at that time, the crime rate was 1,471/100,000 people. By 1992 the rate had risen to 4,326/100,000 based on 447,215 reported cases. Then a slight (about a 10 percent) decrease occurred. In 1995, however, the trend was again upward, reaching a peak of 502,036 recorded crimes. In the last year (1996) from which we have complete data, 446,050 offenses were reported.

The lion's share of crimes committed are offenses against property which, according to 1996 statistics, accounted for 78.4 percent of all offenses. In the same year, crimes against public order (a miscellaneous category for street crime, document forgery, etc.) represented 11.4 percent, traffic offenses 4.4 percent, and criminal acts against persons (murder, manslaughter, and assault) 3.1 percent of the total.

As mentioned earlier, criminal offenses against property constitute the large majority of criminal acts. In 1993 and 1994 fewer offenses of this kind were committed than in the first years of the political transformation—307,396 and 287,095, respectively. In recent years, however, property crimes increased to 365,235 in 1996. Some criminologists have argued that the earlier, temporary decline could be attributed to the termination of many insurance contracts because of rising prices. Thus, people were less interested in reporting since they were aware of the low probability that their property would be recovered via police intervention (*Rendőrségi Évkönyv*, 1994). Thefts represent the most frequent property offenses; in 1994, 135,620 were recorded. In 1996, thefts amounted to 76.7 percent of all property crimes. A growing organizational sophistication and professionalism among criminals have made the task for criminal justice personnel in this field more and more difficult.

The growth in the number of criminal suspects has proven to be less dramatic than that for crimes. In 1985, criminal proceedings were instituted against 85,766 people suspected of having committed offenses (805/100,000 population); by 1992 their numbers amounted to 132,644 (1,283/100,000 people). As with the number of crimes, a small decrease occurred in 1994 when 119,494 (a rate of 1,163/100,000) suspects were recorded, while in 1996 their number rose to 122,226 (a rate of 1,182/100,000).

Although the basic pattern of criminality has not shown radical changes, there are tendencies and emerging new dangers worth mentioning. As was stated earlier, the overwhelming majority of offenses have been committed against property. Within this category, though, there are criminal acts which necessarily or eventually include the use of violence (e.g., robbery). All in all, violent and riotous crime is expanding both in terms of numbers and by its contribution to overall crime. The level of violent crime increased by 20 percent during recent years, and its total crime contribution to the overall crime problem has exceeded 6 percent.

In 1994, 313 murders and manslaughters were registered, an increase from 298 killings in 1993; in 1996, the number stood at 271. Although these numbers are not enormously high by international comparisons, these are disturbing trends to be noted. It is not the only the numbers themselves but rather the motivations and methods of committing murder which are causing serious concern for criminal justice authorities. In earlier years, revenge, hatred, jealousy and covering up other crimes had been typical motives of killing. Now, there are increasingly material motives and considerations for these crimes and more and more are premeditated and involve firearms. Relatively new phenomena—fights among foreign criminal gangs and the hiring of professional killers—add to the growing menace of this kind of criminal activity. Hold-ups represent another dangerous category of crimes against life and property. Offenders committing such crimes more and more often kill their victims without giving them a chance to save their lives by satisfying the criminals' demands. There were 8,464 assaults causing serious injuries in 1994 and 7,486 in 1996. The number and proportion of these offenses are likely to increase, which indicates that aggressive behavior among offenders has become more common.

Sexually motivated crimes have remained relatively stable, with rape cases averaging between 400 and 500 for a long period. In 1994, 436 such offenses were reported.

In recent years (1992–96) about 20,000 crimes related to the economy were reported. This number represented a decline compared to the early 1990s, partly due to the liberalization of regulations affecting the economy. Criminal Code provisions typical of a socialist, state-controlled economy, such as prohibitions against speculation (seeking to make a profit) and criminalizations of violations of economic discipline (such as union organizing work) have been removed. As a consequence, and naturally so, no one has been charged with these offenses, which were part of the earlier economic crime pattern. At the same time some spectacular cases of fraudulent manufacture and labelling (for example, of paprika and wine) shocked the public and revealed underworld involvement in the fight for markets and economic positions (Egységes rendõrségi 1984, 1991, 1992, 1993, 1994, 1995, 1996). Products of inferior and debased quality could be sold at lower prices and thereby push competitors who observed regulations out of the market.

While trends and patterns of criminality are, of course, not to be underestimated, it is my thesis that the increase in crime compared to the communist era is not as dramatic as reflected in the official statistics. The number of reported crimes cannot be used as the basis for evaluating the actual security situation before and after the political change. Dictatorships are always built on the fear and insecurity maintained by those in power, in order to keep people dependent and subservient. Large numbers of people had been deprived of their lives, freedom and/or property under communism without any possibility of seeking justice. Arbitrary proceedings took place contrary to law, disguised in legal formalities (such as the "show trials") or in compliance with formally valid phrases which could be interpreted in any way desired, such as "dangerous to society" or "in the interests of socialist co-existence." Such terms were used in legal codes but courts and administrative agencies also referred to them as principles which were not to be questioned.

Institutionalized insecurity was counterbalanced by a mythologized safety achieved through the people's reliance on an apparently resolute paternalistic power. Citizens did not even have to bother with such "residues and impacts of capitalism" as criminality because crime was expected to disappear under socialism. Statistics and real information on crime were hardly ever published. Compared to other socialist states, Hungary has had a somewhat better reputation in this respect, but a highly selective information system concerning crime data and knowledge of police activities existed here as well. This meant that although figures about reported crime were available to the public, data concerning measures (e.g., identity checks, arrests, etc.) applied by police remained confidential. All information about the personnel, internal organization and budget of the police was considered a state secret. Even the theoretical journal of the Ministry of Interior, which did not contain any secret data, was "for internal use only", that is, accessible only for employees of the Ministry, including police officers and for other designated persons (party officials, prosecutors, loyal academics, etc.). The media, which were strictly controlled by the party, relied almost exclusively on police sources when reporting about crime and investigations. Decisions on the release of information were made by the party-directed Press Department of the Ministry of Interior. Many significant criminal cases were not covered because of prohibitions based on political considerations (e.g. the suspect was from a friendly socialist state, an offense was committed against a public institution of high authority—which would be an admission that some people did not grant legitimacy to the system—, or, most frequently, the possibility of disturbing the tranquillity of society). Generally speaking, informing the public was not a priority in the course of press administration. Instead, decisions were based on considerations of how such information could be used for criminal investigation or the promotion of political interests.

Insecurity also was built into part of the criminal legislation. Undefined and ambiguous concepts and the criminalization of victimless deviance allowed wide discretion by the organs of criminal justice. In addition, criminal behavior could be dealt with to a large extent outside the criminal justice system Formal and tolerated informal diversion played an important role in the system. Formal diversion included adjudication by social courts, and informal diversion could take the form of making a deal with a suspect to recruit him or her as an informant. There were, at the same time, spheres of socio-economic life where institutionalized crime (mostly corruption, bribery) reached a level beyond any hope to be recorded and cleared.

Thus, the assumption of lower crime and greater security under socialism is not supported by analyses built on facts taken from a broader scope than pure criminal statistics. To put it more generally, facts suggest that public safety in former socialist countries did not approach the excellent situation declared by official statements. Studies looking beyond the surface of rhetoric tend to support the results of Maria Los' (1988, pp. 285–286) analysis:

> the repeated claim made by various western and eastern authors that there is less crime in communist countries than in capitalist nations does not appear to be particularly well founded, even in light of the data reported within the communist countries themselves.... The claim that crime and deviance have been significantly reduced due to party efforts to exercise total control over the society is also rather unconvincing.

If the thesis that dictatorships provide more security turns out to be inadequate, arguments for the restriction of freedom in order to gain greater safety can also be challenged. Nevertheless, claims that extended police powers and limited human rights are necessary to achieve law and order are still on the agenda in Hungary.

The Minister of the Interior expressed his views on the supposed contradiction between liberty and police powers in his opening speech delivered at a 1992 conference. The growth and expansion of criminality required that society give priority to fighting crime even at the sacrifice of democratic ideals (Boross, 1992, p. 10). This approach has been embodied in the 1994 Police Act. If higher crime is the price of more democracy and freedom, too high a loss of security has to be counterbalanced by restrictions on liberty. The problem with this argument is that the transformation of the political system was premised on the need to free society from an oppressive and secure police state. Now, the powers of the police are increased in many fields without, at the same time, strengthening or even developing the democratic accountability of law enforcement.

At the same time, there is no proof of the validity of the argument. Even the slight decrease of recorded crime during recent years does not justify the policy of waging war against evil-doers by strengthening the police in all their aspects. The usual statement that 5,000 new recruits contributed to the decrease of crime is misleading because the police have the task to detect as many crimes as they can. Consequently, more police officers should have recorded more criminal offenses; an increase in police personnel should have increased the number of known crimes and decreased the number of hidden ones. Since the number of actual crimes committed is fixed but hidden in any society at any point in time, finding and reporting more crimes will increase the known number of crimes even though the actual number of crimes stays the same. Thus, apart from the unanswered question as to the nature of political transition in terms of the balance between personal liberty versus police powers, theoretical arguments do not underpin the law and order philosophy.

In sum, there is no evidence for the thesis of greater security under a communist system. This is not to deny that, according to official statistics, patterns of criminality have changed. There is certainly more street crime, which is largely attributable to the fact that a growing number of people have left their workplaces and homes because of unemployment and rising costs of living. Street crime is, apart from some spectacular actions by criminal gangs, deviance by the working and lower classes. In addition, a change in economic crimes has been triggered by new opportunities for acquiring and legalizing huge amounts of money. Of course, the process of privatization and transitions in legal regulation are accompanied by contradictions and loopholes which also contributed to the emergence of new criminal opportunities. One example is the return of local, formerly state-owned enterprises to local government ownership. This was a correct policy, but it proved impossible to properly evaluate the worth of specific pieces of property which were returned. This situation led to many illegal bargains being struck among negotiators, and the use of experts and mediators who contributed little yet were paid a lot.

Another factor which changed patterns of crime was the opening of borders, with the inevitable expansion of free movement not only of persons and ideas but of criminal activities as well.

Migration

Hungary is situated on the crossroads of Europe. Throughout its history problems of different forms of migration have been unavoidable. Following the occupation of Hungary by the Turks, for example, Empress Maria Theresa in the eighteenth century encouraged thousands of Germans to settle and cultivate land in the country, mainly in the southern areas. German descendants make up a sizable minority population even now. Problems arose from this situation. After World War II, the descendants of German settlers were accused of having collaborated with the Nazis and many were expelled from Hungary.

But even during the period of communist dictatorship freedom of movement was less curbed than in most other totalitarian regimes. In the late 1980s, cautious steps were taken to loosen border controls with regard to people, most of them of Hungarian origin, escaping from Ceaucescu's Romania. Since 1989 Hungary has been a party to the Geneva Convention on Refugees although it imited the scope of the Convention to Europe; only refugees from European countries are given refugee status. It can be stated that not only was Hungary prepared to enter a new Europe in which traditional borders had become more symbols of national sovereignty claims rather than effective barriers to the movement of goods and people, but it was ready to actively participate in creating the new Europe.

In 1995, there were about 4,000 refugees under the terms of the Geneva Convention and up to 6,000 persons granted refuge on humanitarian grounds (mainly war escapees from the states of former Yugoslavia) living in Hungary (Office for Refugees and Migration, personal communication, 1995).

There is another category of non-Hungarians causing problems to the authorities. These are the illegal aliens staying in the country without appropriate justification and in violation of the rules of entry, sojourn and exit. Since Hungary has concluded visa-waiver agreements with a number of states and no general obligation to register with official agencies exists for aliens, infringements of immigration laws do not require special skills while control proves to be difficult and energy-consuming. Illegal aliens also arrive across the "green border," by crossing outside border control checkpoints and ports of entry, and thereby avoid any passport control. According to police estimates, about 50,000 to 55,000 illegal aliens are in the country (Szabó-Kovács, 1992 p. 68). Most of them do not have lawful means to support themselves, therefore the danger that they may become involved in criminal activities is obvious.

A generalized sense of xenophobia does not exist in the country. There were, however, some criminal attacks against foreigners. Skinheads and other extremist groups, though not significant in number, cause much harm to the reputation of Hungary.

In spite of all the difficulties and the huge mass of foreigners coming to Hungary, the proportion of non-Hungarians among offenders is fairly low and relatively stable. Foreign delinquents constitute about 4 to 6 percent of all offenders over the last five years.

Of course, quantitative indices alone do not paint a comprehensive picture of the difficulties and dangers posed by the illegal activities expanding across all of Europe and beyond. Internationally organized crime involving both Hungarian and foreign offenders is a growing concern for our police. The so-called Balkan Route is a well-known traditional route for smuggling drugs from eastern and southern countries to Western Europe. The illegal traffic in other goods, and even in people, is a prosperous business for many criminals as well.

Legislation to deal with these problems includes the Alien Policing Act of 1993 which gives enormous powers to all agencies dealing with this problem. Section 28, paragraph (4) provides that "the Police may supervise the observance of the rules of supervision of aliens in this Act, and for this purpose, are entitled to enter private flats, and/or private areas" (*Idegenrendészeti törvény*, 1993). The police can enter any place, including private dwellings, at any time without warrant to check on compliance with even purely technical administrative provisions. There is wide scope for discretion in making decisions concerning the entry, stay and exit of aliens. Violations of law can be punished by detention. Uncertain immigration status results in a compulsory stay in designated areas which are similar to detention areas. One of these facilities was closed recently because it was used in practically the same way as detention facilities but did not meet the legal requirements for detention centers.

I am convinced that this law-and-order reaction is not the appropriate answer to the serious problems caused by the movement of people. International cooperation has to be strengthened in order to prevent mass migrations triggered by wars, famine and other inhumane circumstances. This is of paramount importance. Furthermore, international police collaboration should be developed further while taking into consideration basic human rights, including the freedom of movement.

PROFESSIONAL CHALLENGES

Politicization

At the beginning of the transition to democracy a new type of police officer was to be educated. This was a central demand of the opponents of the communist system. The strong mistrust of law enforcement was totally understandable as many members of the opposition had been treated arbitrarily by the police.

In spite of all the hatred accumulated against the communist police, the new government decided to keep its organization basically unchanged. This rather sudden change of attitude was basically due to the situation in which the freshly established institutions of public power found themselves. From the perspective of those in power, institutions such as the police were not enemies any more but potential tools for achieving political goals. Having disbanded the Workers' Militia (a voluntary armed force established to defend the achievements of socialism) and having written into law strict limits on the internal use of the armed forces, it appeared impossible to consolidate the changed political order without the support of the police. The restrictions on the domestic use of the armed forces which have been written into the Constitution stipulate that the armed forces can only be employed in times of emergency when the capacity of police to maintain order would not be sufficient.

Police leaders offered loyalty and active support for the establishment of a democratic system. The government actually had no choice because it was impossible to create a totally new police force within a short time without the active assistance of law enforcement personnel. The resulting collaboration was the survival of the organization and the omission of major political purges.

The State security services have been separated from police and blamed for most of the communist party's political interference in citizens' lives while the crime-fighting image of the "ordinary" police has been emphasized. In an effort to whitewash the existing police rather than reform it, new and partially contradictory slogans have been used to describe the "new" force. At some times democratic values (citizen-friendliness), at other times forcefulness and the courageous actions of the police have been emphasized in order to prove the changed character of the institution.

Under these circumstances it has proven to be difficult to set firm goals for police training and recruitment. The head of the responsible Ministry unit could only state a generalized requirement and goal, which was a platitude as well as impossible to meet: "The future belongs to the policeman who has universal knowledge in every field of the police work" (Korinek, 1992, p. 13). If police officers have to know everything, no requirements for specific skills and knowledge can be stated. The authorities directing the police revealed their incompetence and inability to produce viable operational guidelines by such statements. In practice, depoliticalization and higher levels of education were the priorities in recruitment and training.

Under the communist regime, all aspects of state activities were controlled by the party. At the same time, certain fields of exercising public power received special attention from the party. Particular arrangements ensured exceptionally direct contacts among government units and the corresponding central organs of the ruling political organization. As opposed to normal party direction, agencies engaged in highly sensitive fields had to report directly to party officials, irrespective of their formal position in the state hierarchy.

One of these fields was policing, including intelligence activities. It was the political mission of the police to serve the party. The overwhelming majority of police officers were members of the party. The resolutions of the communist party were implemented as legal norms.

In the course of political transformation, the depoliticalization of the police became one of the most repeated slogans in political declarations and mission statements. The Constitution categorically prohibits party membership or any other political affiliation for members of the force. Section 40/B, paragraph (4) provides that "professional members of the armed forces, the police and the civil national security services may not be members of political parties and may not engage in political activities" (*Alkotmány*, 1949). Thus, on the level of the individual officer there is the guarantee that policing and party politics are kept at a distance. "Ordinary" police and political intelligence agencies are also fully separated. All these regulations culminate in the stipulation of section 2, paragraph (3) of the 1994 Police Act that "in fulfilling their tasks, the police will proceed uninfluenced by political parties" (*Rendőrségi törvény*, 1994).

Abuse of Power

Another challenge is an increase in cases of police brutality. These had not been a major issue for a long time. Recently, however, a number of cases shook the Hungarian public. Instances when the police beat up innocent citizens or caused injuries to suspects during interrogation have been highlighted in the mass media. One county police chief resigned because of allegations of brutality, and disciplinary and criminal proceedings have been launched against other police officers on the same grounds. There are instances of brutality which are serious breaches of law and good faith, even if statistics do not reveal a frequent misuse of physical force by law enforcement officials (see theAppendix for specific cases).

There are several offenses which can only be committed by officials of the state or local governments. Some of them closely relate to the sphere of law enforcement. These are: ill treatment during official procedure, forced interrogation and unlawful detention. Prosecutors seem to be reluctant to investigate allegations of this nature. While refusal to investigate in other crimes, due to a lack of reasonable suspicion, does not exceed 8 percent of the reports submitted to the police, the refusal rate is well above 35 percent (and in recent years even frequently above 40 percent) concerning complaints about the three offenses. There were 1,269 cases related to police abuse reported in 1995 (Kőszeg, 1997, p. 4).

According to statistics published by the Hungarian Weekly *HVG* (August 5, 1995), in 1994 there were 798 reported cases of assault by officers on citizens and suspects while they were carrying out official duties. The offense of assault can be committed by any state or local authority official but in practice all investigations of such allegations concern police-citizen encounters. Of these complaints, 254 were not substantiated enough to trigger a criminal proceeding and 467 investigations were terminated at a later stage. Criminal charges were brought in only 77 cases.

Many prosecutors in charge of investigations of the police claim that the high rate of refusals is to be attributed to the habit of experienced criminals who often blame the police for ill treatment or forced interrogation in order to explain changes in their statement at various stages of the investigative process. This might be true in a number of cases. But the opposite is also true—victims report only a fraction of all ill treatment or forced interrogation cases. During the Hungarian Helsinki Committee's police custody monitoring program, 106 persons detained by the police answered questionnaires used by the monitoring teams. Twenty-three (22.3 percent) of the respondents stated that physical ill treatment had been applied against them in the course of arrest or during detention. There were accounts in a number of cases where already handcuffed persons suffered beatings. No complaints, though, were filed by any of the respondents (Kőszeg, 1997, p. 5).

It is also true that politics and legislation have contributed to the increased self-confidence of the police, a confidence that has not been matched by higher levels of professional knowledge and skills. At the beginning of the transformation of the political system, the opposition attacked law enforcement harshly, charging the police with oppressive activities and unconditional subservience to the will of the communist party. Officers became reluctant to take any action because of fears of being accused of political prejudice. Indeed, emerging political parties and movements threatened to "blow up" the whole organization of the police and to dismiss all politically biased police officers. Opposition forces wanted to radically reform the organization of the police. The national police would retain criminal investigation powers while normal uniformed policing would be performed by units reporting to local governments. As mentioned, this expectation was not met once the new government took power. The official reasons were an increase in crime and the need to keep political transformation on a peaceful course.

But a shocking experience forced the government to modify its policies concerning the maintenance of public security. During October, 1990, a blockade-protest paralyzed practically the whole country. Taxi drivers, joined by masses of people demonstrating against a government decision to raise the price of petrol to unprecedented levels, blocked all major transport hubs and ports of entry. The police were unable to cope with the situation. Some chiefs spoke out for non-intervention referring to the citizen-friendliness of their service. The Minister of the Interior was removed following the event and the government gave up emphasizing the slogan of citizen-friendliness. The new appointee to the post, Peter Boross, did not at all hide his "macho" perception of policing. Since then, increasing the numbers and resources of the police has been the goal of government policies for preservation of public security. The 1994 Act on Police (*Rendőrségi törvény*, 1994) gave enormous powers to law enforcement without counterbalancing them with adequate institutions of accountability. These developments are undoubtedly factors leading to the increased self-esteem of police and, sometimes, to the use of unacceptable force.

RESPONSES TO CHALLENGES

International Influence

As a consequence of the central location of Hungary at the crossroads of Europe, international cooperation has always been vitally important for the Hungarian police. Hungary was among the founding nations of Interpol.

Under the communist regime, contacts with the world were prohibited or limited to interactions with other socialist police forces, mainly with the Soviet Militia. However, similar to migration policies, a cautious opening began in the early 1980s. In 1981 Hungary rejoined the international criminal police organization (Interpol). Bilateral cooperation with Austria was established by an agreement concluded in 1979. This agreement became the model for a series of agreements regulating direct police collaboration (Nyíri, 1995, p. 6).

With the opening of the borders and the lifting of restrictions on travel, the development of international police relations became an urgent necessity. A series of government-to-government agreements (typically concluded by ministers of the interior or justice, depending on who controlled the police) have been signed since the iron curtain was lifted. These agreements create the legal framework for direct cooperation in the prevention and repression of drug trafficking, the illegal trade in firearms, the fight against terrorism and organized crime. Such agreements are in force with Germany, Italy, Ukraine and with many other countries. The Police Act of 1994 (*Rendörségi törvény*, 1994) explicitly promotes common activities among different national forces by stipulating that, based on international agreements, a Hungarian police officer may exercise his/her powers outside the country while members of foreign services are allowed to perform their duties on Hungarian territory.

Encouraged by these agreements, the Police Force of the Hungarian Republic has established and maintains direct working relations with law enforcement agencies of other countries. In addition to contacts among National Police Headquarters there is an increase in cooperation among police units at lower levels and their foreign counterparts. Each Hungarian county police force has, for example, a Dutch counterpart and within that framework not only official but friendly personal ties have developed. An agreement signed by the Hungarian National Commissioner of Police, Mr. Sándor Pintér, and the Dutch Police Chief, Mr. Joop de Wijs, encouraged this development. The document stipulated an exchange of officers for study tours and the sharing of comprehensive information about each other's activities and problems. In addition, it called for similar agreements and collaboration on the part of other territorial units of police. By the end of 1993, each Hungarian county police force had a Dutch partner and substantial cooperation has since developed among the partners, covering practically all areas of policing but focused mainly on the fight against organized crime, drug-related offenses and terrorism.

Even prior to this agreement, the Warnsveld Training Center in Holland had provided substantial support to the Hungarian police in developing training programs. In addition, a project for the modernization of the Hungarian police had been partly financed by the Dutch Government. In 1992, Csongrád County in Hungary concluded an agreement with the Dutch Utrecht Region concerning cooperation in policing.

An effort, originally initiated by the Austrian and Hungarian Ministries of the Interior, to train police officers from different countries of Central and Eastern Europe in a common setting—the Middle European Police Academy—has been especially succesful The Academy offers an excellent opportunity for participants to learn about each others' systems of policing and to establish and maintain the close personal contacts which are sometimes so necessary for successful cooperation. The Academy does not have a permanent site but travels among the eight participating countries. In Hungary, training sessions are normally conducted at the facilities of the International Law Enforcement Academy (ILEA) in Budapest. Training helps officers gain insight into the organization and activities of different forces. Participants are experienced mid- to senior-level police officers. Criminal investigations and intelligence work are stressed in the curriculum but occasionally other aspects of law enforcement (e.g., border guard functions) are studied as well. The teaching language is German.

Another international training institution is the International Law Enforcement Academy (ILEA) (colloquially called the FBI Academy) which provides training for officers from Central Europe and the Newly Independent States (the former Soviet Union republics). This study center was opened in 1995 in Budapest following an agreement between the USA and Hungary. The director is an American police officer (currently an FBI agent) and the deputy director is Hungarian. The Academy's curriculum is modelled on the National Leadership Course offered at the FBI Academy in Quantico, Virginia. The eight-week-long curriculum was modified to reflect the interests and needs of participating countries. Teaching is conducted mainly by police from US federal law enforcement agencies and European police forces. Participants are mid- to senior-level officers in their respective forces. ILEA also offers short-term training in specific topics of interest, e.g., undercover operations.

Without questioning the professional competence of the teaching staff, it is doubtful whether the US style law enforcement can be successfully implemented in Hungary or Europe. The training emphasizes successful methods of fighting crime, while constitutional guarantees and the democratic attributes of the police forces of the US (e.g., a decentralized system of organization) are given less attention. Hungarian police officers now tend to imitate the tough "stop and frisk" methods practiced in the US. In Hungary, people rarely carry firearms and there is generally no threat of being attacked by the persons who have been stopped by the police. Within the context of the Hungarian legal system, such actions by police officers basically contradict the requirement of proportionality. American officers emphasize, based on their own experience, the importance of demonstrating control and power during police interventions. Thus, an aggressive attitude toward suspected criminals has been reinforced by American examples. It is worth mentioning that the Budapest Police Force has maintained friendly ties with the notoriously "hard line" Los Angeles Police Department for a long time.

As well, American colleagues do not say much about institutions that they dislike, such as civilian review boards or control by local governments. As a result, references to law enforcement practices in the USA usually neglect differences in the constitutional setting and the protection of rights to those in Hungary. The Hungarian police are over-centralized and there is no civilian review board in the country. But if tough law enforcement methods are conveyed in training without, at the same time, discussing means for the proper control of police operations, no desirable balance between police measures and guarantees of individual freedoms can be achieved.

Administrative Reforms

Local Autonomy

In spite of declarations and efforts to depoliticize the police, some problems of political involvement by police remain unsolved while new ones have emerged. Policing could not be removed from the political arena because many problems faced by the police proved to be inseparable from politics. One example is the control of mass demonstrations which require political knowledge in order to cope with them effectively. The complete separation of the police from political influence proved to be impossible.

There is separation at the lower levels of the force. Neither local police units nor individual officers are given the right to maintain political contacts. Considering, however, the pyramid-like organizational structure of the police, the most delicate point is, quite naturally, the connection between the commander of the whole police system (the National Commissioner) and the government. Since the Minister of the Interior is empowered to issue instructions concerning particular police operations, serious doubts are raised about the reality of depoliticalization. The authority of the Minister to direct police activities covers any field of policing, including investigations. The only limitations on this power are stipulated by section 5, paragraphs (2) and (3) of the Police Act of 1994. According to paragraph (2), "the Minister of the Interior may not take over any case from the authority of the police by way of concrete instruction, nor may he prevent the police from performing their duties." Paragraph (3) provides that "individual instructions by the Minister of the Interior to the police have to be addressed to the National Commissioner of Police"

(*Rendőrségi törvény*, 1994). In practice this means that the Minister can order an investigation but he or she is not entitled to prohibit the police from launching one. The first contingency constitutes no takeover of the competence of the police because they would still have to act under the relevant legal regulations, while the prohibition of an investigation would prevent them from performing legal duties. These provisions ensure the operational autonomy of the police. So far, no case in which the Minister issued specific instructions concerning police operations has been reported.

Training

Higher standards have been introduced into the recruitment and training system of police. According to the regulations in force, no one may perform police duties without appropriate knowledge and skills. The entry requirement is a grammar school education. In Hungary, grammar school education is twelve years of formal education—eight years in elementary school and four years in grammar school. In addition, passing a maturity examination at the end of grammar school is also an entry requirement. Recruits receive two years of training at the police academy before they are allowed to exercise police powers. New laws and regulations concerning police activities are taught in special courses and tested in examinations.

A higher educational institution, the Police College, serves the needs of law enforcement. In addition to police officers, the College trains members of the Customs Service, Prison Service, and Border Guard. A three-year course of studies provides university-level training in the relevant subjects. The College itself is not a new establishment; it has been in existence since 1971. During the political transformation the curriculum was modified; ideological topics were abolished and more practical knowledge and skills were added.

The institution was, in the beginning, open exclusively to active law enforcement officers. In the Hungarian rank structure, the term *officer* designates a member of the force who is at least a lieutenant. The promotion system is the same as the military's. To achieve a higher rank members must come up through the in lower ones. Officers must have a higher education degree and special conditions may be attached to certain positions. A cautious change took place allowing limited possibilities for lateral entry into the College. Applicants having a grammar school education could be admitted without having previous police service experience. Now the requirements for admission are the same for outsiders and for serving officers. As a result, the overwhelming majority of students are civilians, and some "fresh air" has entered the rather conservative structure of the police. Keen competition (there are generally 10 to 15 times more candidates than places) ensures a good applicant pool from which to select. On the other hand, the chances of being promoted from the rank and file are necessarily decreased. As a corrective measure, in-service training is provided for members of the force who are making an effort to rise within the organization.

Despite all innovative efforts, it has been extremely difficult to reform the attitude of police staff. Ironically, enhanced training contributed to the survival of traditional policing methods which, in turn, causes problems for the constitutional state. Instructors in practical subjects were drawn mainly from the active service and had gained their experience and knowledge while working under the communist system. Thus, the emphasis on practice-oriented training inevitably led to the conservation of some older methods and approaches.

To counterbalance this deficiency, the Institute for Police Research and Management Training was founded in 1994 as a civilian institution independent from the police. Its goal was to disseminate up-to-date information based on comparative and domestic research to the police leadership. It was hoped that police chiefs could and would integrate theoretical knowledge and actual policing practice. In this way, a progressive perspective could be introduced into a policing system, for its hierarchical structure would ensure the distribution of new ideas and knowledge. If the chiefs themselves are aware of and believe in the significance of progressive methods and constitutional practice, their instructions, as they permeate the organization, will necessarily convey the appropriate message to all ranks and positions within the force.

Legislation

The democratic transformation of the country has taken place within the framework of legal statehood (*Rechtsstaatlichkeit*), the German equivalent of Rule of Law. The notion of *Rechtsstaat* embraces the formal aspects of institutions of public power while requiring at the same time full adherence to constitutional principles. Security and the reliability of legal relations are key elements of a *Rechtsstaat*.

Yet the process of replacing a totalitarian system by a democratic one faces almost irreconcilable contradictions. Legal security would require respecting all rights derived from the laws of the dictatorship, yet at the same time those laws have not complied with the substantial principles of *Rechtsstaatlichkeit*. The difficulty is this. After political change, those entitled to decide upon rights and duties based on previous legislation either have to part with the security aspect of *Rechtsstaatlichkeit* by admitting that substantively wrong law may not be enforced in a legal state, or they have to admit—at least in an implicit manner—that under communism there existed a legal system some of whose elements are worth preserving in a democracy.

Of course, each of the answers to the problem leads to injustice. Rejecting the validity of former rules would inevitably cause damage to those honest citizens who acted in good faith in order to acquire rights under socialism without having any chance to choose other ways of achieving their goals. On the other hand, accepting the validity of the socialist legal system could lead to the preservation of all unfair privileges acquired by the communist *nomenclatura* (such as houses or hidden bank accounts) as well as discrimination against certain social groups. According to the experience of the emerging democracies, no general answer to this problem can be given. The applicability of laws from the preceding system depend on their quality and on the degree of continuity in the political arena. The Hungarian Constitutional Court has resolutely emphasized the formal, legal security aspect of *Rechtsstaatlichkeit* by rejecting any retroactive interference with acquired rights. At the same time, it has repealed inherited rules when they do not comply with the principles or provisions of the constitutional order *ex nunc*. In this manner, the police functioned on the basis of old regulations for several years after the first free elections, even as many aspects of the legal environment changed during the same period.

A complete revision of the Constitution took place in 1989. In addition to enhancing guarantees for human rights, the Parliament adopted a new chapter of the basic law dealing with armed forces and police. The new constitutional provisions prescribed that all regulations concerning police organization and activities needed to be codified by an Act of Parliament by a specified majority of votes. Article 40/A, paragraph (2) provided that

> the fundamental function of the police is to safeguard public security and defend internal order. The enactment of the law on the police and the detailed

rules connected with national security require two-thirds of the votes of the Members of Parliament present.

Another important principle in the amendment stated the firm separation of police from the armed forces and strictly limited the latter's involvement in public order maintenance. Article 40/B, paragraph (2) stipulated that

> the armed forces may be used only in times of an emergency situation promulgated in accordance with the provisions of the Constitution, in case of armed action aimed at the overthrow of the constitutional order or at the seizure of absolute power, furthermore in cases of violence committed with arms, or the use of force in a manner endangering the safety of the life and property of citizens on a mass scale, and only when the deployment of the police is not sufficient (*Alkotmány*, 1949).

The prohibition of political activities referred to earlier was introduced only in 1993, though the text from 1989 already authorized legislation to restrict the political activities of police officers.

The comprehensive constitutional reform enacted by the outgoing communist Parliament (based on the political agreement between the party and the opposition) and other acts of the transitional legislation made significant progress toward establishing a firm system of constitutional values in the first years of democracy. This generalization, however, does not apply to regulations dealing with police activities. As was mentioned, the government preferred having a powerful law enforcement organization rather than limit police powers within a strict framework of constitutionality. The preservation of totalitarian norms served the hard line attitude of the administration well for a long time. It is no wonder, then, that the 1994 Act on Police (No. XXXIV) raises serious questions on whether it complies with elementary constitutional standards (*Rendőrségi törvény*, 1994). The following discussion will point out some of the problems.

In general, two classical legal principles are undermined by this piece of legislation. One is that of the division of powers, the other the priority of human rights over the operational and power interests of the state. One example is the special "plea bargaining" provision which disregards both the division of powers and the priority of human rights. Section 67, paragraph (1) of the Act introduces a specific negotiation arrangement that raises serious doubts as to basic principles of presumption of innocence, right to appeal, and so on. The text is as follows:

> With the approval of the prosecutor, in order to obtain information, the Police, promising the refusal or termination of the investigation, may enter into an agreement with the perpetrator of a criminal offence, if the interest of the enforcement of criminal law to be served by the agreement is higher than the interest attached to the criminal prosecution of the case by the state (*Rendőrségi törvény*, 1994).

Of course, it is highly questionable whether one can identify a perpetrator of a criminal offence unless an investigation is started first. In addition to human rights problems this provision challenges the right of the courts to decide upon guilt or innocence.

Introducing the parliamentary debates, the Minister of the Interior explained the theoretical background of the bill. He stated as a principle that "the degree of freedom guaranteed by a legal state belongs only to those having respect for the law" (Parliamentary Records, 1993). In other words, all manner of perpetrators can be deprived of their "*Rechtsstaat* degree of freedom," that is, of their basic human rights. This approach is, of course, contrary to the general perception that basic rights and freedoms are essential attributes of human existence. A further problem in the Act is the definition of those categories of people who should fall under the presumption of innocence. In the speech quoted earlier, the Minister solved this dilemma by stating that the police proceed only against those infringing the law, while citizens complying with it would be left in peace.

Section 3, paragraph (1) of the Police Act provides that

> the Police shall be an armed policing agency performing duties in crime prevention, criminal law enforcement, state administration, and general law enforcement. The central organ of the Police with a nation-wide competence shall be the National Police Headquarters (NPHQ) (*Rendőrségi törvény*, 1994).

There is no clear linkage of the stated functions to different procedural systems. Criminal procedure is regulated by the Criminal Procedure Code. The Code of Administrative Procedure regulates activities (e.g., issuing personal identity cards, driving licenses, etc.) defined as administrative in nature. No specific procedures exist for crime prevention and general law enforcement. The legislature failed to decide what procedural norms should apply to those activities. Crime prevention is most closely tied to prosecution (and deterrence), although the Code does not apply specifically to preventive measures. In short, the police can implement the same measures for both the prevention and detection of crime. Needless to say, there is a big difference between prevention and repression and it should be reflected in distinctive procedures. If the police have a power, such as intercepting communications, for both prevention and detection, the distinction between an actual violation of law versus the mere anticipation of an infraction as the justification for police action vanishes. Under the rule of law nobody should be exposed to such an invasion of their privacy without at the very least being suspected of having committed a crime.

State administrative duties can be extended beyond the traditional scope necessary for maintaining public security. According to accepted views of the police role in Hungary, only that authority may be vested in them which is closely related to public security. Issuing a license for the possession of a firearm is a legitimate function of the police. Issuing personal identity cards, though, crosses the accepted line between the protection of order and purely administrative functions.

The Police Act repeats the text of the basic law on the police which stipulates that their role is to protect public security and maintain internal order. The responsibility for keeping internal order raises a new problem, namely the fact that the police also supervised intelligence agencies (until 1989). Political and "ordinary" policing arrangements changed several times even during socialism. In 1949 an independent state security organization was created and placed under the government but in practice was directed by the highest party leadership. The 1956 uprising revealed the extremely arbitrary character of that security apparatus. During the consolidation of the socialist system following the suppression of the 1956 revolution, police and state security organizations were integrated in order to achieve an acceptable level of transparency. In reality, however, the political police remained independent within the formally united structure. Under the general direction of the Minister of the Interior one deputy minister was responsible for ordinary policing and another for state security. Deputies of county police chiefs were heads of the state security apparatus within the jurisdiction of the force but they could be instructed from above without even notifying the county chief.

This rather independent security and internal order subsystem was an example of direct party interference in policing. Political interest superseded the demands for public order or criminal investigation. At the same time, within this structure, the implementation of politically motivated decisions could be easily disguised as "normal police operations." Members of the underground opposition, for example, were often investigated and detained for "public security" purposes in order to prevent them from attending important meetings or pre-election gatherings. In 1990, the police and national security agencies were again separated. The structure changed, yet the constitutional provisions authorizing the previous arrangement remained. The constitution still contains the phrase "internal order" in reference to normal policing even though, originally, that phrase denoted state security policing, a function no longer performed by the normal police.

The specific tasks of the police include the detection and investigation of crime, the prevention of criminal activities, decision making in cases of petty offenses (qualified as administrative infractions) and policies contributing to their prevention, alien policing, and performing duties related to producing, selling and using certain devices and substances which are dangerous to public security. The police are responsible, as well, for traffic control and the maintenance of order on public premises. VIP protection is also within the scope of police activities. A new task is the licensing and supervision of private security agencies. The police take part in enforcing particular supplementary punishments and measures imposed by courts of justice in addition to the main sentence. A supplementary measure is, in principle, independent from sentencing purposes. The confiscation of dangerous objects, for example, does not seek to punish the convicted but to protect the community from potential harm. In addition to imprisonment, courts can impose such sanctions as the suspension of political rights, the withdrawal of driving licenses, the exclusion from certain municipalities or the suspension of the right to exercise a profession. Special duties arise in times of emergency. In sum, the police have a wide scope of authority both for law enforcement in a narrower sense and for decision-making in the interests of public security.

The Police Act provides some principles to ensure that the potential misuse of power will be limited. Section 2, paragraph (1) declares that

> the Police shall provide protection against acts which imminently endanger or harm the life or bodily integrity of persons or the security of property and shall provide information and help to persons in need of such assistance. The Police shall respect and protect human dignity and they shall guard human rights (*Rendőrségi törvény*, 1994).

Other provisions of the Act seek to guarantee important constitutional values. Any kind of torture or cruel, inhumane or degrading treatment is strictly prohibited. The principles of proportionality in measures taken by the police and the need to use only minimum force govern the conduct of police activities.

Realizing these guarantees is nevertheless difficult under the present situation in which procedural constraints on law enforcement actions are left obscure and undefined. Also, institutional control over the police has been weakened in order to strengthen their powers against the "criminal underworld." There are, in addition to the stipulations designed to safeguard human rights, other highly questionable provisions in the Police Act as well. Officers must, for example, enforce even unlawful instructions by their superiors unless the behavior would constitute a crime. However, many orders can infringe upon elementary procedural rules without actually being a criminal act.

In addition, the obligation to comply with instructions of doubtful legality extends to those encountered by members of the force. Section 19, paragraph (1) of the Police Act requires that

> everyone shall submit to a Police measure and shall obey the instructions of a Police Officer if the measure is aimed at the implementation of provisions set forth in statutes, unless otherwise provided by an Act or an international treaty. In the course of Police action, its lawfulness shall not be challenged unless it may be seen as manifest (*Rendőrségi törvény*, 1994).

In other words, police interventions to enforce statutory provisions do not have to be legal in all their aspects. This raises problems. For example, during a stop and search a police officer may not seize objects which are not suitable for attack or can create danger. However, if he or she does, persons confronted do not have the right to protest unless they have exact knowledge of the specific provisions of the law which were violated. It is not enough to know that they are innocent. The seizure of the object cannot be challenged at that time. It may turn out that the officer had acted properly—for example, he or she had seized the object in order to help in detecting a crime—but the officer may also have been legally in the wrong in her/his actions.

Some provisions of the Act concerning different police measures and the use of coercive means also reveal contradictory tendencies which, in turn, reflect competing interests involved in the drafting of the Act. Most provisions and statements of powers are rather loosely formulated to enable the police to do their job without major constraints. The police can deprive individuals of personal liberty without formal charge or judicial supervision for up to twenty-four hours, and in certain cases for up to seventy-two hours. People can be detained for eight hours, which can be prolonged by another four hours, while the police decide what to do with them. The use of firearms is permitted in a wide range of situations, some of which do not involve imminent threat to the life or bodily integrity of the officer or others. Section 54, paragraph (h) of the Police Act authorizes the police to shoot if necessary, among other goals, "to apprehend, or to prevent the escape, of the perpetrator of an offense against the state ... or against humanity" (*Rendőrségi törvény*, 1994).

Secret methods for collecting information are also authorized and regulated by the Act, granting the police wide discretion in crime prevention and detection initiatives. They may use informants or secret agents, and can prepare (fake) documents to run an undercover enterprise. For example, the police can produce a false identify card for an undercover officer to prevent revenge against the officer after the completion of the police operation. An undercover enterprise is a police operation disguised as a business. A typical example is a secondhand store which has been set up to catch criminals selling stolen merchandise. Surveillance and entrapment activities which do not cause injury or damage to health can be carried out as well. All these actions can be implemented without a warrant, solely by the judgment of the police. One of the members of parliament pointed out during the debate on the bill that it places a higher value on effectiveness, on the exercise of powers by the police, than on procedural protections or the avoidance of potential violations of law by police (Parliamentary Records, 1993). More extensive surveillance and secret intelligence gathering are subject to warrants issued by the courts. With such a warrant, police may secretly search a private residence; and record their observations by technical devices (e.g., video cameras); they can observe and record events taking place in a private residence, and they can intercept any communication in order to know its content or to use it as evidence.

The protection of the confidentiality of data was also established by this Act, though it also provides the police with far-reaching powers to prevent public knowledge and to invade the privacy and rights of persons involved. Banks, tax authorities and telecommunications services are required to satisfy the information needs of law enforcement if that information is deemed necessary for the detection of crimes punishable by at least two years of deprivation of freedom. Such information can be requested by the police only with the prior consent of the prosecutor.

Lack of accountability is another cause for concern. The Act does not establish any forum to keep watch over the exercise of the far-ranging powers of the police. The Act includes provisions for criminal or administrative proceedings against abuses of power by the police. Complaints against the police, except those arising out of confidential matters, are dealt with by the organization itself. Complaints must be submitted within eight days of the incident to the police unit directly responsible for the event. In case of disagreement with the decision reached, appeal lies to the superior police unit for a final decision. This means that omissions cannot be challenged properly, and higher police officials may have a stake in the case and decide, in some cases, that the actions leading to the complaint have been in conformity with the instructions of the superior unit, hence not a violation of citizens' rights.

As mentioned, at the time it was finally adopted, the Police Act did not introduce structural changes because practically no significant interest group urged a radical reorganization of the police. The former enthusiasm of many local authorities had faded because they had been given new responsibilities without appropriate financial support. They feared that taking over policing tasks under such circumstances would cause problems for them. The police themselves argued in favor of retaining the centralized system. This formulation of police needs was quite natural for it was advanced by high-level officials who were mostly interested in retaining power and social status. Politicians also recognized the advantages of central police management. The public became persuaded that the growing menace of internationally organized crime could be fought effectively only by a hierarchically organized force.

The return to the organizational principles of the totalitarian dictatorship has been crowned by the 1996 Service Relations of Officers of the Armed Organs Act (Szolgálati törvény, *1996*). Working conditions, legal status, and social insurance of officers of the military, border guard service (which belongs to the armed forces while at the same time performing law enforcement duties), prison service, state and local fire brigades, customs service and financial guard, civil defense and national security agencies have been given a unified set of regulations by this Act. Unconditional obedience and all external signs of militaristic structure, once so vehemently rejected by the first democratic government of 1990, are again the underlying principles of administering the "armed organs of the state," including law enforcement. Police officers are soldiers in terms of the criminal law, liable for special military offenses in addition to ordinary ones.

The National Headquarters directs and controls the whole policing network. A National Commissioner appointed by the Prime Minister is the head of the police. His or her two deputies, called Directors General, have respective responsibilities for the two basic fields of police work—criminal investigation and public security (uniformed) policing. Control and direction is exercised as a rule—but not exclusively so—through county headquarters. Some special units, such as the Airport Police or Central Riot Police, report directly to National Headquarters. There are nineteen county police forces in Hungary and the police for the capital, who have the same status as county police. Local police stations, usually serving several municipalities, represent the lowest level of the pyramid.

The Challenge of the Media

The police were one of many institutions, such as the party, the army, the Soviet (local council) system, etc., which represented and demonstrated the power of the communist, totalitarian regime. Journalism critical of their actions was forbidden. It was, therefore, a shocking experience for the basically unchanged force when newspapers began to attack their general behavior and specific actions.

Reporters found ways of gaining information about tensions and disagreements among police chiefs. They also interviewed many rank and file members of the force who did not refrain from criticism of their superiors. There were, no doubt, instances when the disclosure of information by the media interfered with the effective investigation of criminal offenses.

These developments led the police leadership to the conclusion that public communication concerning law enforcement should be strictly centralized and controlled. A unit within the National Headquarters is now responsible for giving out information concerning police activities and important cases. Due to competition among the media, however, journalists are not satisfied with official and uniform accounts of events. Many of them rely on additional sources and frequently question the authenticity of police statements.

To counterbalance what they saw as the undesirable influence of the newspapers and other media, the police resorted to direct means of informing the public about their activities. National and county headquarters maintain newspapers which echo police perspectives on policing, although legally they remain independent. There are also programs on the electronic media which are directly supported by the police. Publishers, together with the leadership of the force are, of course, interested in reaching large sections of the population. Wide exposure has been achieved by the weekly *Zsaru* (*Cop*) which is one of the most popular journals in the country. Its spectacular success, though, can be partly attributed to the inclusion in its contents of bloody crime stories, telephone-sex advertisements and television program schedules along with information on policing.

To sum up, the relationship between police and the press is full of conflicts. Police did not manage to adapt their communication policies to the special needs of the public press and journalism, while the latter often disregard the reasonable interests of investigations and the protection of public safety. Focusing on issues of democracy, one can argue that both the police and media tend to be similar in their attitudes. Though done for different reasons, both groups are prone to be prejudiced against suspects. Neither of them is really interested in waiting for the final judgment, the decision of the court. The presumption of innocence and individual rights are often violated by statements referring to the guilt of suspects even before the trial. Waiting for the outcome would lessen the attractiveness and novelty of a story for the press, while, for the police, the guilt of suspects is an organizational and professional given—the police, in their mind and rhetoric, do not arrest the innocent.

CONCLUSION

The police in Hungary survived profound and extensive political transformations basically unchanged. In addition, some new policing institutions and regulations tend to be even more oppressive than those which existed under the totalitarian system. Though one cannot deny that massive efforts were undertaken to improve many aspects of policing, still there is no real progress in terms of democratization. Indeed, it is the openly declared goal of the Government that police effectiveness and efficiency should be strengthened and space for discretionary action be expanded and that human rights and other democratic values be pushed into the background.

No doubt, many arguments of police hard liners seem to be convincing. But notwithstanding the fact that continuity with the past is usually disguised behind democratic phrases, let me put the question very simply. Is any institution to be rejected just because it originated in the socialist era? Taking into consideration the evolutionary character of Hungary's political transformation, the answer is: surely not. It has to be noted, as well, that western commentators had not found the formal features of socialist policing to be totalitarian or distinct from those in developed democracies.

Other than referring to the dependence of Hungarian policing on the guidance of the communist party, critics from traditional democracies often came to the conclusion that "the police role in Hungary ... is not at all that different from policing in other European nations" (Ward, 1984, p. 32). Rubin goes even further stating that, apart from some specific aspects, "the Marxist–Leninist concept of 'public order in the Socialist society' does not differ, in many respects, from public order concepts of western democracies" (Rubin, 1980, p. 23). These arguments support the idea of the essential neutrality of policing. The same organization with the same powers can serve different political systems. Many examples could illustrate surprising survivals of law enforcement structures even during earthquake-like transformations. The reasons for such continuity, in my opinion, must be the relative independence of the police. They not only enforce the law but they maintain the social order as they define and perceive it.

Although formal aspects of policing appear to be neutral in political terms, they have to be fitted into the framework of the existing state structure. The Hungarian police should have been decentralized but not because decentralization is inherently more democratic. Indeed, centralized law enforcement systems can serve democratic values. However, since the whole administrative system in Hungary has been decentralized, keeping the old police hierarchy will inevitably lead to tensions within the state structure.

A similar argument applies to laws and regulations. It is true that one can consider how far basic rights should be limited for the sake of police efficiency. But such a balancing of effectiveness versus rights must not be divorced from the existing constitutional structure. In sum, I think that the most serious problem in the administration of the Hungarian police is that issues have been dealt with independently of the total constitutional setting.

There is little hope for substantial democratic changes within policing in Hungary. Since the opportunity for major reform which existed during the political transformation was not seized, it seems rather unrealistic to expect that constitutionalists can now achieve a breakthrough in creating a law enforcement system that satisfies the requirements of the Rule of Law. People have been persuaded that the growing menace of crime can only be fought successfully by strengthening the police in all aspects of their organization and powers. Politicians cannot openly express their reservations about police behavior without increasing the probability of losing votes. The rhetoric of the law and order lobby seems to be persuasive enough to keep the old police structure in place but without fitting it into the new system of constitutional values and guarantees.

However, several factors still can be relied on to support democratic improvements in the field of law enforcement. The most important and convincing is the need to harmonize Hungary's legal institutions with those of the European Union in order to gain admission. Hungary's joining the Council of Europe was preceded by a thorough examination and at some points modification of Hungarian law.

Admission to the European Union will raise questions about the compatibility of Hungary's policing system and practices with those in the other countries of the Union. Practical police cooperation in several fields will be an essential aspect of participation in the activities of the Union; data protection and other requirements have to be met by the Hungarian police if they wish to be accepted as compatible partners.

Another factor favoring legislation which will comply with Rule of Law principles is the preparation of the new constitution for the country. Hungary did not adopt a new constitution during the political transformation. Formally, the Stalinist 1949 basic law (*Alkotmány*, 1949) is still in force, but with substantial amendments. The drafting of an entirely new constitution is now on the agenda and provides an opportunity to evaluate and revise the basic regulations on police organization and activities. The institutions and procedures essential for a critical oversight of policing (e.g., the administrative or judicial review of all policies concerning human rights) can be written into the constitution.

Research can play an important role in arguing for democratic reforms. Merely referring to universal human values does not convince those being pressured by the law and order coalition. Proving that many of the tough measures being advocated have not achieved desired results can be more persuasive. Investigations into the realities of policing and its impact on society are, therefore, vitally necessary for the democratic development of law enforcement. Of course, research has to be conducted without prejudice. Due to the apparent essential similarities among different policing systems and keeping in mind important differences in socio-legal environments, international comparison is one of the most promising methods of police research and means to promote change.

APPENDIX

CASES OF ALLEGED POLICE BRUTALITY AS REPORTED BY AMNESTY INTERNATIONAL

In September, October, and December, 1993 Amnesty International called on the Hungarian authorities to initiate impartial and independent investigations into reported cases of torture and ill-treatment by law enforcement officers.

Alleged Ill-treatment of Roma (Gypsies) in Ujpest and Orkeny

On May 6, 1993, Kalo Istvan and members of his family were allegedly ill-treated by police officers in Ujpest, a suburb of Budapest. Kalo Istvan and his wife, Kalo Istvanne, were arguing with each other while returning to their home when they were approached by two police officers. One police officer reportedly asked Kalo Istvan why he had called him [the police officer] crazy and then hit Kalo, his wife and Kalo Anita, their 14-year-old daughter, with a rubber truncheon. Kalo Istvan tried to take hold of the truncheon to prevent further beating, but after the other officer reportedly took out his gun, the Kalo family fled into their apartment. Later six or seven officers broke down the door of the apartment and started to beat the family, including 16-year-old Kalo Maria and 12-year-old Kalo Julia, as well as a neighbor, Nani Gyulane, and her son, Nani Aladar. Having beaten Kalo Istvan all over the body they took him by the feet and dragged him unconscious through the courtyard into the street. He was taken to a police station, but was released the same day. He was treated in the 4th District Hospital for a concussion, facial, head, chest, arm and leg contusions and extensive skin abrasions which he suffered as a result of the beating and other ill-treatment.

Dozens of Roma were ill-treated by police officers on the 21st of May, 1993, in Orkeny, a town some fifty kilometers south of Budapest. About twenty police officers came to search homes in the Roma neighborhood after a German national was robbed in a nearby motel. The incident reportedly began when two plainclothes policemen checked Radics Marton's identity card. They then reportedly twisted his arm behind his back, threw him against a car and handcuffed him. His wife, Radics Martonne, and neighbors came out of their houses and the police reportedly started to beat them. Shortly afterwards a special police unit of around 100 officers arrived in Orkeny. According to accounts given to an Amnesty International representative, they used force indiscriminately, beating and spraying with tear gas the Roma in the neighborhood. At least nine people were injured. Lakatos Laszlone fainted and was taken to a hospital after an officer beat her, ripped off her tracheotomy tube and sprayed tear gas into her face. Feher Peterne, who was five months pregnant, was also beaten and sprayed with tear gas. She later required medical treatment for injuries suffered as a result of the ill-treatment and lost her child. Thirteen-year-old Radics Krisztián was taken to the police station in Dabas together with eight men. The boy was locked up in a cell from 9 PM until 1:30 AM when his mother, Radics Martonne, found him unconscious. He had reportedly been beaten by the police. The bruises on his body were described in a medical certificate issued after the incident (Amnesty International, 1993).

The case of Mohammed Radwan

Mohammed Radwan, a Jordanian national, was arrested on October 14, 1993, in Budapest on charges of possession of and trafficking in illegal substances. Police officers reportedly pulled a hood over Mohammed Radwan's head, pushed him against a police car, and then kicked and beat him. Because of injuries suffered during the assault, Mohammed Radwan's condition deteriorated and he was taken to a hospital in the evening of October 15. His ruptured right testicle had to be removed surgically (Amnesty International, 1993).

Ill-treatment and Violation of the Right to Freedom of Expression—the Case of Herman Peter and Palinkas Jozsef

Amnesty International wrote to Minister of the Interior, Konya Imre, on the 15th of March, 1994, urging him to investigate promptly and impartially reports that Palinkas Jozsef and Herman Peter were beaten and otherwise ill-treated by police officers in Szarvasgede on January 19, 1994. Amnesty International was also concerned that Herman Peter, an activist of the Green Alternative, who was campaigning against the decision of the Szarvasgede Council to approve the construction of a medical waste incinerator, was subsequently held in detention until the 28th of January, apparently for exercising non-violently his right to freedom of expression. Amnesty International's concern was described in a report published in April (Amnesty International, 1994).

REFERENCES

Alderson, J. (1979). *Policing freedom*. Plymouth: Macdonald and Evans.
Alkotmány (1949). *XX. törvény az Alkotmányról* [Act XX. on the Constitution]. Budapest: Miniszterlnöki Hivatal (Prime Minister's Office).
Amnesty International. (1993). Amnesty International concerns in Europe, May–October. Internet: AI Index: EUR 27/01/94.
Amnesty International. (1994). Hungary—ill-treatment and violation of the right to freedom of expression. Internet: AI Index: EUR 27/01/94.
Berkley, G. (1969). *Police behavior in a democratic society*. Boston: Publisher.
Boross, P. (1992). Megnyitó beszéd [Opening speech]. In *Társadalmi változások, bûnözés és rendõrség, nemzetközi konferencia* [International Conference on Social Changes, Crime, and Police]. Budapest: ELTE–RKI (Eötvös Loránd University–Police Research Institute).
Concha, G. (1988). A rendõrség természete és állása szabad államban (Nature and position of the police in a free state). In *A magyar közigazgatás-tudomány klasszikusai 1874–1947* (Classics of Hungarian public administration science 1874–1947). Budapest: Közgazdasági és Jogi Könyvkiadó.
Egységes rendõrségi és ügyészségi bûnügyi statisztika, 1984, 1991, 1992, 1993, 1994, 1995, 1996 (Uniform police and prosecution criminal statistics, 1984, 1991, 1992, 1993, 1994, 1995, 1996). Budapest: Ministry of Interior and the State Prosecution Service.
Fosdick, R. B. (1969). *European police systems*. Montclair, NJ: Patterson Smith (Reprint of the 1915 edition).
Fõvárosi rendõrségi törvény. 1881. XXI. törvénycikk a Budapest-fõvárosi rendõrségrõl (Act No. XXI on the Police of the Capital, Budapest). In *Törvények és rendeletek tára* (Collection of laws and regulations]. Budapest: Ráth Mór.
http://www.hvg.hu/plusz/vilag/index.htm (Home page of the weekly magazine *HVG*).
HVG (1995, August 5). *Heti Világgazdaság* (Weekly World Economist). Budapest.
Idegenrendészeti törvény (1993). *LXXXVI. törvény a külföldiek beutazásáról, magyarországi tartózkodásáról és bevándorlásáról* (Act No. LXXXVI on the entry, stay in Hungary, and immigration of foreigners). Budapest: Miniszterelnöki Hivatal (Prime Minister's Office).
Kmety, K. (1900). *A magyar közigazgatási jog kézikönyve* (Handbook of Hungarian public administrative law]. Budapest: Politzer Zsigmond.
Korinek, L. (1992). Modernization of police. *Rendészeti szemle* (Special issue in English). Budapest: Belügyminisztérium.
Kõszeg, F. (1997). A rendõr és a polgár (The police officer and the citizen). *Belügyi Szemle, 2*, 35, 3–16.
Los, M. (1988). *Communist ideology, law and crime: A comparative view of the USSR and Poland*. Houndmills, Basingstoke: Macmillan Press.
Nyíri, S. (1995). A rendõrségek nemzetközi együttmûködése (International cooperation of police forces). *Belügyi Szemle, 4*, 3–12.
Parliamentary Records (1993). *Országgyülési jegyzõkönyv as Országgyûlési 1993. Október 5-I ülése* (Parliamentary Records, Session of Parliament, October 5, 1993). Budapest: Office of Parliament.
Reith, C. (1943). *British police and the democratic ideal*. London: Oxford University Press.
Rendõrségi Évkönyv (Police yearbook) (1994). Budapest: ORFK (National Headquarters of Police).
Rendõrségi törvény, 1994 (1994). *évi XXXIV törvény a rendõrségrõl* (Act No. XXXIV on the Police, 1994). Budapest: Miniszterelnöki Hivatal (Prime Minister's Office).
Rubin, F. (1980). The policing system of Hungary. *The Police Journal, 53*, 20–29.
Skolnick, J. H. (1975). *Justice without trial: Law enforcement in democratic society*. New York: John Wiley and Sons.
Szabó-Kovács, J. (1992). A jelenkori népvándorlás kriminogén hatásai (Criminogenic effects of the contemporary migration of people]. *Rendészeti Szemle, 12*, 66–71.
Szikinger, I. (1989). Rendészet és önkormányzat (Policing and local self-government). *Állam és Igazgatás, 39*, 1057–1064.
Szolgálati törvény (1996). *XLIII. Törvény a fegyveres szervek hivatásos állományú tagjainak szolgálati viszonyáról* (Act No. XLIII on Service Relations of Officers of Armed Organs). Budapest: Miniszterelnöki Hivatal (Office of the Prime Minister).

Tomcsányi, M. (1929). *Rendészet—közigazgatás—bírói jogvédelem* (Policing—Public administration—Protection of rights by the judiciary]. Budapest: MTA (Hungarian Academy of Sciences).

Trojanowitz, R., & Bucqueroux, B. (1990). *Community policing: A contemporary perspective*. Cincinnati: Anderson Publishing Co..

Ward, R. H. (1984). Police and criminal justice in Hungary. *Police Studies, 7*, 31–34

Westley, W. A. (1970). *Violence and the police*. Cambridge, MA: MIT Press.

6

The Challenges of Policing Democracy in Poland†

EMIL W. PŁYWACZEWSKI

INTRODUCTION

Poland (*Polska*) lies at the physical center of the European continent. The total area of Poland is 120,727 square miles (312,683 square kilometers); its population is 38.6 million. Its capital is Warsaw. Its current frontiers, stretching for 2,198 miles (3,538 kilometers) were drawn in 1945 and presently touch six countries. In the west, Poland borders Germany along the Oder (*Odra*) and Lusatian Neisse (*Nysa Luzycka*) rivers. In the south, the borders mainly follow the watershed of the Sudeten (*Sudety*), Beskid and Carpathian mountain ranges which separate Poland from the Czech and Slovak Republics. In the northeast and east Poland touches the borders of Russia, Lithuania, Belarus and Ukraine. The Baltic coast forms the northern frontier.

Polish society, since World War II, has been transformed by two great movements: the growth of a dominant, urban, industrialized working class and the continuing drift of peasants from the rural areas into towns and cities. The aging agricultural population which remains working the land has experienced great distress. Emigration has been a permanent feature of Polish life for most of the last two hundred years, and roughly one Pole in three lives abroad.

In 1990 Poland experienced rapid political change. The changes which have happened so far have not, however, altered Poland's absolute need to grapple with the uncertainties of further political and economic transitions. Also, in 1990, the Polish government decided to move the Polish economy toward a free market system.

In 1990 the parliament passed a new Police Act which sought to create a police organization which would be adequately prepared to fight against crime within the new democratic framework. The national police force currently is about 90,000 strong, mostly male (females are very rare in the uniformed police) and controlled and commanded by the Ministry of Interior Administration. The national police headquarters is organized along three broad divisions: Criminal Police, Traffic and Prevention Police, and Administration. Besides the police, the Ministry of the Interior also controls the Border Guards and the Fire Brigade.

THE CONCEPT OF DEMOCRATIC POLICING

Democratic political transformations initiated in Poland in 1989 had an influence on police structure and functioning. Transformations occurred in all spheres of public life and altered inter-relations among the state's institutions.

The 1992 Provisional Constitution, and regulations of the 1952 Constitution which continue in force, catalogue the capital principles of the Polish Republic's political system and establish Poland as a democratic state governed by law. The principle of the rule of law, introduced to the 1952 Constitution on the strength of a December 21, 1989 amendment, affects the functioning of the police and other state security agencies. The new Polish Constitution was enacted on April 2, 1997.

The new principles of policing, along with effective administrative control over the police, meet the needs of a democratic society. These changes had become necessary by the thrust and direction of political reforms, including those within the police, and by the need to cooperate with other states (especially European ones) in the maintenance of internal security and the successful prevention of transborder crime, in particular that of an organized nature.

Democracy, obviously, has both advantages and disadvantages. Among the latter are the inefficiency of institutions, corruption and other forms of social pathology. All the same, to paraphrase Winston Churchill's words, though democracy is the worst of all possible systems of government, no-one has ever come up with a better one.

In a democratic state, the role of the police is to provide protection against internal and external threats to the state and to secure the peaceful functioning of society. On the other hand, the fundamental responsibility of the state with regard to the police is to maximize their capacities for the prevention and control of crime and other undesirable phenomena, and to minimize the risk of the police being used in conflict with principles of democracy.

Charles de Montesquieu's model of government with its basic tripartite divisions of powers and functions strongly influenced the drafters of the constitution and helped change the understanding of executive power, which at present is wielded by the President and Government of the Polish Republic. Montesquieu's model also contributed to changes in that part of government administration (which includes the police force) called the special administration.

The police were made responsible to the Minister of Interior Administration, who is their supreme authority within the state. As a member of government, the Minister is responsible for enforcing all statutory tasks in the field of public safety and order. Of some importance for the functioning of the Polish police is the fact that the force falls under the authority of one of the so-called presidential departments. The appointment of the Minister of Interior Administration lies with the President, while the nomination is by the Prime Minister. The Chief Commissioner is the head of the force and is constitutionally responsible to the Tribunal of State, which is a special judicial organ that deals with cases concerning infractions of the Constitution committed by the President of the Republic, members of the government, the General Inspector of Fiscal Control, the Attorney General, heads of central administrative departments and members of the National Broadcasting and Television Committee. Moreover, taking into account that both the Minister of Interior Administration and the Chief Commissioner of the Police are supervised by parliament, it can be stated that the present structure of the police in Poland meets all requirements necessary for a democratic state and the rule of law.

ORGANIZATIONAL CHALLENGES

When considering the position of the police in a democratic society, three issues need to be examined: the legal status and role of the police, control over the police, and the political disengagement of the police.

Legal Status and Role of the Police

In a democratic society, the authority and basis for police actions should be among the most important legal acts passed by the legislature. In Poland, police work is regulated by the Police Act which was adopted on April 6, 1990 (*Journal of Law*, 1995c). Under this Act, the police have been separated from other organs of internal affairs and administration. According to section 1 (Galster, Szyszkowski, Wasik, and Witkowski, 1996), the police are a uniformed and armed force serving society and are responsible for public safety and order. Given their roles, their activities must be evaluated by the standards of effectiveness in protecting the people's life, health and property as well as by the observance of human rights and freedoms (which, in Poland, are sometimes violated by the police).

One problem with the notion that the police are servants to their democratic society arises when a police officer refuses to execute orders and commands which he or she feels are against the law. Democracy calls for the absolute observance, irrespective of the circumstances, of the principle that decisions which are arrived at democratically must be carried out. The question is whether the police are to be allowed to intervene in society using their own discretion in such a way that their actions may endanger the democratic legal order. The only answer to this question is that democracy can be protected by the police only when they are controlled in accordance with democratic principles and legitimate laws are faithfully executed by them.

Control over the Police

Control over the police is one of the fundamental features of public life in democratic states. External control is perceived as one of the main factors which guarantees conformity to constitutional principles by law enforcement agencies. The forms and scope of control over the state's administrative agencies (including the police force) are defined by constitutional laws and acts and statutes of lower order. In Poland, the police are controlled by parliament, the Chief Board of Supervision, courts, prosecutors, internal control agencies, the Ombudsman, as well as by society in general through public control institutions regulated by law (such as local administrative agencies).

Moreover, influence on the police is exerted by district courts and local administrative agencies and authorities, for it is they who approve candidates for the posts of police commandants of regional (*voivode*) and district headquarters and of police stations. Every year, police commandants must present reports about their work, and they must inform their *voivodes*, their regional and local authority councils, about the state of public order and safety of the area. In the case of threat to public order or major disorders, reports and information must be submitted immediately to the oversight bodies mentioned. The police must accede to every request for information at other times. This requirement establishes democratic control over the police and promotes a dialogue between them and civil authorities.

Political Disengagement of the Police

The principle of political disengagement of the police is explicitly expressed in the Police Act. In accordance with this principle, police officers cannot belong to any political party; on joining the police they automatically cease to be members of a party. They must execute only orders issued by constitutional state agencies and they cannot submit to party discipline. In addition, some restrictions have been imposed on police officers as regards their involvement in non-police associations. They have to inform their superiors about their membership in national associations. To become members of foreign or international associations and organizations, they have to obtain the permission of the Minister of Interior Administration or another authorized person. However, political disengagement is not total. Police officers enjoy the right to vote and to belong to trade unions.

The question can be raised whether or not existing legal regulations provide adequate protection against police officers' involvement in political activity. The answer is yes. Further restrictions would violate the constitutional principle of the equality of citizens.

The police exist by the authorization of the community which they represent. In a democratic country their role is to protect society and the state against internal threats and, in some situations, external dangers. On the one hand, the police must reflect the democratic nature of the society they are part of; on the other hand, the society should have confidence in the police, support them and be ready to bear the costs of their maintenance and accept the price and tradeoffs for being safe. This postulate, that the police are part of society but are also partly autonomous, requires maximum openness by the police. Their structures and procedures must be clearly visible in order to ensure that the promotion of public safety and the development of police policies are subject to the legal control and accountability mechanisms of the competent state authorities.

OPERATIONAL CHALLENGES

Crime

Common Crime

The ongoing transformation of society and state in Poland has changed the nature and structure of crime. New forms of crime and danger have appeared. However, the comparison of current crime patterns to those of the 1980s may result in misleading conclusions. Criminal policy of that time was strongly influenced by political and ideological factors and criminal statistics were subject to manipulation. A steady increase in the number of reported crimes in the 1980s coincided with a slow decline in the legitimacy and functioning of the political structures of the Polish People's Republic. According to the Computer Science Bureau of the Police High Command (KGP), the crime rate in the years 1980–1989 increased by 53 percent, from 946/100,000 to 1,446/100,000 people. The increase was caused, among other factors, by an acute crisis in the operation of prosecution agencies which were then involved in political arguments with the opposition, as well as by a growing demoralization of administrative authorities.

But it was in the years 1989–1990, at the time of the shift from a socialist to democracy-based market economy, that the most rapid growth of crime occurred. Crimes jumped by 61.3 percent, from 547,589 in 1989 to 883,346 in 1990. The corresponding increase in the crime rate was from 1,446/100,000 to 2,317/100,000, to 2,526/100,000 in 1995, with a slight decrease to 2,325/100,000 in 1996. The clearance rate decreased, continuing a downward trend observed since the beginning of the 1980s. In the years 1980–1989, the clearance rate decreased from 77.5 percent to 55.8 percent, and in 1980 to 40.1 percent, but rose in 1995 to 54.7 percent and 54.4 percent in 1996. The numbers may not be completely reliable (Siemaszko, 1996, pp. 220–227), but they do reflect the efficiency of prosecution agencies, which at present are in a deep crisis due to numerous problems of both organizational and financial natures and to the inadequacies of current regulations.

Although after 1990 state policy on prosecuting and punishing criminals became relatively stable in comparison with that of the 1980s, the state was faced with new types of crime, especially organized crime, brought about by the new political and economic situation (Pływaczewski, 1996b, p. 19). Following the rapid growth of crime in 1990, the situation stabilized somewhat at a steady, but small, upward tendency. In comparison with 1990, in 1995 the number of crimes on record increased to 974,941, that is by 9.4 percent, with a small downward fluctuation in 1993, when the crime decreased by 3.3 percent compared to 1992. A similar small decrease by 7.9 percent in comparison with 1995 was observed in 1996 when 897,751 offenses were recorded. These fluctuations do not exceed criminological norms and should not be interpreted as downward trends. Optimism could be expressed only if the tendencies were to continue from year to year.

It is evident, however, that some unfavorable phenomena in the crime structure intensified. There has been a considerable increase in crimes against life, health and property, and their perpetrators have become more violent, aggressive and ruthless to victims. To give an example, in the years 1988–1995 the number of murders more than doubled—from 530 to 1,134. But 1996 did not see a further increase in murders. The above data also include murders committed by juvenile delinquents; their number grew from 17 in 1990 to 36 in 1996.

Of much concern is the growth of homicides by firearms, which is undoubtedly connected with the opening of borders and the increasing number of weapons smuggled from western countries (Palka, 1996, p. 35). Police statistics show that between 1960 and 1990 none or just a few cases of murder by firearm occurred each year. It was only in the years 1990–1995 that the numbers increased to a total of 337 cases, with clearance rates in particular years varying from 57.7 to 72.5 percent, for an average clearance rate over the whole period of 58.8 percent. For people having connections with the criminal world, it is no problem to buy a gun for a reasonable price. Weapons are more and more commonly used to commit crimes. For example, in 1995 there were 4,570 crimes which involved the use of firearms, including 1,464 robberies (in 1991 there had only been 167 robberies). This is a clear rise in the kind of crime which is most dangerous to people. At the same time, in 1995 the clearance rate for armed robberies (excluding car thefts) was only 38 percent, the smallest clearance rate for all types of crimes, according to the KGP.

As far as common crime is concerned, the figures are as follows: rapes—1,660 cases in 1989 with a steady growth to 2,267 in 1995; bodily injuries—8,588 in 1989 and 19,371 in 1996; assaults—2,988 in 1989 and 11,575 in 1996. There was also a steady and extremely worrying leap in extortion and robberies from 9,067 in 1989 to 26,852 in 1995. In 1996, a slight decrease (to 26,257 cases) in these crimes was reported for the first time, which appears to be the result of proper detention policies for this type of crime which were introduced in 1995. Perpetrators of these crimes had been punished leniently for some time, as prosecution agencies and courts had been too restrained in using detention, while other preventive measures proved ineffective.

The greatest fluctuations, among the total number of crimes reported each year, are seen in cases of thefts of private property and burglaries. The former were first included in the statistics as a separate entry in 1990, with 169,365 cases reported. In the following years the figures fluctuated continuously, reaching 211,602 in 1995 and falling to 157,479 in 1996. However, the biggest drop was recorded in 1992 (to 137,510 cases), which also seems to be a consequence of the greater use of the detention policy mentioned above. In comparison to the theft of private property, in 1995 there were 53,819 cases of public property theft or roughly one quarter that of private property theft, which gives some indication of economic relations in Poland and the scale of threat to property.

The years 1989–1990 saw the biggest rise in burglaries; they doubled from 218,581 to 431,056, but fell gradually to 299,975 cases reported in 1993, and then increased steadily to 305,703 in 1996. No reasonable explanation for such a big fluctuation can be given. One can only suspect that after 1990 some crimes were not reported to the police because the possibilities of detecting crimes were seen as decreasing (especially in cases of petty offenses, as police do not have enough time to deal with them), resulting in less and less confidence in the efficiency of the police.

A real problem currently is the deepening demoralization of minors which is reflected, among other indications, in the growing number of crimes they commit. In 1990, 60,525 juvenile offenses were recorded. The number increased to 82,551 in 1995, while in 1996 it dropped to 70,073. This drop should be interpreted as an indication that the new emphasis on prosecuting and punishing juvenile delinquents was beginning to have an impact. However, more general conclusions cannot be drawn about the efficiency of this policy in the long run, all the more since the drop did not occur in all kinds of juvenile crime. Although it is typical of young offenders that they commit mostly burglaries and theft (the biggest decrease in 1996 was observed in these offenses), there has been a horrifying leap in violent crimes such as murders, bodily injuries, batteries, muggings, robberies and extortions committed by juveniles. Between 1990 and 1996, bodily injuries increased from 741 to 2,527; batteries from 198 to 1,340; muggings, robberies and extortions from 1,463 to 14,656; and 163 cases of rape were reported in 1996. In addition, the mass media report more and more aggressive behavior in schools. The reasons why juvenile crime is rising are known, yet there are no effective methods of preventing the decline of morality and subsequent delinquency among minors (Wójcik, 1995, pp. 73–75).

Economic Crime

Poland is still a developing country with few firmly institutionalized socio-economic structures. This has led to a further rise in common and organized crime and, in particular, in economic crime (Górniok, 1994, pp. 109–121). The origin of economic crime can be traced to human nature and existing systems of values. Socialism did not develop values which would support the reduction of the ever-increasing crime problem. The socio-economic system introduced after 1989 gave rise to new aspirations. With no possibility that aspirations could be fulfilled, due to the lowered standard of living which followed the initial introduction of capitalism, crime started to expand rapidly.

According to B. Hołyst (1994, pp. 220–221), crime is also affected strongly by the elimination of the information barrier between the western and the so-called underdeveloped countries. The depiction of high standards of living and welfare in western countries offered by mass media easily reaches the young and results in their rejecting all authorities and moral principles. Generally, such a cultural phenomenon is typical of all developing countries.

Lowered standards of living, awakened aspirations and a decline in moral standards are conducive to the spread of economic crimes. The lack of adequate legal regulations and economic instability tempts lawbreakers to commit economic crimes by their potential for easy and quick profits. In 1990 there were 78,337 economic crimes. In 1992 the number increased to 108,051, and in 1995 it went down to 83,893.

It is worth noticing that imperfections in the legal system contribute to the existence of the so-called shadow economy and to crimes in the turnover of capital, economic transactions, foreign trade, banking, tax and customs systems, commercial law and intellectual property protection. Actually, prosecution agencies are rarely informed about these crimes. It is estimated that about 70 percent of shadow economy incomes are generated in legal companies, while about 30 percent are derived from activities beyond state control (Bednarski, 1992, pp. 49ff, Dziemidowicz, 1996, pp. 70–73).

There has also been a considerable growth in activities which were declared illegal and criminal by the provisions of the 1994 Economic Transactions Protection Act. In 1996, there were over twice as many such violations reported compared to 1995. Moreover, after the year 1989, the ailing legal system led to a number of large-scale economic crimes which in turn weakened the state's new democratic system and the state-regulated economy. Such crimes are usually committed by bank clerks, customs officers, businessmen and, sometimes, top officials of the state and bank administration (Hołyst, 1994, pp. 184–185). It can be taken for granted that the "black number" of economic crimes is high, presumably even higher than the number of unknown common crimes. On the basis of fragmentary surveys, it is estimated that about 70 percent of all economic crimes are not reported to law-enforcement agencies (Siemaszko, 1996, p. 205).

Organized Crime

Common and economic crime is closely connected with organized crime which grew rapidly in the 1990s. The existence and growth of organized crime had been mainly ascribed to the social and economic conditions created under the totalitarian system, but the new political and economic reality has forced people to revise some of these myths (Hołyst, Kube & Schulte, 1996). According to estimated data by the KGP's Bureau of Organized Crime, there are about 300 organized criminal groups in Poland "employing" a total of about 6,000 professionals, while an additional 300,000 people earn a living from professional criminal activity. These groups are often connected not only with state authorities and administration but with law-enforcement and criminal justice agencies as well.

Opinion polls carried out in 1995 showed that 35 percent of Poles considered the fight against organized crime to be one of the major problems Poland had to cope with. Media surveys of the 1996 showed an increase to 44 percent, a percentage even higher than the fear of unemployment. Other surveys conducted in 1995 pointed out that 60 percent of respondents regarded paying racketeers for "protection" a common event and an inevitable cost in running a private business. In B. Hołyst's opinion (1996b, pp. 49–50), organized criminals form an independent social class satisfying their own and other people's demand for illegal services, and they make fortunes from their activities.

Existing criminal groups are of an international nature with respect to both territorial scope of their activity and members. Typical activities of criminal groups are car theft and trafficking; the production of and traffic in drugs; money laundering; arms trade and smuggling; alcohol and tobacco smuggling; extorting ransoms and crimes of revenge; customs and foreign currency abuse; banking and capital turnover; night life and white slavery; and computer crime. Foreigners contribute much to these activities. In particular, there is a high degree of involvement on the part of citizens of the Commonwealth of Independent States and other republics of the former USSR (Pływaczewski, 1993, pp. 92–95). Occasionally there are also acts of terrorism.

To complete the picture of organized crime, some statistical data should be cited. In 1991 there was only one case of drug smuggling, while in 1995 there were sixty-nine. The growing number of detected cases of amphetamine production and smuggling indicates that Poland has become a major producer of this so-called "Polish heroin." Not only does Poland produce drugs, it is also a transit country. Currently, Polish youth can consume all the drugs which are available elsewhere in the world. Liberal regulations are responsible for the fact that drug traffickers are to be found even in primary schools. The mere possession of drugs is not yet a crime or penalized, which makes prosecution in this field extremely difficult (Serdakowski, 1996, pp. 195–205). But since the drug market in Poland is just being created, this is the time to introduce proper preventive measures.

Apart from the above, Poland is under threat from a plague of car theft. In 1988 the figure for stolen cars was 4,173; in 1989, 26,389; and in 1995, 54,807, while the clearance rate was as low as 11.5 percent. The main factor hindering car theft prosecution is the relatively easy legalization of ownership (through false documentation) and the relatively unhampered transportation of stolen vehicles. A comparison with other countries shows that Poland is a transit country and the place where cars are stolen rather than where they are sold (Pływaczewski, 1997).

Owing to the lack of proper preventive regulations, Poland is thought to be a good location for money laundering (Pływaczewski, 1993, pp. 150–151). Money laundering is an essential support for organized crime. There can be no doubt as to its occurrence in Poland. This appears to be proven by the fact that some businessmen involved in organized crime make a great deal of money in a very short time. The black or unknown number of this kind of crime may reach 100 percent; practically all incidents remain hidden and very few cases of this type are brought to court. For example, in 1995, bank managers informed prosecutors about thirty-two cases of alleged money laundering, but no report was to be found in the public prosecutor's office. This supports the argument made by Pracki (1997, p. 10) that the initial allegations could not be substantiated.

It should also be stressed that there are more and more instances of counterfeiting, both of Polish and foreign currency, as well as insurance fraud. It is estimated that about 200 to 300 different insurance frauds are committed in Poland every year, with the loss of about 200 to 300 million new *zloty*, which is one-fourth of all insurance premiums (Hołyst, 1996b, p. 51) There is also a striking increase in computer crime, especially in computer piracy. Because the police are not trained well enough to deal with this kind of crime, the detection rate is minimal compared to the scale of the phenomenon.

To sum up, according to public and scholarly opinion, the expansion of crime after 1989 is a result of the system's political and economic transformations. However, the growing threat of crime cannot be accepted as the price of democracy and freedom.

Migrations

Issues related to foreigners constitute one of the biggest problems within the context of unification processes in Europe. Changes within the European Union and efforts by post-communist countries to join have added a new dimension to the migrations of people. More possibilities for traveling exist for inhabitants of former Eastern Bloc countries, and for many reasons: tourism, family, business and economics. Such travel can lead to permanent migrations. Moreover, these countries, especially the former USSR republics and Poland, are used as a transit route by refugees and emigrants from Asia, Africa and the Near East who are heading towards Western Europe and other countries, such as the USA or Canada.

Poland plays a major role in these migrations because of its size, geographical location and desire to join the European Union, as well as its political transformations (processes of democratization) and pace of economic development. So far, Poland has mainly been a transit country but it is becoming a destination country as well. Examples are the growth of new ethnic groups, e.g., Armenians, Vietnamese and Chinese, as a result of the legal and illegal inflow of foreigners into Poland (Sklepkowski, 1996, pp. 215–221).

Migrations through and into Poland are a new social phenomenon which can have a great and long-lasting impact, as has been the case in West Europe. Immigration in Western European countries resulted in the permanent settlement of many ethnic minorities which, in turn, brought about considerable consequences of a political, economic, social and criminal nature (Korybut-Woroniecki, 1993, p. 86).

Poland as a Transit and Destination Country for Illegal Migration

The years 1989–1991 saw a substantial decrease in emigration and Poland, once perceived as one of the main sources of politically and economically motivated migrations, became itself a transit and destination country for illegal migrants. In the years 1990–95, Poland became the destination for a massive and varied influx of people. According to figures for 1994, illegal migrants from 89 countries arrived in Poland. The gradual tightening of immigration regulations for people seeking economic betterment (rather than political freedoms), introduced by many countries, brought about specific counter-reaction on the part of increasing numbers of possible migrants and also resulted in organized crime developing an interest in migration. Stricter rules generated new ways of crossing international borders, such as smuggling people by air or hidden in other means of transport, crossing borders in unguarded places or using forged, expired or other people's documents (Siemaszko & Woźniak, 1995, pp. 100–103).

The majority of legal and illegal migrants arriving in Poland treat it as a transit country on their way to the West, in particular to Germany. At present, transit migration is comprised mainly of refugees who claim, falsely or correctly, that they are looking for asylum, of people who pretend to be (temporary) tourists, and of people who enter Poland illegally, often with the help of organized smuggling groups. So far there have been more emigrants leaving for the West than legal immigrants arriving from the East. This is the consequence of the fact that the majority of migrants, with and without visas to other countries, enter Poland legally but then cross the border to the West illegally.

The growing interest in Poland as a destination country by people from Eastern and Southern Europe, Asia and Africa can be seen by a gradual increase in the number of applicants for permanent residence (about 3,000 a year) and for work permits (about 15,000 a year).

In addition, every year more than a million people prolong their visit in Poland after they enter on a tourist or limited stay visa, in a kind of temporarily delayed migration. Many of them are interested in staying here permanently or in going to the West. The ongoing political unification of Europe and, in particular, the future inclusion of Central European countries in the European Union, may intensify migration to Poland with the intent to stay there temporarily. Once Central European countries, including Poland, are admitted to the Union, migrations from Poland to and through Western Europe will be easier to accomplish. Therefore, a considerable inflow of foreigners to Poland can be expected before it joins the Union (Sklepkowski, 1996, p. 171).

When Poland is admitted to the Union, its eastern borders will become the border of Europe, which will undoubtedly increase border crime in the region neighboring the former USSR countries. Such crimes will be encouraged by highly organized criminal smuggling groups operating on the eastern border, which are becoming increasingly professionalized in their activities. Having connections with the mafia and enormous funds at their disposal, these groups make use of perfectly forged documents or transit visas gained by false pretenses from consular offices of the former USSR countries, to organize the smuggling of individuals and groups across the border.

Of some importance for the efficiency of these groups are a good knowledge of the border and attempts, sometimes successful, to corrupt the border services of particular countries. Signs of organized smuggling across the border are also noticeable on the western border of Poland, to a lesser degree on the southern border, and at seaports and airports, where the threat seems to be on the increase.

Threats from Illegal Crossings of the State Border in 1992–1995

Table 6.1 shows the number of people stopped by the Border Guard in the years 1992–1995. During this time for various reasons, about 60,000 people a year were not given permission to enter Poland. According to estimates by the Border Guard, border protection is effective at an 80 percent level. However, this estimate does not apply to people who legally enter but continue to stay in Poland. So far, the Polish immigration system cannot fully monitor foreigners staying on in Poland.

Among already noticeable side effects of transit and permanent migration and temporary migration with the intent to stay are: vagabondage; work without permit by foreigners; serious crime (robberies, racketeering, homicides by hire, etc.); crimes committed by and targeted at foreigners; the organized smuggling of tobacco, alcohol, electronic equipment, stolen cars, works of art, precious metals, drugs, etc.; forgeries of financial documents; money forgeries (legal tenders of many countries are counterfeited in Poland); forgeries of passports, visas and other travel documents; illegal trade at marketplaces; money laundering; the organized smuggling of migrants and nightlife-related crime (Sklepkowski, 1996, p. 150). One of the most serious problems, both a consequence and sign of illegal migration, is organized white slavery, which accompanies the spread of illegal sex enterprises and occasionally the traffic in children.

Moreover, migration processes are conducive to the creation of a social and economic infrastructure (starting legal and illegal firms and smuggling routes, etc.) for international organized crime. Migration-related phenomena, including pseudotourism, are commonly considered to have been one of the main factors which significantly contributed to the growth of crime in the years 1990–1995, both in Poland and in other European countries. This growth is observed mostly in organized crime as a result of an expansion of ethnic criminal organizations, including supranational and international ones. The development of criminal groups and a supporting infrastructure is also affected by the permanent inflow and settlement of criminals from the Commonwealth of Independent States and Asia together with

Table 6.1. Number of persons stopped at the border

	Outside border check points	At border check points	Coming from Poland	Entering Poland
1995	11,900	11,028	15,626	30,934
1994	3,667	3,288	2,672	2,647
1993	12,291	11,394	15,906	31,635
1992	3,306	2,922	2,392	1,946

At particular sections of the Polish border:

	with Russia	with Lithuania	with Belorussia	with Ukraine	with Slovak Republic	with Czech Republic	with Germany	at the sea border	at Okęcie Airport Warsaw	Total
1995	79	627	144	336	850	1,793	11,243	116	409	15,597
1994	82	503	357	252	634	1,635	10,225	221	407	14,316
1993	59	245	278	220	508	1,393	15,154	166	275	18,298
1992	66	190	222	295	–	–	30,387	130	223	33,581

the organized criminal activities typical of these countries (Płilwaczewski & Sklepkowski, 1995, p. 78).

The threats posed by immigration and crimes committed by and against foreigners are not all of one kind. Important are immigration-related criminal offenses, and economic and organized crime which in combination, although not each to the same degree, influence the level and internationalization of the shadow economy. In this context, one should recognize specific police duties which result from the role the police play among the many other institutions responsible for public order and safety protection. In Poland there is no immigration police as yet (in the future, this responsibility will be assumed by the Border Guard) and Polish law concerning foreigners still lags far behind international standards. There exists, as well, a lack of financial resources for creating and maintaining a sufficient number of refugee facilities and detention centers for illegal immigrants. Moreover, the police do not have enough funds to finance accommodation and food for temporarily detained immigrants, which sometimes reach large numbers. There have been groups of Asians as large as a hundred persons or more who have been temporarily detained.

In some cases, arresting a large group of immigrants almost completely disrupts the functioning of the police unit concerned. The main reason for this is the requirement to carry out a number of specified activities which are necessary to begin deportation procedures. Yet the police cannot always detain illegal immigrants because of the lack of space in detention centers. In such cases, released immigrants are issued with an order to leave the country from the *voivode* (chief official of an administrative district), with which they do not usually comply.

A separate issue is the frequent inability to communicate with immigrants as they often quite deliberately speak rarely used languages in their contacts with the police. In addition, it is worth stressing that organized illegal immigrants receive advice from their lawyers who are familiar with the rights immigrants enjoy and take advantage of inefficient legal procedures in order to avoid expulsion or deportation.

Among the basic Polish police responsibilities in regard to immigrants are detecting and fighting migrant-related crime, protecting immigrants against Polish and foreign offenders as well as providing victims of crimes with assistance, enforcing legal procedures with particular emphasis on human rights and arranging for humanitarian aid for immigrants if necessary.

The Polish immigration system is subject to continuous change due to the growing threats posed by illegal migration. The responsibilities of the police in this area now and in the future depend on changes in the immigration system and the role assigned to the police within it. Now, despite various difficulties, the police carry out their tasks in accordance with the law and human rights.

RESPONSES TO CHALLENGES

International Influences

In the 1990s Poland has reached an inglorious position among other countries with respect to crime figures. First of all, it has become a transit country for international drug trafficking along the so-called Balkan and Asian routes. Poland is also the main smuggling route to the East for cars stolen in Western Europe. One can observe a recent escalation of criminal activity by international groups in the Baltic Sea countries. Drug production and trafficking, money laundering, trade in arms and radioactive materials, and the smuggling of stolen cars, cigarettes and alcohol are all problems which police forces in this part of Europe are faced with more and more often.

Due to its geographical location, Poland plays a crucial role in the development of these criminal phenomena. The number of foreigners suspected of committing crimes in Poland has been steadily increasing for a few years now.

As Table 6.2 shows, in 1991 there was a dramatic increase (by 250 percent) in the number of foreigners committing crimes in Poland and the numbers have remained at a high level since. In 1994 the crime rate for foreigners went up by as much as 30 percent. According to 1994 data, nearly half the offenders were citizens of Russia (22 percent) and the Ukraine (26.5 percent). Other criminals were citizens of Germany (13.6 percent), Romania (5.3 percent), Lithuania (3.5 percent), Bulgaria (3.4 percent) and the Czech Republic (2.7 percent).

Given this situation, individual actions undertaken by the police forces of different countries to fight the criminal world cannot be very effective in the long run. What is needed is transnational police cooperation at different levels and a search for new structural solutions with respect to the functioning of international police organizations. It is also important that legal standards in the area of extradition, money laundering and drug crime become harmonized. However, the matter which is of utmost urgency is that a uniform internal security strategy throughout Europe be worked out. Rainer Schulte (1995, p. 12), the director of the Police Executive Academy (*Polizei Führungsakademie*) in Münster-Hiltrup, Germany, has outlined the need for cooperation in three fields of European police relations. He advocates that cooperation be encouraged within the contexts of the European Union and the Schengen agreements, and that bilateral cooperation be cultivated among the countries of Europe.

Existing boundaries do little to restrain the expansion of contemporary crime and that makes crime one of the major threats to the social and economic stability of the continent. Limiting police cooperation only to the Schengen and European Union partners would be a mistake, particularly when the rapid growth of criminal groups in post-communist Europe makes it necessary for these countries to get involved in the process of creating a uniform European internal security strategy.

One can argue that, despite numerous legal barriers, the objective necessity to fight transnational crime has contributed to the harmonious development of police cooperation in all fields of social and economic life. Most successful has been cooperation in fighting organized crime. A conference on drug trafficking and international police cooperation was held in Berlin in September 1994. Among its participants were ministers of justice and internal affairs from European Union

Table 6.2. Foreigners suspected of committing crimes in Poland in the years 1976–1994

Year	Total number of suspects	Suspected foreigners	Their percentage of the total number
1976	231,677	820	.35
1982	251,231	216	.09
1985	310,888	386	.12
1989	220,265	957	.43
1990	273,375	719	.26
1991	305,031	2,402	.79
1992	307,575	3,575	1.16
1993	299,499	3,010	1.01
1994	388,855	3,983	1.02

member countries and representatives from Central Europe, including Poland. The conferences issued the so-called "Berlin Declaration" which emphasized the importance of intensified initiatives to be undertaken against the illegal traffic in drugs, radioactive materials, immigration and money laundering (Pływaczewski, 1996c, pp. 87–88). The ratification of the UN's 1988 *Convention Against Illicit Traffic in Narcotic Drugs and Psychotropic Substances* and of the Council of Europe's 1990 *Convention On Laundering, Search, Seizure and Confiscating of Proceeds From Crime* had authorized international police cooperation in such areas as transborder surveillance, controlled purchase, undercover operations and money laundering prosecutions.

The last issue was of particular interest to participants of a subsequent conference of the Council of Europe, which took place in Strasbourg in December 1994. Poland was represented by the Minister of Justice and officers from the Police High Command Bureau for Fighting Organized Crime. Analyzing the "East Wash" Project (a program against money laundering initiated by the Council of Europe), it appeared that in 1989 only 17 cases of money laundering had been reported in Central and East European countries; by 1994 the number had increased to 400. Of much concern was the fact that only 15 percent of suspected money laundering cases were reported by banks. The reason is that banking laws in Central and Eastern Europe still have not yet been adapted to the changing economic situation, particularly with regard to bank secrecy regulations. This analysis resulted in the proposal to exchange information among national institutions which in different countries mediate between banks and financial institutions and law enforcement agencies in regard to the investigation of transactions with money laundering features.

Another example of a contribution to international cooperation by the Polish police is their involvement in regional collaboration with members of the European Union and other countries. This kind of activity is part of the movement toward full European integration. In 1993, the first conference of the Baltic countries was held in Borgholm, Sweden and was attended by representatives of law enforcement agencies from Belorussia, Denmark, Estonia, Germany, Norway, Lithuania, Latvia, Poland and Russia, and observers from Ukraine, Holland and the Council of Europe. Conference participants dealt with the following issues: illegal production of, traffic in and smuggling of drugs; illegal migration of people, or even whole social groups, trying to cross borders with West European countries as their destination; tobacco and alcohol smuggling; illegal traffic in stolen vehicles; illegal transfer of capital, including money laundering; illegal traffic in weapons and radioactive substances; and money forgery.

During subsequent meetings of the Baltic group, including one held in 1995 in Warsaw, and considering the rising crime rates in the Baltic Sea countries, participants agreed to enhance direct contacts among their law enforcement agencies and to carry out regular crime analyses.

Meetings of representatives from the criminal police in Poland and the former USSR countries, also attended by observers from thirteen West European countries, resulted in the setting up of the Police Computer Data Bank (abbreviated BIK) located in Warsaw, which gathers and provides criminal information for the police forces. Other joint projects which were agreed upon dealt with carjackings, homicides, acts of terrorism and kidnappings for ransom.

Apart from being involved in international initiatives, in the 1990s Poland has entered into a number of bilateral agreements with respect to transnational policing matters. According to Schulte (1995, p. 17), quick intervention by the police, particularly in case of regional incidents, is possible only because of such bilateral agreements. Among the direct results of these agreements were the establishment of liaison police officer posts in signatory countries, whose work provides the foundation for further international police cooperation. At present, liaison officers representing German, American, French, Swedish, British, Russian and Ukrainian police forces are based in Poland.

Most advanced is Polish-German cooperation. In 1991, both countries entered into an agreement to fight organized crime, terrorism and illegal migration, and (a provision added later) cross-border crime.

The success of European integration and an internal security strategy will depend largely on the awareness and professional training of police officers. At this time, a number of regional and bilateral initiatives have been started which seek to promote the exchange of experience-based knowledge about methods and forms for fighting crime, enhancing police–public cooperation, or other aspects of public order protection. In 1992, the Middle European Police Academy (MEPA) came into being, originally as a partnership between the Austrian and the Hungarian police. MEPA was joined a year later by the Czech Republic, Slovakia, Poland, Germany and Slovenia, and in 1996 by Switzerland. Its aim is to train police officers in the prevention of and fight against organized crime. Academy member countries are extremely interested in cooperation in various areas, including training, because of the threat posed by international criminal groups operating in this part of Europe. Another interesting initiative aiming to integrate post-communist police forces are conferences of police academies' commandants. So far four such meetings have taken place: in Münster (Germany, 1993), in Prague (Czech Republic, 1994), in Vilnius (Lithuania, 1995) and in Bucharest (Romania, 1996).

The Law

The socio-economic situation in Poland after the year 1989 has brought about far-reaching changes in the origins, quantity, pattern and quality of both economic and common crime. The changing crime problem necessitated alterations in almost all branches of law, most of all in the substantive criminal law and rules of criminal procedure. Changing the law, obviously, does not mean that criminal law is thought to be omnipotent. Law reforms simply reflect the conviction that the law ought to be *ultima ratio* rather than *prima ratio*. Of no less importance was the fact that in many cases the existing criminal law proved to be dysfunctional (Tarnawski, 1996, pp. 19, 24). In addition, Polish penal law had to be brought in line with operating standards defined in the Universal Declaration of Human Rights, the European Convention for the Protection of Human Rights and Fundamental Freedoms and the International Covenant on Civil and Political Rights (Hofmañski, 1995).

Thus, on the threshold of a newly born democracy, the Polish police faced unknown challenges connected with controlling, detecting and fighting an increasing crime problem, in particular that of economic organized crime. The lack of adequate legislation, especially in the beginning, prevented the prosecution of non-penalized conduct (acts not considered crimes in law earlier). Prosecution was hindered by the incoherence of existing legal regulations, e.g., there were no proper regulations with regard to the new economy which resulted in a large number of large-scale crimes.

To handle the developing crime problem, the legislators of the Third Polish Republic, in the period between June 1989 and June 1996, introduced a number of changes in the existing criminal law. Among others, eight new crimes were defined in the specific part of the criminal code and stricter penal sanctions were introduced in relation to ten existing offenses. On the strength of other legislative enactments, seventy-six new crimes were spelled out, and stricter sanctions were introduced in relation to fifteen offenses after rewording some elements of definition but without changing the core of the crime itself. As a consequence, eighty-four new crimes were added to the law. The changes, since they criminalize conduct considered legitimate earlier, have aroused criticism from some legal scholars (Filar, 1996, p. 55).

Apart from introducing a number of penal regulations to deter criminal threats and challenges to the new economic system, new regulations regarding economic relations were introduced to create social, moral and political conditions to prevent or eliminate criminal economic behavior. These include, among others, a law imposing excise duties and the obligation to place tax stamps on spirit and tobacco products; amendments to the economic activity and tax obligations acts; a tax on goods and services; a new fiscal control act whose aim it is to improve the efficiency of fiscal control, introduce new policies and powers facilitating the fight against the shadow economy, and to assist the police in their fight against economic crime; the introduction of copyright and related rights regulations, which also create new criminal offenses for twelve different types of products, including video and audio piracy. Such violations constitute a big part of the shadow economy.

Unfortunately, these laws are largely ineffective because of the growing pauperization of the society. Such crimes are rarely reported to the police. Computer crimes pose similar problems, a situation made worse by the fact that the police are not prepared to prosecute them. In order for their prosecution to be effective, new methods and operational techniques have to be developed (Hołyst, 1996a, p. 242; Adamski, 1994; Jakubski, 1996, pp. 48–49). Thus, despite the existence of regulations penalizing such conduct, the efficiency of prosecuting this kind of crime is very low, a deficiency which is condoned by society's indifference as far as the prosecution of economic crime is concerned. In order to limit economic crimes, the legislature is working on other legislative initiatives, such as a tax regulations act, a general customs inspector act, an economic court registry act, and banking law and customs law together with related regulations. The advantage of these regulations is that the creation of non-penal provisions targeted against economic crime is in line with remedies outlined in the Council of Europe Recommendations (Council of Europe, 1981, pp. 8, 37; for further details see Górniok, 1994, pp. 163–167).

Undoubtedly, the above policies will either directly or indirectly affect police work and make the prosecution of criminal and economic offenses more effective. Of the utmost importance in this respect was the passing of the economic turnover protection act (Journal of Law, 1994), although it happened relatively late (on Oct.12, 1994) and after the most serious large-scale crimes had been committed. Under this act the following crimes can be prosecuted: 1) the crime of punishable mismanagement which protects trust-based legal relations (section 1); 2) the crime of economic fraud connected with a) public auctions, to ensure their proper organization and progress (section 2), b) obtaining a loan, credit guarantee, grant, or order for public works by means of any fraudulent behavior (section 3), obtaining compensation under insurance contracts by fraud (section 4), laundering money derived from organized crime activities connected with the traffic in drugs or weapons, counterfeiting or extorting a ransom (section 5); and 3) crimes against creditors—these crimes were, in fact, reestablished as they had been included in the criminal code of 1932 (section 6–10) (Górniok, 1995, p. 8; Ratajczak, 1994, pp. 19–20).

Although some of these acts, if they met certain conditions, had already been criminalized under the existing criminal code, financial law and the commercial code, an unquestionable advantage of the new formulation of criminal liability is that they take into account the specific elements of these acts and are based on lessons drawn from criminal justice experience, including the achievements of foreign countries. The new criminal law and regulations equip the police with fundamental and ultimate authority in their fight against economic crimes, including those committed by organized crime.

The two-year history of the act shows that the regulations have been applied more and more often with increasing efficiency and better equipment and resources. The majority of cases, so far, have been tried under sections 1–3 of the act, whereas no convictions under section 5, that is money laundering, have been reported. This omission is extremely disadvantageous from the standpoint of criminal policy—this type of crime continues to thrive despite the existence of regulations penalizing such activities. The reason seems to be that the statutory regulation of money laundering focuses on intent and future consequences, which makes it difficult to substantiate allegations with evidence. The regulations have also been criticized, justifiably so, for applying only to some forms of organized crime (expressed in section 5[1]) (Pływaczewski, 1996a, p. 130). Moreover, the enforcement of section 5 has been hindered by the fact that prosecuting agencies have no direct access to banking data owing to rigorous principles restricting the flow of information and protecting secrecy, a restriction which hampers investigation procedures and undermines the financial security of banks.

Thus, Polish legislation in this area is not yet in line with section 18 of the Council of Europe Convention, which provides that no party shall cite banking secrecy as the basis for the refusal to cooperate in disclosing and seizing property derived from money laundering. It should be pointed out that, at present, the Ministry of Interior Administration is participating in the drafting of the Prevention of Illegal Activity Proceeds Legalization Bill initiated by the Ministry of Finance. It is expected to include the long-awaited liberalization of regulations concerning police access to secret bank data (now such data are only available to financial control agencies and partly to the public prosecutor). The establishment of a State Financial Data Agency should increase the efficiency of detecting money laundering cases as well as improve criminal prosecutions.

Another regulation requiring additional comment is section 4 of the act, since it does not protect the Insurance Guarantee Fund. The act does not punish a person who causes an incident in order to obtain compensation from the Fund. Such a staged incident can be prosecuted under section 205 of the criminal code, but with some limitations, for example that compensation had already been paid out. Furthermore, the new Insurance Act of June 8, 1995 (*Journal of Law*, 1996), introduced a number of new criminal regulations concerning various forms of insurance activity, since the previous regulations had proven to be of no use (Bentkowski, 1996, pp. 57–62).

It seems that the provisions of the Acts discussed above should successfully deter amateurs in search of quick and easy profits. Current and planned amendments to the provisions of the Acts will seek to eradicate criminogenic factors which, according to the casual models and criteria proposed by Lernell (1965, p. 179), support economic crime. The aim should be the creation of laws and provisions which are approved by the whole society. This would be one indication that democracy is being achieved.

The old criminal substantive law, that is, the criminal code, did not include many new regulations which respond to the latest threats posed by crime. On the whole, all recent amendments (*Journal of Law*, 1995b) seek either to mitigate or aggravate criminal liability, in particular for the gravest types of crimes, such as organized crime. In the context of fighting organized crime, essential for the police are the amendments to sections 267–277 of the criminal code, which complement the former notion of a criminal bond with the notion of an organized criminal group. That notion, which includes all forms of cooperation and complicity, may prove to be an efficient tool in the fight against organized crime, even at the stage of preparations for an offense regardless of whether its aim has been attained.

In addition, section 276[3] of the criminal code introduces criminal liability for persons who start or head a criminal band or organized criminal group—a completely new, aggravated form of a crime. Some justified controversy concerning these provisions arise from the explicit assumption that the "crime organizer" is the "instigator" (the person who wants another person to commit a crime and induces him or her to do so) which may cause serious inconsistency in criminal justice. To prevent inconsistencies, it will be necessary to penalize all forms of preparation and activities involved in organizing criminal groups (Daszkiewicz, 1995, p. 34.) so that organizers cannot avoid punishment, as may have been the case so far. As in the past, section 277 of the criminal code includes regulations conditioning the rupture in solidarity ties among criminal group members. That is, whoever voluntarily withdraws from participation in a criminal group and makes known to competent authorities the planned commission of a crime, and thereby makes possible the prevention of that crime, shall not be subject (unless that persons started or headed the criminal group) to a penalty for an offense specified in article 276. The court may apply extraordinary mitigation of the penalty for an offense under the same article. However, these regulations appear to be used rarely by prosecuting agencies and courts largely because there is little cooperation between them.

Worth mentioning is also section 202 of the criminal code (illegal seizure of property on a large scale), which protects the economic activity not only of socialized economic units, as has been the case so far, but also of other legal persons, natural persons and other organizational units which do not enjoy the status of legal persons. This provision establishes the long-advocated equal protection of all property by the criminal law.

These changes, in combination, are undisputed and precious innovations in the war against the most dangerous crime. It seems that for unknown reasons the police do not take full advantage of sections 276–277 of the criminal code. This may be due to the fact that it is more and more difficult to infiltrate criminal bands and organized criminal groups because of procedural restrictions on the police and the increasingly impermeable criminal world.

More attention should also be given to some of the latest amendments to the code of criminal procedure introduced on June 29 and July 6, 1995 (*Journal of Law*, 1995a), which can directly affect police efficiency in their fight against the most dangerous crimes. Section 65[1](1) of the code of criminal procedure provides the legal basis for direct identification parades (police line-ups), with the identifying witness kept out of sight so that the person to be identified does not see or recognize the witness. This procedure is of great significance for the investigation process and to victims and witnesses. This arrangement for presenting suspects for identification avoids the potentially intimidating influence of the person presented on the witness stand, decreases witness stress and thus increases his or her reliability. Another advantage of this method is the possibility of carrying out the identification parade with the participation of anonymous or unknown witnesses, a provision which was introduced into the criminal code by section 164a (Hofmański, 1994, pp. 31–32; Wasek, 1995, pp. 81–87).

The institution of an anonymous or incognito witness was introduced into the criminal proceedings in order to conceal the witness' personal data and other information given in statements. Such information could reveal the witness' identity and endanger his or her own life, health or property, or that of close relatives. The concealed information is available exclusively to the public prosecutor or court and, if need be, to the police officer who is conducting preparatory proceedings. However, section 164a is not perfect as it does not completely eliminate the possibility of identifying the anonymous witness. There is high probability of this happening since the defendant or defense lawyer can move to admit the evidence given by a witness presumed to be taking part in the trial as an incognito witness. The judge cannot dismiss such a motion as he or she is bound by section 155 of the code. Although section 164a does not ensure full protection for witnesses, it has been used so far in several cases concerning different forms of organized crime, e.g., racketeering.

In addition to section 164a, section 173[3] of the procedural code expands powers of the police and other law enforcement agencies as regards "ordinary" witnesses. It increases the protection of witnesses by the possibility of reserving information of his or her address exclusively for the prosecutor or court, in case of a reasonable belief that a threat of violence exists against the witness or next of kin in connection with the proceedings. Unfortunately, this regulation does not conceal the witness' place of work. This makes his or her protection insufficient, as it is easy to establish a workplace address or even threaten him or her there (Waltoś, 1993, pp. 42–44).

Among other significant amendments to the code of criminal procedure are: 1) detention awaiting trial, when there is reasonable suspicion that the defendant may commit another crime against life, health or public safety, when the crime is punishable by at least an eight-year-long imprisonment, and in particular when the defendant previously threatened to commit such a crime—section 217[3]; and (2) a new preventive measure which prohibits the defendant or accused person from leaving the country, which may be combined with confiscating documents needed to leave the country or a ban on issuing such documents—section 235a. These, together with other subsidiary preventive measures, should ensure the proper course of criminal proceedings.

From the standpoint of combating particularly dangerous crimes, momentous regulations were introduced into the Police Act on July 21, 1995 (*Journal of Law*, 1995c). These regulations fully recognize the importance of certain operational police activities in detecting crimes and their perpetrators. These activities have not been treated adequately in the provision of the code so far, as they were often wrongly identified with intelligence gathering activities (Hanausek, 1993, pp. 475, 477). The controlled purchase (section 19a) and delivery of drugs (section 19b), introduced by the amendments, are examples of these operational activities.

The above-mentioned techniques of police operational work are most of all used in cases of particularly dangerous crimes. Controlled purchase can consist in secretly buying or receiving things either derived from crime (subject to confiscation), or which, according to the law, cannot be produced, possessed, conveyed or traded, as well as in accepting or giving material profit. On the other hand, controlled delivery can be used in order to supply documentary evidence for such grave crimes as traffic in explosives, weapons or drugs. Moreover, these activities can be carried out to identify people involved in crimes or to seize the object of crime, and, according to the code of criminal procedure, the techniques aimed at gathering information that eventually may be used as evidence in criminal proceedings (Waltoś, 1996, p. 359).

The introduction of these two methods was necessary and long awaited. However, there are still no adequate rules governing the way they are to be carried out, supported with documents and transferred to the realm of criminal proceedings. The incompatibility of these regulations with the provisions of the code presents procedural problems for the police. According to Hanausek (1993), the rules which govern the introduction of evidence derived through these new methods should not be developed in criminal proceedings on an ad hoc and case-by-case basis, but should be introduced by means of a definite procedural decision issued by the body conducting the proceedings—for example, the prosecutor's decision to recognize specific results of such police activities as evidence. Still, it is the so-called controlled bribe which poses the biggest problems. The present regulations do not guarantee the efficient prevention and detection of bribery. The prosecution of bribery, in which both parties are liable to punishment, is hindered extensively if not made impossible, without special provisions in the law (Jaroch, 1996, pp. 359–363). It would be desirable, for example, to include an amendment whereby the person who provides information about giving or taking a bribe would not be prosecuted.

In addition, section 19 of the Police Act expands the circumstances under which the police can use technical means to secretly obtain information and record clues and data of evidential value. In connection with these changes, the police and, in some particularly justified instances, persons other than police officers have been authorized to carry and use documents which prevent their identification when carrying out these activities (Daszkiewicz, 1996, p. 317).

The final change in law requiring some discussion is section 20[2] of the Police Act. It empowers the police to take, collect and use for detection and identification fingerprints, pictures and other data concerning people suspected of intentional crimes or prosecuted on indictment who are trying to hide their identity or are unidentified. This section expands the procedural, forensic and operational powers of prosecuting agencies and provides the authority for setting up and maintaining necessary identification data files, e.g., photo and fingerprint records, and for using them in criminal proceedings (Kudrelek & Lisiecki, 1996, pp. 18–21).

The new legal regulations resulted from new threats faced by Polish society. There arose much controversy as regards their interpretation, procedures and ethics. Nevertheless, they are of extreme importance given the new (criminal) reality. At present, the legislature is developing new solutions concerning auto theft, financial transactions, and property protection. On June 25, 1997, the legislature enacted the Law on the Crown Witness (*Journal of Law*, 1997) which seeks to break down the solidarity of criminal groups by providing special protection for witnesses who inform on criminal groups of which they were members. All of these changes should equip the police with more efficient tools and means in their fight against the crime wave.

Individual Professionalism

Police ethics are derived from social ethics in their broadest sense. The basic criteria for evaluating police conduct are part of the moral heritage of the whole society. The ethical standards of the police have to be in congruence with general moral norms. One cannot judge as good something which is commonly regarded as evil. However, though police and societal ethics are based on the same basic norms, this does not mean they are identical. Police ethics are also influenced by the characteristics of the job and by the position and role of the police in a democratic state and society. Moral standards of the police are defined taking into account their tasks, methods and means. Thus, the specific nature of police ethics embodies a different hierarchy of socially approved values, which is reflected in such traits as wearing a uniform, an organizational structure based on hierarchy and submission, special equipment, weapons, the power to use reasonable force against citizens, availability for duty at any time, service loyalty, a special attitude to law, team work, discipline, the secret nature of many police activities, frequent contacts with the criminal world (which may bring about unfavorable consequences for the police), the necessity to resort to violence or a statutory ban on getting involved in politics.

Police ethics embodies the moral notion that police methods must be subordinated to the superior values, tasks and objectives assigned to the police by the state and society. Moral standards, for example, require that police officers respect standards of human dignity when enforcing the law. Law often does not always provide clear guidelines as to proper conduct, which means that police officers must exercise their discretion in conformity with moral principles.

Police ethics also regulates relations within the force by encouraging solidarity, mutual assistance, consideration for recruits and loyalty towards superiors. In practice, few cases of violating these standards have been observed.

Of much greater importance are norms regulating relations among the police, society and state agencies. These norms define the core traits expected of police officers. The nature of police work requires that police officers exhibit such qualities as bravery, firmness, adherence to principles, determination, politeness, sensitivity to harm, tolerance, good manners, impartiality and honesty. Operational work requires shrewdness, the ability to mislead the enemy and the capacity to act in secret. Because these required traits are many and diverse, it is impossible to state only one and unchanging profile of the good police officer. The changeability of circumstances sometimes calls for completely different attitudes when dealing with citizens and offenders: humanity and strictness, tolerance and uncompromisingness, trust and criticism. The choice of a given attitude depends on the situation. However, irrespective of the situation, the police officer should always obey the law and respect human dignity.

Disengagement from politics is a crucial aspect of police ethics, for it ensures the realization of the most precious procedural and moral principles underlying the work of law enforcement agencies, namely impartiality, broad-mindedness, and an authentic humanitarianism free of political, religious or racial preferences.

The notion of police ethics—the basic list of theoretical principles of conduct—does not generally arouse reservations among police officers themselves. However, essays, written by police officers under conditions ensuring full freedom of expression, repeatedly reflect opinions which emphasize a distinction between theory and practice, for example in the principle of political disengagement. All police officers approve of it, yet they have much doubt as to the possibility of its full application in practice. Although bribery and corruption in the police are utterly inadmissible in theory and regulations, some officers justify their occurrence by low police salaries; the morally justified dominance of crime prevention over prosecution is questioned in practice because the fight against crime requires repression rather than prevention. In police work, the "ideology of power and fight" prevails over an "ideology of service"; maintaining an objective attitude during investigation is called into question since criminal procedure places the police on the side of prosecution. And, finally, moral guidelines for the police which call for effective yet totally lawful performance are perceived as inconsistent.

Police Training Systems

The proper fulfillment of statutory police tasks rests, largely, on having good personnel—on their recruitment, attitudes and, most of all, their training (Miciński, 1994). Improvements in the efficiency of police training and the structure and methods of teaching are continuously being sought, often by lessons drawn from the police experience of other countries, e.g., France, Germany, Great Britain and Holland.

Police training is carried out on a number of levels and consists of the following stages:

Stage 1, basic training: Its main objective is to develop simple skills and practical techniques for performing basic police duties. Graduates of this type of training, which is carried out in police training centers and is given to all recruits, occupy the lowest posts.

Stage 2, special training (medium level): Special training qualifies police officers to perform tasks within particular branches. Police schools, the Police Training Center among them, which carry out this type of training equip their graduates with medium-level professional qualifications.

Stage 3, higher professional training (higher and post-graduate studies): Professional training is carried out at the Higher Police Training School in Szczytno, which is a higher vocational school granting a first (university level) degree. Although the school's main objective is to train police officers for their future job, the school also provides educational services for departments in the Ministry of Interior Administration, as well as for the national education system.

Apart from updating knowledge, skills and abilities which reflect progress in policing theory and practice, the school provides its students with the necessary formal qualifications essential for fulfilling the role of commissioned officers and police managers. In addition, the school offers post-graduate studies addressed to university graduates specializing in various fields of activity (in this case connected with the police work and the organization of the force). Post-graduate training automatically leads to promotion (Fiebig & Pływaczewski, 1995, pp. 46–48).

Professional Development

The professional development system deals with demands addressed to the Higher Police Training School (WSPol) and other police schools by field police units and the Police High Command through its Personnel and Training Bureau. Such demands take into account principles and needs of personnel policy (promotion, creating reserves) and the level of training carried out in particular schools. In addition, professional development training takes place in those police units which have training personnel and specialists posted to them.

One natural consequence of socio-economic transformations in society can be traced by changes in the police curriculum. At present, much emphasis is placed on both theoretical knowledge and practical skills. In connection with the new concept of policing (that the police provide a service), the aim of training is to develop attitudes which are desirable from the citizens' point of view, that is, professional relations between the police and the public in general and between the individual police officer and a member of the public.

Corruption in the Police

Similarly to other countries, Polish criminal law penalizes the crime of bribery, which is defined as accepting material or personal profit, or the promise of such, conditioning the execution of one's duty on receiving a profit, or demanding such a profit, as well as accepting a profit or its promise for performing an act violating the law. If a person commits the above-mentioned crime when holding a post involving particular responsibility, or if he or she accepts material profit of great value, or a promise of such, he or she is liable for at least three years' imprisonment (Spotowski, 1972, pp. 88–122).

Although crime statistics do not indicate the number of bribery cases which involve police officers, research indicates that they constitute about 18 percent of the total number of reported cases. Police officers are at particular risk of being tempted by illegal profits owing to the specific nature of their service. They have the power and opportunity to affect the interests of various groups and individuals.

Corruption is induced by various factors deeply rooted in history, culture, economy and psychology. In Poland, the pathologies created within social life, the economy and the state by communist ideology contributed substantially to the growth of corruption. Today, corruption is also ascribed to the new and complex socio-economic situation, deficient values and moral standards cherished by some officers, and an inefficient system of internal control within the police (Kojder, 1995, pp. 42–46).

Particularly vulnerable to corruption are the following areas of police work (Cichorz & Chmielewski, 1996): traffic control (omitting the performance of duties in connection with traffic offenses); control over foreign trade (traffic in cars and taxable goods) and the criminal branch (investigation into frauds and crimes against property).

In 1996, there were 104 recorded cases of accepting material rewards and 28 cases of demanding bribes by police officers. A total of 115 police officers were involved in these cases, which is about 0.1 percent of police personnel. Similar corruption rates were reported in previous years.

At present, the police function is under great pressure from the public which is manifested, among other indications, by the wide publicity given by the media to offenses (often petty ones) committed by police officers. However, it should be stressed that the publicity given to these perpetrators has not changed public opinion about the force. People seem to understand that the number of revealed corruption cases in a force of more than 100,000 police officers, although being reason for concern, does not discredit the force as a whole.

On the other hand, in order to improve the efficiency of policing and to avoid the police becoming subject to more external control, mechanisms of internal control are being introduced in the police with the aim of protecting officers against dangerous factors. Specifically, particular attention is paid to the motivations of people intending to join the police (Miciński, 1994, p. 9).

Moreover, a change in the remuneration system was introduced in 1997 which allows police officers to obtain additional income by performing specific work commissioned by government and other tasks performed off duty. This will reduce the attractiveness of other sources of income, especially illegal ones. Because accepting bribes does not result only from a police officer's wish to grow rich but also from his or her thoughtlessness and naivete, training has an important role to play in this respect by increasing awareness, shaping right moral attitudes and eliminating undesirable ones.

Media

Liaison activities of the police implement tasks defined in the Police Act. Their aim is to safeguard the right of the public for access to information; help shape public opinion on the police; ensure conditions for public control over police activities; and carry out other statutory tasks of the police by the use of the mass media.

These tasks are fulfilled mainly by information officers posted to Chief Commissioner's and *voivode* commandants' offices. It is their duty to coordinate liaison activity and, in particular, to provide the press with information about incidents; respond to criticism of the police; cooperate with the local and national press; help in organizing contacts between journalists and particular police units; and instruct police officers in how to give information at the scene of incidents.

In addition to police unit commandants and their liaison officers, all other police officers enjoy the right to make comments (on behalf of themselves) to the mass media. Polish press law (January 28, 1984 Act) provides that every individual, in accordance with freedom of speech and the right to express critical opinions, may provide a journalist with information, and nobody may be liable to suffer a loss or be blamed for providing the press with information, as long as he or she has acted in accordance with the law.

As far as the police are concerned, limitations on making public comments result mainly from the necessity to keep state or official secrets, as regulated by the act of December 14, 1982. Moreover, formal restrictions imposed on the freedom to make public statements are a consequence of the provision of the Police Act which states that a police officer cannot give information about a citizen to any person or institution other than the court or public prosecutor, if the information was obtained during an investigation. In addition, the information can only be used for the sake of prosecution. The criminal code also forbids unauthorized publication of information on preparatory proceedings before the first-instance or closed hearing (Majer & Misiuk, 1997, p. 84).

Yet, while keeping in mind the aforementioned restrictions, police officers are often obliged to cooperate with representatives of the mass media and to follow appropriate procedures in these contacts. First of all, officers performing their duties at the scene of an incident of public concern should provide a journalist with basic facts concerning the incident, at the journalist's explicit request and on seeing his or her identity card. Police officers should also enable the journalist to fulfill his or her duties even as prevention units restore public order. Journalists must remember, however, that their safety cannot be guaranteed by the police (Wachowski, 1995, p. 72).

The formal regulation of police–journalists relations through the above-mentioned acts obviously does not ensure full harmony among representatives of these two professions, due to the conflict of interests between the journalists' urge to reveal the so-called "whole truth" and the requirements of the police work. The strength of the conflict varies depending on the kind of media involved.

In Poland, there are two public TV channels, one commercial channel, two encrypted commercial channels and a network of local broadcasting stations. Channel one on public television regularly broadcasts a police program called "*Gliny*" ("Cops"), which is produced with the participation of police officers. The program seeks to show day-to-day police work and how strenuous it really is. Although it is on the air as late as 11 p.m., it is very popular with viewers. Other police information, usually following failures or spectacular achievements of the police, and often presented or commented on by police officers themselves, is of an incidental nature and is included in news or journalistic programs. In order to acquire skills useful when speaking in front of the TV camera, some of the officers have undergone training organized by Polish television.

Radio coverage of events is carried out by four national public radio stations and numerous private national or local ones. Like television, public radio deals with police issues only in cases of widely reported crimes. Of much interest are also significant changes in police personnel, new powers and responsibilities given to them, as well as the activity of police trade unions, particularly when they appeal to public opinion. A regular part of the radio news is police information on road conditions, especially in rush hours, and traffic accidents and stolen cars.

The printed press, the analysis of which here includes only the most popular national newspapers, devotes the most attention to police matters compared to the other media. The most important impact on public opinion is exerted by three morning newspapers: *Rzeczpospolita* (wrongly regarded by some readers as a government paper), *Gazeta Wyborcza* (founded in 1989 by former political dissidents and now representing the views of the political opposition), and *Trybuna* (linked to *Trybuna Ludu*, the paper of the former ruling Polish party, founded in 1944).

Among these papers, *Rzeczpospolita* has a reputation for presenting police issues in the most objective and reliable way. Police material is found in a four-page daily addition entitled *"Prawo na codzień"* ("Law for Every Day"), which is mainly devoted to law-enforcement issues of various natures. Among these are legal acts which have passed through the legislative process and brief accounts of the most important criminal incidents. In addition, the supplement includes analyses of threats posed by various kinds of crimes, interviews with top police officers, reviews of police reports, and news concerning police forces from other countries.

At present, as it is linked with the ruling coalition, *Trybuna* presents a positive attitude towards the police. It prints the greatest number of articles on successful actions of law enforcement agencies, as well as information on easy-to-understand crime prevention techniques.

Compared to *Trybuna*, *Gazeta Wyborcza* is much more critical of the police. It regularly prints short accounts of criminal incidents and more detailed articles, often including information obtained from spokespersons from the Chief Commissioner's office. In addition, it often quotes comments by anonymous police officers, especially commissioned ones. The most critical articles usually appear on the front page, and their ironic or malicious tone can be easily discerned, for instance when police officers do not want to reveal details concerning inquiries they are conducting. Were it not for the neat style and lack of vulgarity, fragments of some articles would be suitable for the gutter press rather than quality papers.

The gutter press in Poland is best represented by two afternoon papers—*Super Expres* and *Expres Wieczorny*. Hunting for cheap sensations, their reporters commonly question the motives of the police or the purpose of their activities. Reporters often claim that, instead of catching criminals, the police "prey on people" or "participate in celebrations." Articles give terrifying details of crimes and their headlines are often tactless or even vulgar.

On the basis of this necessarily cursory analysis of the press one can notice a clear distinction between newspapers which seek to show an objective image of the police and those which choose the easier way and try to create a negative picture of the force. Objective writing is typical of newspapers with an established position and regular readership, as seen in their circulation figures. The gutter press offers something quite opposite. As its only chance for survival in the market is either maintaining or increasing the number of copies sold, it prints sensational news to attract the highest possible number of readers. It can be stated that those newspapers and journalists who maintain the best relationships with the police present an objective picture of the force.

CONCLUSION

At the beginning of the 1990s, democracy spread to vast new territories, mainly as a result of the fall of communist governments in Eastern and Central Europe (Brand & Jamróz, 1995). On adopting democracy, people in these countries seek answers to the following questions: What is democracy? What principles should democratic society be based on? What rules should a democratic state obey? Other questions may concern rights enjoyed by an individual in a democratic society, the status and role of political parties, as well as the status and tasks of law enforcement agencies, the police occupying a special position among these.

The time of transformation in Poland witnessed great organizational changes which had objective consequences in many areas of social life. The police had to adapt to the changes in their organization, legal regulations and professional awareness; they had to keep up with societal changes and, consequently, had to undergo a transformation themselves. In 1990, the new management of the Ministry of Interior Administration realized that the Civic Militia, the armed force of the Polish United Workers' Party, would have to be transformed into a modern, apolitical police in a very short time if Poland were to become the democratic country it aspired to be after 1989. It was not an easy task for, on the one hand, the people's confidence in the new police had to be gained, and, on the other hand, the police had to be taught their new roles in the new constitutional order, that of public servants under the law and protectors of society.

Despite the objections of various political circles and public opinion at the time, the decision was made to introduce into the newly created police profound organizational changes, including shifts among managerial staff and the dismissal of discredited people who were unable to work within the new framework. This policy resulted in a shortage of manpower at all levels and reduced police efficiency, which was lowered further by a lack of senior officers to occupy managerial posts in the basic organizational units of the police.

At the same time, the widely supported political, economic and social transformations necessary for the development of democratic freedoms and a free-market economy brought about a rapid increase in crime and changes in its structure and forms. The most dangerous development was that of organized crime of both national and international natures, which started to resemble traditional mafias. The dynamic and practically undisturbed development of organized crime resulted from, among others factors, wrong political decisions which reflected a blind faith that social processes would lead to the automatic eradication of this form of crime. Yet decisions taken after 1989 hindered the effective reorganization of law-enforcement and criminal justice agencies.

The rapid rise in crime was accompanied by an even greater sense of threat. Fear of crime is dangerous in itself as it leads to more and more repressive, and thus less efficient, criminal justice policies. Fear of crime is aroused not only by real threats but also by the mass media and politicians. In particular, it is politicians willing to gain power who eagerly stress the fact that people in control now are helpless when challenged by increasing crime. Such rhetoric is quite common, not only in this part of Europe, and results from a misunderstanding of the situation and the failure to react to crime in appropriate ways.

It can be assumed that the increase in crime after state borders have been opened is a normal consequence of the lessening of state control over the movement of people. In addition, communism had eroded informal public control. In a communist country everything was controlled by the orders of superior authority. Now, social control institutions are in a more or less serious crisis, while citizens and local communities do not face up to new challenges since they think that crime prevention is a task to be handled exclusively by an anonymous state bureaucracy. Thus, more publicity should be given to one of the basic theses of victimology, namely, that citizens should take care of their own life, health and property, and that it is the obligation of the state to create proper conditions for that. However, people generally limit themselves to criticizing the state of safety rather than cooperate with law-enforcement agencies and criminal justice personnel.

The unwillingness of people to engage in such cooperation cannot be blamed only on the "heritage" of the previous system. The idea of protecting the individual's rights, a position consistently advocated and put into practice since the beginning of the 1990s, sometimes assumed extreme, even pathological, forms. On the one hand, there occurred an excessive increase in the rights and protections of perpetrators in all stages of legal proceedings; on the other hand, the interests of victims and witnesses of crimes were neglected and they were exposed to various forms of aggression and threats from perpetrators and their supporters, as well as to the so-called "secondary victimization" at the hands of representatives of law-enforcement and criminal justice agencies. Thus, there developed a drastic imbalance in the rights of victims and witnesses versus those of perpetrators of crimes. This was one of the reasons for the creeping paralysis of criminal justice. The unwillingness to cooperate with the police resulted in a lack of witnesses in preparatory and court proceedings, which made it impossible to examine cases and bring alleged criminals before the court. This situation led to the introduction of the institution of the incognito witness and to work on drafting a crime witness act.

Yet, the initiatives came too late in the face of a growing and extremely dangerous phenomenon of "autojurisdiction": citizens themselves began to administer justice. A new criminological category came into being—crimes due to settling accounts—resulting from, among other reasons, the calculation that a debt could be regained much faster, cheaper and more effectively in this manner than by means of court proceedings. Generally speaking, Polish courts are in a very serious crisis as the number of cases to be tried increased rapidly with the introduction of a democratic system based on a free-market economy while, at the same time, proper initiatives to respond effectively to these new challenges to the finances and personnel of courts were not pursued. Trials take a very long time and there are big problems with executing judgments. Without further and substantial aid provided to Polish courts, it will be difficult to regard Poland as a *Rechtsstaat* (Siemaszko, 1996, pp. 4–25).

The Polish police must cope with even greater problems in trying to ensure the safe functioning of the society. The basic tasks of a democratic state with regard to the police are to maximize their capacities to prevent and fight crime and other socially harmful phenomena and to minimize the chances that their powers are used in conflict with democratic principles. At the same time, Polish crime has been subject to the process of democratization as well—criminals can be found in all social classes. Criminal regulations in force are not adequate to tackle new types of crime (especially of an organized nature) and new criminal methods. Hence, in the last few years Poland has suffered from a total lack of preparation by the police to fight the transformations in economic and other types of crime after the free-market economy was introduced. The preparedness of the police was further weakened by the neglect among new decision makers of the existing criminological and criminalistic prognostic research on the new types of crime (Kołecki, 1992, pp. 27ff). This neglect led to an inefficient appreciation and knowledge of present forms and methods of economic and financial offenses.

In sum, the issue remains unresolved of how to design police training and work in such a manner that the public can feel assured that the police are for the society and not against it. One should also bear in mind that an increase in public safety and order depends on overcoming the following three barriers: legislative barriers, which can hinder the introduction of the legal solutions necessary for effective crime prevention; economic barriers, which can reduce budgetary resources for law enforcement and criminal justice agencies; and barriers in social awareness, particularly when, in the process of shaping public opinion about suggested solutions, there is a tendency to offer only one-sided presentations of limitations, regulations and prohibitions. It is mainly in the field of social awareness that one can observe the escalation of the potential conflict between the rights and freedom of citizens and their safety. But unless the police are equipped with adequate powers to enable them to fight successfully against crime, the legal protections offered to criminals will be too strong.

The efficient protection of human rights during police interventions depends not only on the congruence of Polish law with international standards (Hofmański, 1995; Rzepliński, 1994), but also on the ability of citizens to initiate control of police activities through the prosecutors and courts, as well as on the right to obtain compensation for loss or injury suffered. In all these aspects, the Polish legal system offers significant possibilities for redress, an ability supported by the institution of the Ombudsman.

NOTE

† This chapter benefited immensely from the editorial assistance of Dr. A. Misiuk, Director of the Institute of Police Research at the Higher Police Training School at Szczytno.

REFERENCES

Adamski, A. (Ed.). (1994). *Prawne aspekty nadużyć pope»nionych z wykorzystaniem nowoczesnych technologii przetwarzania informacji* [Legal aspects of computer-related abuse]. Toruń: "Dom Organizatora" TONiK.

Bednarski, M. (1992). *Drugi obieg gospodarczy* [Secondary economic circulation]. Warsaw: Wydawnictwa Uniwersytetu Warszawskiego.

Bentkowski, A. (1996). *Nowe przepisy karne dotyczące przestępstw ubezpieczeniowych* (New criminal law regulations concerning insurance crime). *Państwo I Prawo* (State and Law), 11, 57–62.

Brand, P. & Jamróz, A. (Eds.). (1995). *Democracy yesterday and today*. Białystok: Temida.

Cichorz, T. & Chmielewski, R. (1996). *Komunikat z badania ankietowego pt. Korupcja w organach ścigania I kontroli* (A survey report on corruption within prosecution and supervision agencies). Unpublished report, Szczytno: Higher Police Training School.

Council of Europe, Legal Affairs. (1981). *Economic crime*. Strasbourg: Council of Europe.

Daszkiewicz, W. (1995). *Zorganizowana grupa przestępcza* (Organized criminal group). *Prawo I Życie* (Law and Life), 50, 34.

———. (1996). *Proces karny. Cześć ogólna* (Criminal proceedings—general part). Poznań: Przedsiebiorstwo Wydawnicze "Ars boni et aequi."

Dziemidowicz, Z. (1996). *"Szara strefa" w Polsce* (The shadow economy in Poland). In W. Pływaczewski & J. Świerczewski (Eds.), *Policja Polska wobec przestępczości zorganizowanej* (Polish police and organized crime). Szczytno: Higher Police Training College.

Fiebig, J. & Pływaczewski, W. (1995). *System szkolenia Policji w WSPol. Stan obecny I perspektywy* (The police training system at the Higher Police Training Academy—the present and future). *Policyjny Biuletyn Szkoleniowy* (Police Training Bulletin), 1–2, 42–51.

Filar, M. (1996). *Polityka kryminalizacyjna III Rzeczypospolitej* (The criminalizing policy of the Third Polish Republic). *Państwo I Prawo* (State and Law), 11, 46–56.

Galster, J., Szyszkowski, W., Wasik, Z. & Witkowski, Z. (1996). *Prawo konstytucyjne—zarys instytucji w okresie transformacji ustrojowej* (Constitutional law—Outline of institutions in transition). Toruń: Wydawnictwo Uniwersytetu Mikołaja Kopernika.

Górniok, O. (1994). *Przestępczość gospodarcza I jej zwalczanie* (Economic crime and the fight against it). Warsaw: Wydawnictwo Naukowe PWN.

———. (1995). *Ustawa o ochronie obrotu gospodarczego* (Economic Transaction Protection Act, with commentary). Warsaw: Wydawnictwo AWA.

Hanausek, T. (1993). *Niektóre problemy wykrywania sprawców przestępstw w Ňwietle przyszłej kodyfikacji* (Some problems of detecting crime perpetrators in the light of future codification). In S. Waltoś, (Ed.), *Problemy kodyfikacji prawa karnego* (Criminal law codification problems). Kraków: Firma Wydawniczo-Reklamowa "PERFEKT".

Hofmański, P. (1994). *"Świadek anonimowy" w procesie karnym?* (Anonymous witness in criminal proceedings?). Przeglad Policyjny (Police Review), 2, 20–35.

———. (1995). *Konwencja Europejska a prawo karne* (The European Convention and criminal law). Toruń: TNOiK Publishers.

Hołyst, B. (1994). *Kryminologi a* (Criminology). Warsaw: Wydawnictwo Naukowe PWN.

———. (1996a). *Kryminalistyka* (Criminalistics). Warsaw: Wydawnictwo Naukowe PWN.

———. (1996b). *Przestepczośf zorganizowana I jej implikacje* (Organized crime and its implications). In W. Pływaczewski & J. Świerczewski (Eds.), *Policja Polska wobec przestepczości zorganizowanej* (Polish police and organized crime). Szczytno: Wydawnictwo WSPol.

———, Kube, E., & Schulte, R. (1996). *Przestepczość zorganizowana w Niemczech I Polsce* (Organized crime in Germany and in Poland). Warsaw, Münster, Łódź: Polskie Towarzystwo Higieny Psychicznej.

Jakubski, K. J. (1996). *Przestepczość komputerowa—zarys problematyki* (Computer crime—Outline of the problems). Prokuratura I Prawo (Prosecution and Law), 12, 34–50.

Jaroch, W. (1996). *"Lapówka kontrolowana"—prawne I praktyczne moóliwości zwalczania lapownictwa* (Controlled bribe—Legal and practical possibilities of fighting bribery). Monitor Prawniczy (Law Monitor), 10, 359–363.

Journal of Law (1994). Issue 126, entry 615 (November 30).

———. (1995a). Vol.89, entries 443, 444.

———. (1995b). Criminal Code and Other Laws Amendment Act of July 12, 1995. Issue 95b, entry 475.

———. (1995c). 1990 Act. Vol. 30, entry 179 with further amendments (April 4).

———. (1996). Issue 11, entry 62.

———. (1997). No. 114, item 738.

Kojder, A. (1995). *Korupcja* (Corruption). In J. Jasiński & A. Siemaszko (Eds.). Crime control in Poland. Warsaw: Oficyna Naukowa.

Kołecki, H. (1992). *Policyjno-kryminalistyczna problematyka wspólczesnej przestepczości ekonomiczno-finansowej w Polsce* (Police and criminalistic issues of contemporary economic and financial crime). Poznań: Polski Dom Wydawniczy "Lawica."

Korybut-Woroniecki, A. (1993). *Migracyjne problemy* (The problems of migration). In *Polska w Europie* (Poland in Europe). Warsaw: Ośrodek Studiów Miedzynarodowych, Vol. 10, 86–89.

Kudrelek, J. & Lisiecki, M. (1996). *Nowe uprawnienia kryminalistyczno-operacyjne policji* (New forensic and operational powers of the police). Jurysta (Jurist), 5, 18–21.

Lernell, L. (1965). *Przestepczość gospodarcza* (Economic crime). Warsaw: Wydawnictwo Prawnicze.

Majer, P. & Misiuk, A. (1997). *Policja a spoleczeństwo* (Police and the public). Szczytno: Wydawnictwo WSPol.

Miciński, R. (1994). *O nowych zasadach przyjecia do służby w policji* (On new principles of police recruitment). Policyjny Biuletyn Szkoleniowy (Police Training Bulletin), 1–2, 9–18.

Palka, P. (1996). *Zabójstwo przy użyciu broni palnej* (Homicide with the use of firearms). Jurysta (Jurist), 10–11, 34–35.

Pływaczewski, E. (Ed.). (1993). *Proceder prania brudnych pieniedzy* (Money laundering procedure). Toruń: "Dom Organizatora" TNOiK.

———. (1996a). *"Pranie brudnych pieniedzy"* (Money laundering). In W. Pływaczewski & J. Świerczewski (Eds.), *Policja Polska wobec przestepczości zorganizowanej* (Polish police and organized crime). Szczytno: Wydawnictwo WSPol.

———. (1996b). *Wezlowe problemy przestepczości zorganizowanej w Polsce* (Key issues of organized crime in Poland). In W. Pływaczewski & J. Świerczewski (Eds.), *Policja Polska wobec przestepczości zorganizowanej pod* (The Polish police and organized crime). Szczytno: Wydawnictwo WSPol.

———. (1996c). *Polska Policja na tle miedzynarodowej wspólpracy w zakresie zapobiegania I zwalczania przestepczości* (Polish police and international cooperation in preventing and fighting crime). In W. Pływaczewski & J. Świerczewski (Eds.), *Policja Polska wobec przestepczości zorganizowanej* (Polish police and organized crime). Szczytno: Wydawnictwo WSPol.

———. (1997). *Kradzieóe samochodów. Studium kryminologiczne* (Car theft. A criminological study). Szczytno: Wydawnictwo WSPol.

—— & Sklepkowski, L. (1995). *Tendencje rozwojowe zorganizowanej przestepczości gospodarczej w Polsce* (Developmental trends in organized economic crime in Poland). In S. Lelental & M. Zajder (Eds.), *Kryminalistyczne I prawne problemy wspólczesnej przestepczości* (Criminological and legal aspects of contemporary crime). Szczytno: Wydawnictwo WSPol.

Pracki, H. (1997). *Ściganie przestepstw gospodarczych* (Prosecution of economic crimes). *Jurysta* (Jurist), 10—11, 1.

Ratajczak, A. (1994). *Ochrona obrotu gospodarczego*. Warsaw: Wydawnictwo Zrzeszenia Prawników Polskich.

Rzepliński, A. (Ed.). (1994). *Prawa człowieka a policja. Problemy teorii I praktyki* (Human rights and the police. Theory and practice). Legionowo: Centrum Szkolenia Policji.

Schulte, R. (1995). *Europejska wspólpraca policyjna* (European police cooperation). Prokuratura I Prawo (Prosecution and Law), 4, 11–19.

Serdakowski, J. (1996). *Zorganizowana przestepczoś f narkotykowa w Polsce* (Organized drug-related crime in Poland). In W. Pływaczewski & J. Świerczewski (Eds.), *Policja Polska wobec przestepczości zorganizowanej* (Polish police and organized crime). Szczytno: Wydawnictwo WSPol.

Siemaszko, A. (Ed.). (1996). *Quo Vadis Iustitia? Stan I perspektywy wymiaru sprawiedliwości w Polsce* (Quo Vadis Iustitia? The present and future of criminal justice in Poland). Warsaw: Instytut Wymiaru Sprawiedliwości.

—— & Woźniak, D. (1995). *Niektóre problemy przestepczości zorganizowanej w Polsce. Tendencje rozwojowe grup przestepczych* (Some problems of organized crime in Poland. Trends in criminal groups). In E. Pływaczewski (Ed.), *Przestepczość cudzoziemców. Nowe wyzwania dla teorii I praktyki* (Foreigners-related crime. New challenges for theory and practice). Szczytno: Wydawnictwo WSPol.

Sklepkowski, L. (1996). *Polska jako kraj tranzytowy I docelowy nielegalnej migracji. Nielegalne przerzuty przez granice* (Poland as a transit and target country for illegal migration. Organized smuggling across the border). In W. Pływaczewski & J. Świerczewski (Eds.), *Policja Polska wobec przestepczości zorganizowanej* (The Polish police and organized crime). Szczytno: Wydawnictwo WSPol.

Spotowski, A. (1972). *Przestepstwa sluóbowe* (Public functionaries-related crime). Warsaw: Wydawnictwo Prawnicze.

Tarnawski, M. (1996). *Dysfunkcjonalności w prawie karnym* (Disfunctions in criminal law). *Wojskowy Przeglad Prawniczy* (Military Law Review), 1, 16–26.

Wachowski, I. (1995). *Policja a środki masowego przekazu* (Police and the mass media). *Przeglad Policyjny* (Police Review), 3, 137–144.

Waltoś, S. (1993). *Dylematy ochrony świadka w procesie karnym* (The dilemmas of witness protection in a trial). *Państwo I Prawo* (State and Law), 4, 39–51.

——. (1996). *Proces karny. Zarys systemu* (Criminal proceedings. An outline of the system). Warsaw: Wydawnictwa Prawnicze PWN.

Wasek, A. (1995). *Świadek anonimowy w rzetelnym procesie karnym in Kierunki I stan reformy prawa karnego* (Anonymous witness in criminal proceedings ...criminal law reform). Lublin: Lubelskie Towarzystwo Naukowe.

Wójcik, D. (1995). *Przestepczość nieletnich* (Juvenile deliquency and victims of crime). In J. Jasiński & A. Siemaszko (Eds.), *Crime control in Poland*. Warsaw: Oficyna Naukowa.

7

Challenges of Policing Democracies: the Russian Experience

YAKOV GILINSKIY

INTRODUCTION

The Russian Federation came into existence in 1991 after the break-up of the Union of Soviet Socialist Republics, a collapse of the Soviet empire which was set in motion by the *perestroika* (restructuring), *glasnost* (openness) and *demokratizatsiya* (democratization) policies of President Gorbachev. Under the Soviet Union, what is now Russia had been labelled the Russian Soviet Federated Socialist Republic. Its formal name now is the Russian Federation, and includes twenty-one republics, six *krai*, fifty provinces, one autonomous area (Chukotsk) and two cities under federal administration (the capital Moscow, and St. Petersburg). Following the termination of the Soviet Union, Russia joined with some other former Soviet republics to form the Commonwealth of Independent States (CIS).

Its total area is 17,075,000 square kilometers, an area about twice as large as that of the United States or China. Russia's population grew from 101.4 million in 1950 to 148.4 million in 1994, with 78 percent living in the European sector and 22 percent in the Asiatic sector (West Siberia, East Siberia and the Russian Far East). The population ranks as the sixth largest of the world's countries. In 1994, 47 percent of the population were male and 53 percent female; 75 percent were living in urban and 27 percent in rural areas (*Russian statistical annual*, 1995). The ethnic distribution, according to the 1989 census, was 81.5 percent Russian, 3.8 percent Tartar, 3 percent Ukrainian, 1.2 percent Chuvashians, with all other ethnic groups each representing less than 1 percent of the total population. Life expectancy fell from 63.8 years for males in 1990 to 57.3 in 1994 and from 74.4 years for females in 1990 to 71.1 years in 1994. The death rate rose from 10.4/1,000 in 1986 to 16.2/1,000 in 1994 and continues to rise. The birth rate fell from 17.6/1,000 in 1986 to 9.6/1,000 in 1994. The natural growth of the population (the difference between birth and death rates) grew at a rate of 7.2 percent in 1986 but fell by 6.6 percent in 1994 (*Moscow News*, 1995; Shkolnikov, Mesle & Vallin, 1995).

The economy of the Soviet Union and Russia was, from 1917 until 1991, under the control of the state as guided by a socialist ideology and the desire for party control. Since 1991 Russia has sought to shift toward a market economy. The immediate effects of economic reforms have been a sharp rise in inflation and crime, the collapse of the currency, the encroachment of organized crime and speculative, "primitive" capitalism into the economic arena, and severe hardships for large segments of the Russian population.

The Soviet Union, even as it sought to establish a social utopia, was a totalitarian regime from its beginning in 1917 to its collapse in 1989. In common with all types of totalitarianism, the regime in the former USSR encompassed control of all spheres of social life; the regulation of all forms of human conduct; an all-embracing control over the carrying out of orders; compulsory, absolute uniformity and unanimous approval of all actions by the party and the government; and the harshest reprisals, which reached the scale of genocide.

This unique social experiment included an artificial negative selection process whereby for seven decades the leading scientists, scholars, philosophers, workers, peasants, poets and military commanders were suppressed and destroyed, while the mediocrities and drab figures survived and were extolled. Even using official data (which may be understating the numbers), between 1921 and 1953, at least 3,777,380 persons were convicted for political reasons with at least 642,980 of them being sentenced to death. In addition, numerous facts have been revealed about extra-judicial executions on a mass scale—the number of victims is beyond estimation. Between 1934 and 1948 alone, at least 963,766 convicts died in the GULAG[t] camps (Zemskov, 1991). According to the official numbers, since 1941, 210,559 Poles, 949,829 Germans, 575,768 North Caucasus inhabitants (Ingush, Karachaev, Balkar, Chechen and people of other nationalities), and 90,940 Kalmyks were exiled to the remote regions of Siberia. Besides that, 154,258 Letts, Lithuanians and Estonians were exiled in the periods 1940–1941 and 1945–1949 (Zemskov, 1990).

All this eventually led to catastrophe. The system led to the disintegration of production and the economy, as well as a loss of trade skills; the proletarianization and marginalization of the population with no middle class emerging; severe crises in the social infrastructure, including health care, education, culture, communication and transportation; an ecological crisis of vast proportion; political crises; the degeneration of moral and spiritual values; an increase in the extent and forms of deviant behavior (crime, drug addicion, suicide, etc.) and the growth of mafia-type organized crime.

The *perestroika* programs instituted by Gorbachev were a much-needed attempt to save the power structures by way of reform. A similar attempt had been made by Khrushchev (the "Thaw") during the 1953–64 period, the first attempt at reforms in the USSR after Stalin's regime, and had been considered by Andropov. However, every attempt ended with the actual or political death of its propagators and was followed by stagnation.

The constitutional system of Russia after 1991 was modelled on the presidential system, including popular elections for the president and a two chambered legislature. The dominance of the Communist Party was curtailed and, ultimately, the Party itself was dissolved. A new constitution enshrined a list of human rights and civil liberties, established the formal independence of the judiciary from political control, and announced that legal and criminal justice matters would be governed by rule of law notions and standards.

With all due credit to Gorbachev, his reforms turned out to be the most radical although even these—*glasnost*, the multi-party system, the release of those states occupied by Stalin (Latvia, Lithuania and Estonia), the lifting of the iron curtain, and the right to hold private property—did not turn out to be fully satisfactory. All the symptoms of stagnation and catastrophe mentioned earlier remained untreated. Power was continually returned to the *nomenklatura* (the party and bureaucratic elites); corruption took on a monumental nature in all organs of power and establishment; the militarization of economics and politics continued; inter-ethnic conflicts resulted in massive numbers of deaths; nationalist, anti-semitic and neofascist groups organized and met with no resistance (Lacqueur, 1994). The criminal war in Chechnya is terrifying evidence of this neo-totalitarianism. The ever-growing economic polarization of the poverty-stricken majority and the "New Russians"—a criminalized, nouveau riche minority—is a source of real social conflict. The differentiation between the incomes of the 10 percent least prosperous and the 10 percent most prosperous stood at 1 to 4.5 in l991; at 1 to 8 in 1992; at 1 to 10 in l993; and at 1 to 15 in 1994. The Gini Index (the coefficient of income concentration) rose from .256 in 1991 to .346 in 1993 (Makarevich, 1995; also see Zaslavskaja & Arutjunjan, 1994; Zaslavskaja, 1995). The country is also permitting massive human rights abuses, particularly in penitentiary institutions where tyranny and torture dominate (*The white book*, 1994; *Amnesty International Newsletter*, 1993, 1995).

THE CONCEPT OF DEMOCRATIC POLICING

The police of Russia, known as the Militia, was set up a month after the state coup of October 1917 and continues as the Militia to this day. Until 1931, the Militia were under the authority and control of the local organs of power (the local Soviets) and in this respect partly justified its name—the militia, defenders of order organized according to territorial principles and made up of volunteers and non-professionals. However, in the 1930s, the Militia was transformed into a professional police force, its name being retained only for the sake of propaganda and ideological complicity; that is, the "Soviet Militia" had to be seen as different from the "capitalist police" of the earlier periods (Borisov, Dugin & Malygin, 1995; also Conquest, 1968; Juviler, 1976; Shelley, 1996).

The current directives, functions and structure of the Militia are laid down in the Russian legislation, "On the Militia," passed on April 18, 1991. The Militia are under the auspices of the Ministry of Internal Affairs (MVD) of the Russian Federation, as are the internal army, specialized police forces (e.g., railroad, air or river police), the fire safety organs, penitentiary service personnel and others. The internal army is responsible for dealing with internal conflicts, rebellion, disturbances and riots. The directives of the Militia are to provide citizens with personal safety, to stop and prevent crime and civil law-breaking, to solve crimes and to secure civil order and security within society (Art. 2, On the Militia: RSFSR Law, 1991).

The Militia are organized into two main sub-divisions: the Criminal Militia and the Militia for Civil Safety (that is, Public Order) at the local level (Law On the Militia for Civil Safety, 1993). The Criminal Militia include the Detective Service, the Economic Crime Prevention Service, scientific-technical specialists, operational investigators and others who supply material for the criminal Investigation and Economic Crime Prevention Services. The Civil Militia include the Duty Service, the Service for Securing Civil Order, the State (Government) Automobile Inspectorate (GAI), the Security Service, divisional inspectors, temporary detention guards, the prophylactic service which includes the Inspectorate for dealing with juveniles and other departments. The Criminal Investigation Service is a separate unit under the Ministry of Internal Affairs.

The Militia are given far-reaching powers, including the right to enter, without a warrant, the living places and other premises of citizens, the premises of companies, organizations and official departments (excluding those of foreign diplomatic representatives) and to conduct searches on transport facilities and to search the baggage and bodies of citizens (Art. 11, On the Militia, 1991).

The law, "On the Militia: RSFSR Law," regulates the terms and procedures for use of physical force and special methods (rubber truncheons, tear gas, water cannons, armored cars and others) and firearms (Art. 12). Unfortunately, it is often the case that the Militia exceed their powers including violations of the terms and procedures for physical force and other methods.

The Militia employ staff from the ages of 18 to 35. The length of service is defined within each unit according to position and rank. The officers of the Militia are conferred with special ranks, wear uniforms, have the right to hold and carry firearms and travel free on public transport. They are not permitted to engage in commercial ventures or to hold other employment, except in scientific and educational fields. The Militia are not allowed to form or actively support political parties, units or activities. Article 19 provides the bases for discharge from the service (age, state of health, serious breach of discipline, voluntary retirement and others). The size of the Militia forces stood at about 540,000 and that of the internal army at about 278,000 in 1995 (*Everyone's Newspaper*, 1995, No. 51).

The specific principles guiding the actions of the Militia are: lawfulness, humanism, respect for human rights and *glasnost* (openness or transparency). As with restrictions of force, Militia actions often violate all of these principles, with concrete examples of such violations being regularly reported in the Russian and foreign media.

ORGANIZATIONAL CHALLENGES

The governing structure of the Militia is as follows. The entire Militia in the Russian Federation are under the direction of the Ministry of Internal Affairs. The Militia in the Russian Republics are directed by the republican Ministry of Internal Affairs, and in each territory, region, *krai*, city or district they are directed by the heads of administrations or departments, as established by the Constitution and law of each territory. The railway, river and air transport Militia are directed by the heads of administration and departments of internal transport affairs which are located in the Ministry of Internal Affairs of the Russian Federation.

Legal Culture

For centuries, stern authoritarian regimes wielded power in the country and the population is not accustomed to living under conditions of freedom and democracy. The only prior democratic phase in Russia was the period between February and October 1917, when czarist autocracy gave way, only to be replaced seven months later by an even more formidable communist totalitarianism.

In Russia, officially proclaimed and legally fixed principles and norms have always been at variance with reality, including the practice of administering justice, law enforcement and the work of the executive power. Such incongruity existing since time immemorial is reflected in folklore, proverbs, sayings and jokes, such as "laws are all right, but judges are malefactors," or "law is just a pillar; one cannot hurdle it, but one can walk around it." A joke of the early twentieth century (before the October coup) went like this: "There exist four legal systems in the World. The English system—everything is permitted except what is prohibited. The German system—everything is prohibited except what is permitted. The French system—everything is permitted even when it is prohibited; and the Russian system—everything is prohibited even when it is permitted." A joke of the Soviet era (in the 1960s and 1970s) described this scene: "A client comes to a lawyer, presents a legal problem and asks: 'So I have a right, haven't I?' The lawyer answers 'Yes, you have.' The client, 'So I may …' The lawyer interrupts, 'No, you mustn't.'"

This disjuncture and incongruity between law and practice has not been eliminated by the reforms of the 1980s and 1990s. *De jure*, Russia, as the legislative heir to the USSR, is obliged to accept and ratify the basic international conventions and treaties on human rights signed by the USSR, including the International Treaty of Civil and Political Rights (1966) which was signed in 1973, and the Convention Against Torture and Other Cruel, Inhumane and Humiliating Treatment and Punishment (1984) signed in 1987. Article 17 of the Constitution of the Russian Federation states that the Russian Federation acknowledges and guarantees the rights and freedoms of the people and citizens and supports generally accepted principles and norms of international law.

However, *de facto*, human rights in Russia are being constantly and thoroughly abused: the right to life (such as the casualties during the criminal war in Chechnya, the mass killings of soldiers while on military service and other acts), the right to freedom and the inviolability of the person (by illegal arrests, the wide-ranging powers of the special services in conducting searches, telephone taps, entering into premises, etc.), and the freedom of movement.

The internal passport, visa and registration system, the *propiska* system, continues. The *propiska* is a unique juridical institution which requires permission from the police to reside in a place stated in the passport. The *propiska* refers to the residence registration stamp in the internal passport (for instance St. Petersburg, Nevskiy Avenue, House No. 3, Apartment No. 4). The person without the *propiska* (the stamp) does not have the right to reside in a city, to work, to receive many kinds of medical services and to get a pension. The person without a *propiska* is homeless and without support. For guests, there exists a temporary *propiska*. Legally, the *propiska* system has been abolished but it continues in practice, especially in the cities.

Organizational Instability

The role of the police in the present political conditions is also dual. On the one hand, the police are declared to be non-partisan. Party organizations cannot set up shop in police detachments. But the police were forced to be involved in political actions such as during the attempted coup against Yeltsin in 1991, the October 1993 disturbances in Moscow, and in the Chechnya war. They are objects of political manipulation.

The work of the police in the 1980s and 1990s has been adversely affected by the continual turnover of Ministers of Internal Affairs in both the former USSR and in Russia. In the span of thirteen years from 1982 to 1995, ten top officials in the MVD have been changed. One should bear in mind that, in line with the old Russian tradition, each succeeding minister begins with structural and staff changes both in the central ministry and in its local administrations. The tenure of General Fedorchuk has been described in these terms by V. Egorshin, the Chief of the Criminal Investigation Department of the St. Petersburg Headquarters of the MVD:

> Everything old was gotten rid of ... Virtually all workers and chiefs of any rank, the operational staff primarily, were under suspicion ... Their careers were ruined and efficient professionals were fired (Egorshin, 1994, p. 106).

The instability of the administration of the MVD, combined with the general economic, political and social instability, impedes the work of the police. The most efficient members of the police are forced to leave and seek employment as private detectives.

In law, the main goal of the police is to ensure the security of citizens and to protect them against illegal infringements on their person and property (Art. 2, On the Militia). In reality, the main goal of the police is still the maintenance of the state's security and interests and the protection of its administrative structures and of the top officials of the state and municipal institutions.

The police under the MVD remain one of the three armed power structures (together with the army and state security bodies) of the new Russian state. The chiefs of the police are directly subordinated to the President. The State Duma (the legislature—*Gosudarctvennaya Duma*) is not empowered to either appoint or relieve them of their posts. The municipal police depend virtually completely on the local administration. They are financed and provided with office premises and housing by local administrative bodies. The local police completely serve the interests of local administrators (e.g., governors, mayors).

Side by side with political subservience, a tendency which is new in Russia has been manifesting itself more and more distinctly of late: guarding their own interests, besides those of the authorities, has become the primary goal of the Militia. The police stand more and more for their own interests as they see it: security against criminals and the mafia, economic status and political influence. The protection of their own interests is sought by legal, semi-legal (exerting influence on mass opinion, the lobbying of state bodies for rights and financial support, or the intimidation of citizens and authorities) and illegal means (blackmail, mass concealment of crime from the public record in order to create a false impression of the well-being of the state and citizens, ostentatious success in combating crime, and mass-scale corruption, bribery and direct services rendered to the criminal world). Legal protection of their status and economic well-being are provided under Articles 27–32 of the law On the Militia: RSFSR Law (state insurance, social security, material security, provision of housing, free travel passes, etc.) and in the law On the Pension Provision for the Army, MVD Staff and Their Families of February 12, 1993.

This critique does not mean that the police in Russia have no positive achievements in combating crime. One can cite many names of honest and highly skilled police personnel, many crimes which are detected and prevented by the police and officers who died discharging their duties heroically and honestly.

But, regrettably, the social, political and economic situation in Russia throughout the twentieth century has aggravated police misdeeds. The Russian police were implicated in the large-scale reprisals of the Stalin regime. During the Brezhnev stagnation, the police prevented known crimes from being recorded. And the police continue to participate in corruption and lawlessness in contemporary Russia.

PROFESSIONAL CHALLENGES

Abuse of Authority

Violations of human rights and international norms take a variety of forms, including the perpetual disregard of the rights of citizens to make appeals regarding the misconduct of the police, the public prosecutors and the courts; the absolute and complete denial of rights to soldiers during time of military service, even when current execrable conditions lead to mass killings, mutilations and suicide (see *Human rights abuses in the armies...*, 1992); the vulnerability of the population to the corruption of the forces of law and order and other officials; and the abuses of power in penitentiary institutions. Such violations are supported by a firmly entrenched police culture which has not changed simply because law and politics have.

Extremely harsh regimes in prison increase the deprivation of freedom suffered by those awaiting trial or under conditional sentence and contravene human rights. Overcrowding in the isolation blocks housing those awaiting trial compels inmates to sleep in shifts. Bad and insufficient food and unsafe and unsanitary conditions lead to diseases (e.g., tuberculosis) which afflict many. Torture is widespread, in so-called "press cells" for those awaiting trial or under investigation and in torture colonies (such as the notorious White Swan detention centers) for convicts who contravened the regime or have fallen into ill favor with the prison authorities. Periodical mass beating of inmates, called "prophylactic" or "training sessions," are administered by special police units. Life in the institutions for the deprivation of freedom is unbearable and the possibilities for correction and rehabilitation are non-existent.

De jure criminal liability of officials is provided for in cases of deliberately illegal and unlawful arrest (Art. 178, Criminal Code of 1960; Art. 301, Criminal Code of 1996) and for compulsion to give evidence by means of threats, violence or unlawful actions (Art. 179, Criminal Code of 1960; Art. 302, Criminal Code of 1996). The Criminal Code of 1996 came into force on January 1, 1997. But in reality, threats together with psychological and physical violence by a policeman or investigator against those suspected of the commission of a crime or an administrative offense or against arrested suspects are common and hardly concealed. These events are covered in newspapers and legal journals. Some policemen who commit such acts are found criminally liable, but only in rare cases. For example, *Everyone's Newspaper* had been publishing articles for two years about the systematic use of torture by the police in the Mordovian Republic (Russia). As a result some functionaries were penalized and Kossov, the Ministers of Internal Affairs, was demoted. However, Rosin, the first deputy minister, retained his post.

Police often collaborate with the mafia, betraying the interests of their institutions and of their colleagues. As a result, in late 1995, a new Administration for Internal Security was established within the Ministry of Internal Affairs. At the press conference held by the Administration for Supervision over the Investigation of Crimes of the General Office of the Public Prosecutor and by the Chief Administration for Public Order Maintenance (held December 1995), it was acknowledged that "the presence of corrupt workers, sadists, bandits and even organized criminal groups is not anything exceptional, but a phenomenon that can be found in the law enforcement bodies" (*Everyone's Newspaper*, 1995, No. 50). In 1995, 805 police officers were found criminally liable (*Everyone's Newspaper*, 1995, No. 50). But this is a tiny fraction of the number of criminals in uniform.

Stereotyping

Not infrequently, the police, as do politicians and the mass media, tend to attribute the high crime rate to the criminal activities of migrants, foreigners, refugees, individuals of Caucasian nationalities (Azerbaijanians, Georgians, Chechens, Ingushes, and Ossetians) or to other social outcasts (e.g., drug addicts). However, statistical data combined with the findings of criminological research show that the contribution of these groups to the crime problem is relatively insignificant. It is not they who determine the criminal situation in the country.

Thus in 1993 to 1995, crimes committed by foreign citizens or stateless persons accounted for 1.7 percent to 1.9 percent of the total number of solved crimes, with 1.3 percent committed by citizens of the Commonwealth of Independent States (CIS) countries (*State of crime in Russia in 1995*, 1997). (CIS is the official name of the group of states of the former USSR after its disintegration.)

OPERATIONAL CHALLENGES

Crime in Russia

Normal Crime

Trends in crime are assessed using official statistics (*Crime and delinquency in the USSR*, 1990; *Crime and delinquency*, 1993, 1995; *Conditions of crime in Russia*, 1996; *State of crime in Russia*, 1997). Inasmuch as these are incomplete, it is preferable to use those crimes which suffer the least from being underreported or are most likely to become known (e.g., visible crime such as premeditated murder and grievous bodily harm) as indices of the crime problem and of trends. In some cases, the findings of empirical research are used. Unfortunately, in Russia there is as yet no established system for monitoring research (such as self-report studies and victimization surveys) which would allow a more accurate reflection of the state of crime.

Since 1966 there has been a growth in recorded crime in both Russia and the rest of the former USSR. Since 1978 increases in crime have been particularly significant, growing at roughly seven percent annually. However, the beginning of the *perestroika* period was marked by a fall in the level of crime. Since 1988, there has been a yearly increase in the level of crime with some types of crime increasing at a particularly fast pace. (See Table 7.1).

The overall crime rate stood at 987/100,000 in 1984, rose steadily to 1,888/100,000 in 1993, declined slightly to 1,779/100,000 in 1994, rose to 1,857/100,000 in 1995, and declined, again, to 1,756/100,000 in 1998. The decrease in the rate during 1986 and 1987 was due to the influence of the social and political changes brought about by *perestroika* and their effects on the consciousness of the people. Subsequent political and social crises caused the crime rate to rise again, in 1989 by 32 percent and in 1992 by 27 percent.

The increase in violent crime is particularly significant. The rate for premeditated murder stood at 8.5 in 1985, declined to 6.3 in 1987, and then rose steadily to 21.8 in 1994, and stood at 21.4 in 1995 and 20.1 in 1998. The rate for serious bodily harm stood at 19.9 in 1985, declined to 13.9 in 1987, and then rose steadily to 45.7 in 1994, and declined to 30.7 in 1998. (All rates are per 100,000.) Such an explosive growth in the level of violent crime indicates the extremity of criminalization and reflects the severity of the social crisis. The rate of premeditated murder (including attempted murder) was always higher in Russia than in Western Europe. From 1989–90 it not only caught up with but overtook the rate in the US. In 1993–94 it reached the level of some Latin American and Caribbean countries. Consequently, the real incidence rate of murder is higher than the official data indicate. Some regions in Russia are marked by particularly high premeditated murder rates: in 1994, the murder rate stood at 39.0 in Altai, at 28.1 in Tuva, at 37.8 in Khahassiya, at 36.0 in Buryatiya, at 32.7 in Primosk territory, at 40.0 in the Kemerovo region, at 33.0 in the Magadan region, and at 34.2 in the Chitinsk region (all rates are per 100,000).

At the same time, the number of unrecorded murders is increasing, as indicated by the number of persons reported missing, some of whom have fallen prey to crime and may be dead and murdered. In 1990, 13,214 missing persons were not found; in 1994, the figure rose to over 21,000 (Dolgova, 1994, 69; *Izwestia*, 1995), and to about 25,000 in 1996 (*State of crime in Russia*, 1997).

The rate of growth of robbery is also high: from 21 in 1987 to 127 in 1993, to 100 in 1994, to 95 in 1995, and to 83 in 1998 (all per 100,000). Assaults increased 6.5 times, from 3.9 in 1987 to 27.0 in 1993, to 25.5 in 1994, to 25.4 in 1995, and to 26.2 in 1998

Table 7.1 Trends in crime in Russia, 1985–1996

	1985	1986	1987	1988	1989	1990	1991	1992	1993	1994	1995	1996	1997	1998
Total registered crimes (in 1,000)	1,228	1,228	1,286	1,220	1,619	1,839	2,168	2,761	2,800	2,633	2,756	2,625	2,397	2,582
Rate (per 100,000)	989	930	817	834	1,098	1,243	1,463	1,857	1,888	1,779	1,857	1,771	1,629	1,756
Murder, attempted	1,2160	9,434	9,199	10,572	13,543	15,566	16,122	23,006	29,213	32,286	31,703	29,406	29,285	29,551
Rate	8.5	6.6	6.3	7.2	9.2	10.5	10.9	15.5	19.8	21.8	21.4	19.8	19.9	20.1
Grievous bodily harm	28,381	21,185	20,100	26,639	36,872	40,962	41,195	53,873	66,902	67,706	61,734	53,417	46,131	45,170
Rate	19.9	14.7	13.9	18.2	25.0	27.7	27.8	36.2	45.1	45.7	41.6	36.2	31.4	30.7
Larceny-theft	464,141	380,582	364,511	478,913	754,824	913,076	1,240,636	1,650,852	1,579,600	1,314,788	1,367,866	1,207,478	969,800	1,143,364
Rate	324.7	264.4	251.1	327.2	512.1	616.8	837.3	1,110.2	1,065.2	884.4	884.4	818.0	716.3	777.8
Robbery	42,794	31,441	3,441	43,822	75,220	83,306	101,956	164,895	184,410	148,546	140,597	121,356	112,051	122,366
Rate	29.9	21.8	21.0	29.9	51.0	56.3	68.8	110.9	124.3	100.4	94.7	82.2	76.2	83.2
Aggravated assault	8,264	6,018	5,656	8,118	14,551	16,514	18,311	30,407	40,180	37,904	37,651	34,584	34,318	38,513
Rate	5.8	4.2	3.9	5.5	9.9	11.2	12.4	20.4	27.0	25.6	25.4	23.3	23.3	26.2

(all per 100,000). The fall in crime rates since 1994 is not due to a real positive change but to new policies of the police and political leaders which encourage the cover-up of recording and accounting of crimes.

The rate of growth of crime has been particularly high in "outdoor" crimes, those committed in the street, squares or parks, and involving shootings, explosions and armed skirmishes. In 1990, various crimes claimed 43,634 lives in Russia and in 1994 the number rose to 93,872, an increase of 2.2 times in four years (*Crime and delinquency*, 1995, p. 19).

The number of registered street crimes rose from 226,250 in 1990 to 333,682 in 1993. A decrease in street crimes to 283,000 in 1994 and to 248,000 in 1995 were most likely the result of non-reporting and concealment by authorities. This is not the opinion of the author alone, but it is shared by other experts (Dolgova, 1994, pp. 24–27).

In 1995, the share of street crimes (as a percentage of all crimes) was 9.8 percent for all of Russia. In Moscow, street crimes constituted 25.7 percent of all crimes and in St. Petersburg 17.2 percent (*Conditions of crime in Russia in 1995*, 1996). In St. Petersburg, the rate of aggravated assault and robbery, in 1993–94, was the highest in Russia (*Crime and delinquency*, 1995).

The number of reported crimes involving firearms and explosives increased by 4.3 times from 4,463 in 1990 to 19,154 in 1993, but declined to about 18,000 in 1994 (*Crime and delinquency*, 1995). In the period January to November 1995, 1,355 such crimes were reported in Moscow, and 538 in St. Petersburg (*Conditions of crime in Russia in 1995*, 1996).

Organized Crime

The nature of crime has changed in a number of ways. There has been an increase in violent crime remarkable in its cruelty. Organized criminal groups of the Mafia type have been formed. White-collar crime and mass corruption among the organs of state power, administration and law enforcement have emerged.

The theme of organized and economic crime deserves special attention. Here only a few points will be made, derived mainly from published sources (Dolgova and Djakov, 1989, 1993, 1996; Ovchinski, 1996; Gilinskiy, 1994, 1996), published information in the mass media and our own research in St. Petersburg.

Organized crime has been a long-standing feature in Russia. Only its forms, level of organization and working methods have changed. Since the 1920s, there have existed bands or gangs. In the 1930s there emerged *Voryi v Zakone* (Thieves by Law) who were professional thieves and swindlers, well-known in the criminal underworld, who had chosen crime as a permanent way of making a living and adhered to a thief's code of conduct. The 1950s saw the emergence of the *Tzechoviki*, groups within professions of illegal production, named after the word *tzech* (shop or department in a factory). The 1970s saw the merging of the *Tenyeviki* (shady dealer) gangs, named after *tyen* (shade), within state structures and the beginnings of institutionalized corruption.

Currently, three levels of criminal organizations exist. First, there is the criminal group; second, the criminal organization; and third, the criminal society. These Mafia groups divide the business sphere in St. Petersburg and in other regions of Russia among themselves. In general, the growth of organized crime in recent times is one of the consequences of the intensified level of organization in the economic, political and social structures. This is probably a worldwide tendency. As other organizations have grown in complexity and size, so has organized crime.

There exist two levels of racketeering. Organized crime extorts tribute from small street kiosks (which have sprung up in urban areas as private enterprise has taken hold) and commercial organizations by straightforward threats ("black racket") or by indirect, disguised extortion: by "guarding" businesses, by "rendering services" in the field of marketing or by assisting in "joint work" in executing a contract.

But the main fields of activity by Russian criminal organizations are as follows: shady bank transactions using fake documents and letters of advice; fictitious transactions in real estate; the hijacking and reselling of cars; the illegal export of non-ferrous metals; black market transactions, including the wholesale purchase and diversion of humanitarian aid; bribing city functionaries; the production of and traffic in fake hard liquor and drinks; the sale of arms and counterfeiting. Criminal organizations are also beginning to dominate gambling, the agencies for supplying sexual services and the narco-business.

The Mafia display a keen interest in privatization. One of our respondents states that "their goal is to take hold of real estate." Criminal organizations obtain information about upcoming auctions, come to the auctions with armed men and determine who buys what property at what price.

Legal business in today's Russia can no longer function without becoming involved in some kind of criminal activity, such as bribery, tax evasion or working relations with criminal organizations. In the existing environment, legal business cannot avoid becoming criminalized.

By the middle 1990s, a stable criminal community had arisen throughout Russia and its regions. Having a "share" of influence on economics and in politics, criminal organizations have effected a welding of criminal, economic and political power structures. Consequently, we are confronting the criminalization of business simultaneously with the politicization of crime. Russia, in the 1990s, can be generally characterized as a criminalized society governed by a criminal state.

The bosses of criminal associations have found themselves in the spotlight. There are books devoted to them (Dikselius & Konstantinov, 1995); the mass media dwell on them; portraits and interviews with them are published in the press.

The high rate of crime, particularly that of crimes committed in the streets, squares and parks, which frequently involve shootings, explosions and armed skirmishes in public, and the lack of protection of citizens against daily crime have inspired fear and given rise to a "moral panic" (Cohen, 1980).

Fear of Crime

Episodic empirical research of victimizations experienced by the population, which were carried out under the direction of the author by the Sector for the Sociology of Deviant Behavior of the St. Petersburg branch of the Russian Academy of Sciences, in St. Petersburg (in 1989, 1990, 1992, 1993, 1994 and 1996) and in Pskov (in 1992), revealed that the actual number of crimes according to the victims was 10–15 times higher than those shown in the officially recorded data. In St. Petersburg, according to the findings of the victimization surveys, 12 percent of respondents reported having been victims of crime in 1991; by 1994 the figure stood at 26 percent and in 1996 at more than 30 percent. However, officially, the crime rate had decreased by 13 percent in that period.

Research carried out by the All-Russia Center of Social Opinion Studies in August 1993 revealed that 83 percent of all respondents were disturbed by the rise in prices, and 64 percent by the growth of crime. According to the result of the St.Petersburg survey done in 1993, 56 percent of respondents felt themselves to be under threat from aggression and violence, and 42 percent from organized crime. In retrospect, they felt themselves to have been safe until August 1991 compared to the time of the survey. Forty-three percent had felt themselves to be "not absolutely safe" or "absolutely unsafe" before August 1991 with the figure rising to 69 percent at the time of the survey. While 21 percent felt "unsafe" alone at night in their homes before August 1991, the number for 1993 was 37 percent. In 1993, 42 percent percent feared falling victim to the wiles of teenagers, 50 percent to robbery, and 45 percent to murder. Thirty-two percent of women feared being raped (Alfanasjev, Gilinskiy & Golbert, 1995).

According to the results of the 1994 St. Petersburg survey, violent crimes against persons and property was ranked highest as a social problem. At the city level, 64 percent of respondents, and 60 percent of respondents at the community level, ranked violent crime first as a problem needing solution.

Drug-related Crime

The levels and trends in drug-related crimes are presented in Table 7.2.

The numbers are likely to be somewhat higher if one bears in mind that drug crimes are often unreported. The share of crimes committed by persons who were in a state of alcohol intoxication rose from 38 percent in 1990 to 41.2 percent in 1994 (*Crime and delinquency*, 1995, p. 227).

The share of crimes committed by able-bodied persons who neither work nor study has tended to grow constantly, from 17.8 percent in 1990, to 20.2 percent in 1991, to 27 percent in 1992, to 35.9 percent in 1992 and to 41.2 percent in 1994. Especially large is their share of premeditated murders (48.5 percent), grievous bodily harm (46.6 percent), rape (43.9 percent), aggravated assault (62.8 percent), robbery (54. percent), larceny–theft (48.2 percent), and drug related crimes (58.7 percent). (All percentages are for 1994.) This tendency has several tentative explanations. First, the unstable economic situation, unemployment and partial employment (incomplete working days or weeks) incite people to commit crimes for material reasons and as a form of protest, even though they may not always be aware of the latter motivation. Second, it is easier for the police to detect and find criminally liable representatives of the least protected and most unfavored strata of society.

Migration

Prior to *perestroika*, emigratory tendencies were typical of the former USSR, including Russia. During the first wave of emigration (1917–1938), about 3.5 to 4 million persons left; during the period 1939–1948, about 8 to 10 million; and during 1948–1990, about 1.1 million (*The population of Russia*, 1993). All, in all, during the *perestroika* and post-*perestroika* periods (1987 to 1994), about 580,000 to 590,000 persons left Russia (Vardanian, 1995, p. 65). Immigration was minimal as the poor totalitarian countries could hardly attract anyone.

Inter-ethnic conflicts and the disintegration of the USSR have brought a new tendency to the region: the mass migration to Russia from former Soviet republics (the so-called "near" foreign countries) and a growing migration (though not yet considerable in numbers) from "far" foreign countries (Turkey, Afghanistan, Iraq, Vietnam, China, Korea, Cuba, etc.). In addition, economic investment has given rise to a small-scale migration from the developed Western countries.

According to the official data, the number of immigrants in Russia was about .7 to.9 million annually during 1991–1993 and about 1.2 million in 1994 (*Russian statis-*

Table 7.2 Drug-related crimes in Russia, 1987–1995

	1987	1988	1989	1990	1991	1992	1993	1994	1995
Total	18,534	12,553	13,446	16,255	19,321	29,805	53,152	74,798	79,819
Rate (per 100,000)	12.7	8.6	9.1	10.9	13.0	20.0	35.7	50.3	57.7
Drug thefts	823	470	439	413	433	315	475	529	691
Manufacture, sale, carrying, trafficking, and acquisition	15,506	9,527	10,594	13,646	17,036	27,115	49,249	70,420	72,457
Persuade to use drugs	503	194	130	182	187	190	338	613	648
Keeping premises for drug use	444	252	171	206	181	324	499	721	750
Forging documents and prescriptions for drugs	602	248	277	222	296	*	85	162	129
Illegal growing of poppies or cannabis	136	74	3	72	76	91	343	593	666
Breaking stock-taking and storage laws	365	776	793	642	488	813	1,066	1,690	1,886
Percentage of drug-related crime to common crime	1.6	1.0	.8	.9	.9	1.1	1.9	2.8	2.95
Percentage of drug-addicts crime to common crime	.2	.2	.1	.1	.1	.2	*	*	*
Percentage of crime committed while under the influence of drugs	.3	.2	.2	.2	.2	.2	.7	.5	.4

tical annual, 1995; Vardanian, 1995). At that time, the share of immigrants from "near" foreign countries dominated and accounted for 94.2 percent in 1993 and 96.2 percent of immigrants in 1994. Within Russia migration rose from 51,600 in 1991, to 176,000 in 1992, to 430,000 in 1993, and to 810,000 in 1994 (*The Population of Russia*, 1993, p. 67; Vardanian, 1995, p. 59). In most cases, migration was nonvoluntary and caused by inter-ethnic conflicts and discrimination in the "near" foreign states (Chervakov, Shapiro, & Sheregi, 1991; Vardanian, 1995; Vitovskaya, 1993). The sociological monitoring of forced migrants and refugees indicates that the reasons for emigration from permanent places of residence were as follows: discrimination on the basis of nationality accounted for 51 percent to 94 percent, and discrimination on the basis of nationality with regard to the nearest relative accounted for 25 percent to 60 percent of the motives (Chervakov, Shapiro & Sheregi, 1991).

The MVD estimates that the number of foreign citizens from "far" foreign countries residing in Russia is about .5 million. The number of provisional, temporal immigrants (staff of joint enterprises, foreign firms, investors, etc.) was about 130,000 in 1995, including 20,000 Turkish workers (Vardanian, 1995).

But one should also take into account the weaknesses of official policy and statistics. Many refugees have either immigrated illegally or have not been granted the status of refugee. Provisional, temporary immigrants are not registered and included in the state statistics in contrast to international practice. One can expect that the scheduled 1999 census will make it possible to discover the size of the migration.

Yet the total number of immigrants in the country, who have entered during the years of a market economy, does not exceed .7 percent to .9 percent of the overall population. In contrast, in some European states, the share of immigrants reaches 10 percent. On the other hand, because of conditions of economic, social and political crisis in Russia, even this small number of immigrants—small by internationally established measures—gives rise to problems, such as lack of housing, the pressures on the manpower resources market, unemployment among refugees and pressure on the urban infrastructure.

The inhabitants of Russia regard legal, but particularly illegal, immigrants as the main causes of crime and organized crime (the "Azerbaijan mafia," the "Chechen mafia"), higher food prices and other social ills. But as was noted earlier, the actual percentage of crimes committed by immigrants is not great, even though many refugees have to commit offenses in order to survive. Some immigrants are suspected to be involved in criminal organizations. However, the concealment of organized crime and drawbacks of statistics data afford no opportunity to precisely estimate the share and role of immigrants in organized crime.

Crime due to ethnic conflicts in the former USSR has been studied by Luneev (1995). The data he obtained demonstrate that ethnic conflicts in the USSR claimed 95 lives in 1988 and 222 in 1989. In two months in 1990 alone, 293 persons were killed. The number of injured was 2,669 in 1988, 2,824 in 1989, and 1,326 in two months in 1990. Cases of arson numbered 2,422 in 1988, 862 in 1989, and 728 in two months in 1990. Since 1991, the registration of such crimes has been discontinued. After 1991, wars broke out in Azerbaijan, Georgia, Abkhazia, Moldavia, Pridnestrovjjye and Chechnya. Luneev believes that

> the total casualties are unknown. Some day we shall count and feel revulsion. Suffice it to say that in Chechnya alone casualties among Russian servicemen, Dudayev supporters and the civil population have been running into the tens, if not hundreds of thousands (1995, p. 100).

The reaction of Russian authorities, the MVD, and the police to the involvement of migrants and refugees in the commission of crimes is extremely inconsistent. Sometimes they are merely idle. Sometimes individuals of the Caucasian nationality are used as scapegoats and a "witchhunt" campaign is launched. Massive police raids in Moscow, St. Petersburg and other big cities and arrests and beatings of people with Caucasian features and documents are common features. According to stereotypes, individuals of the Caucasian nationality find themselves lumped as suspects with drug users (in Russia, a drug addict evokes the image of a criminal and not that of a patient who needs treatment and social care), refugees, and the homeless. In contemporary Russia, individuals of the Caucasian nationality are treated similar to Negroes in the USA during the epoch of racism (Iljin, 1994). Thus the police commit the grossest infringements of human rights while the real members and bosses of criminal associations, including those with a national or ethnic basis, remain safely beyond the reach of the police.

RESPONSES TO CHALLENGES

International Influences

After Stalin's death, the international ties of the Russian police were gradually reestablished during the Khrushchev "thaw." International contacts have increased rapidly since the mid-to late 1980s, as an element in Gorbachev's *perestroika* reforms. Contacts have progressed along several lines.

First, cooperation under the umbrella of Interpol and other international organizations has progressed at a fast pace.

Second, there has been a growth of contacts through international scientific and research conferences and congresses, such as the International Association of Criminal Law, International Criminological Society, United Nations Congresses for the Prevention of Crime and the Treatment of Offenders (the USSR took part for the first time at the Second Congress), International Center for Criminal Law Reform and Criminal Justice Policy and Penal Reform International.

Third, bilateral and multilateral treaties of cooperation among states and local (municipal) agencies have been enacted. For instance, there are agreements regarding cooperation among the Russian and American police (Egorshin, 1994). The forms of such cooperation are diverse: the exchange of delegations and groups of police officers in various fields such as drugs, patrol operations, penitentiary institutions and expert services; material and technical aid in the areas of transport and technological applications; and all-embracing programs of joint work.

The St. Petersburg and Finnish police have held several conferences, and programs of joint activity have been worked out to deal with problems connected to business travel, the security of companies operating in St. Petersburg, relationships with clients and competitors and relationships with authorities (Aromaa & Lehti, 1995).

Pursuant to special international agreements, programs on the exchange of operational information in joint detective work and for the extradition of felons have been established. (Such cooperation has a long history. Russia had signed agreements on extradition with France, England and Byzantium during the feudal epoch, at the beginning of the Middle Ages.) Current cooperation has become especially important as international organized crime has made rapid advances in such areas as drugs, arms trafficking and the pornography business.

Sometimes, cooperation assumes unexpected and very local forms. For example, the Jail Staff Training Center in Finland published Matti Laine's book *Criminology and the sociology of deviant behavior* in Russian and distributed it among Russian colleagues, primarily in St. Petersburg.

The Middle European Police Academy, established initially by Austria and Hungary and now grown to eight member states, regularly entertains representatives of the Russian police and publishes papers by Russian authors (see for example the contributions in *MittelEuropäische Polizeiakademie*, 1995).

Individual Professionalism

Democracy and a market economy as consequences of the fall of Soviet totalitarianism should have resulted in the growth of professionalism, skills and responsibilities of police personnel. In fact, contradictory and complex processes are under way.

In spite of all the drawbacks and atrocities of totalitarianism, there were highly skilled people in the Soviet militia—primarily those who sincerely believed in the advantages of a socialist society and the prospects for improvement of the human condition. They were personally honest and conscientious. During the Soviet years, a system of personnel training developed and took shape: secondary schools of the Militia, higher schools of the Militia (MVD), the Academy of the MVD, as well as various short-term advanced training courses. The higher MVD schools provided a high level of legal training, special training in operational and detective work and military discipline. Some of the higher schools had their own specializations. The Volgograd higher school trained investigators who worked in the MVD system; the Ryazan higher school educated personnel for penitentiary institutions and the Nizhegorodskaya higher school prepared personnel of the special services for combating organized crime. The Academy of the MVD located in Moscow was designed for the training of senior and higher-level police officials. In addition, some law graduates for universities went to work for the Militia.

Several recent factors have affected the police adversely. These include the disintegration of the totalitarian system, the frequent movement of the heads of the MVD, the slump in the prestige of the MVD, the low salaries of personnel and the lack of social care for the police. Experienced personnel were pensioned off or have left for other employment. A shortage of personnel forced the authorities to hire candidates without testing them properly. Long-term study at higher educational institutions was replaced with short-term courses.

Police professionalism has been affected by police dismissals and disqualifications. According to observers,

> together with skilled and experienced people, the knowledge of the criminal world as well as many reliable forms and methods for combating crime are receding into the past. The lack of a stable core of highly skilled specialists in the Militia is one of the key reasons for the low standards in fighting crime. The situation is aggravated by the corruption which has grown in the law enforcement system and has noticeably impaired the sense of camaraderie indispensable in the operational branch (Gilinskiy, Luneev & Michailovskaja, 1995, pp. 5–6).

Consequently, the following situation has developed. In 1993, 224,500 new personnel were hired by the MVD. The shortage of officers and rank and file reached 78,200 in that year. Over 55 percent of personnel had been in the police service for less than three years. Only 59.8 percent of investigators, the most prestigious position in the MVD, had a higher eduction degree (Milukov, 1995, p. 220). In response,

> striving to make up for the personnel shortage, the MVD is hiring persons who do not meet the necessary requirements ... . An outflow of highly skilled personnel from the leading functional services to commercial and other agencies has not discontinued. In 1993, 4,300 were answerable for breaking the law. Over one-third of the them were punished for withholding crimes from investigations and over 500 personnel in the operations division for collaboration with criminal gangs. The police themselves committed 2,200 crimes (Gilinskiy, Luneev & Michailovskaja, 1995, pp. 5–6).

In 1994, 9,300 police officers were made to answer for various illegal activities, and 2,000 of them were prosecuted. Of these, about 12 percent were charged with taking bribes and about 20 percent were brought up on the charge of exceeding their authority or legal power (Milukov, 1995, p. 72). But these figures are just a tiny part of the enormous numbers of crimes committed by Militia personnel.

Anonymous interviews with police personnel and criminals conducted under the author's supervision by the Department for the Sociology of Deviant Behavior of the St. Petersburg branch of the Institute of Sociology for the Russian Academy of Sciences reveal the scope of corruption and bribery. Extracts from some interviews are given below. A Militia officer (road militia) was asked "Who offers the most money?" He replied:

> last year it was yet not like it is now. But this year [1995] the money is being offered by everybody, ranging from a drunk who did some trifling thing to bandits [gangsters]. You would stop them; they would obey and slip 30,000 rubles or more to you without so much as bothering themselves to get out of the car. Only you'd better not ask if I took the money or not.

The representatives of bandits (member of a criminal organization), when asked about the safety of their criminal activity, gave the following answer: "Do you think they don't know in the Big House [the slang name for the building of the Main Administration of Internal Affairs in St. Petersburg] how I work? You just have to make friends with everybody." And also,

> with us, all were implicated: the residence office, notaries, and a couple of district Militia officers. This machinery is impossible to stand against. It is not only in St. Petersburg. I know smaller towns where everybody is interlaced: the Town Hall, Militia, municipality, bandits. They have already got married into each other's families. This is just mafia, a real family. They control the whole town.

Low professional standards together with the pressure of work have resulted in a number of evils. In the first half of 1994, the number of criminal cases handled by an investigator ranged from 14 to 37; in St. Petersburg, the number varied from 47 to 62 (Milukov, 1995, p. 202). First, there is the deliberate withholding of crimes from registration, especially ones which are hard to solve because of a lack of clues, eyewitnesses and other proofs. Second, unlawful methods of investigation, such as the use of physical and psychological violence against suspects, many of whom fall into the hands of the police by accident, have increased. Tortures used on suspects make the innocent plead guilty while real felons remain at large and continue their criminal activity.

Laws

The transition to a market economy and political reforms in post-Soviet Russia, as well as the new goals and changing roles of the police (Militia) in state and society, are partially mirrored in changes in law.

The fundamentals of democratization are provided by the 1993 Constitution of the Russian Federation and other legal acts. Under Article 1 of the Constitution, Russia is a democratic federal state with a republican system. It is stipulated by Article 2 that "man, his rights and freedom are the supreme value." Observance and protection of the rights and freedom of persons are the duty of the state. The Constitution also embodies the concept of the presumption of innocence (Article 49). Other rights and freedoms are thoroughly elaborated. A jury system is provided for in the area of judicial reform. The Law of the Russian Federation has mitigated some of the sternest articles of penitentiary legislation and liberalized them. The provision of citizens' personal security is declared to be the key goal of the police (Article 2 of the Law on the Militia).

But many of the legislative innovations in post-Soviet Russia remain only as good intentions and are violated in the process of law enforcement (*The white book of Russia*, 1994). The jury was introduced as a legal experiment only in nine regions of Russia in 1994. All in all, 173 cases against 241 persons were heard by a jury in 1994, while the total number of persons convicted by all courts in Russia that year was 924,574 (*Crime and delinquency*, 1995, p. 135). Owing to a very low quality of investigation, 32 percent of the cases heard by a jury in 1994 were returned to the investigating bodies for further investigation, and 18 percent of the accused were found not guilty. In general, in the courts of Russia, only 1 percent of all accused are found not guilty.

The Law of the Russian Federation "On the Imprisonment of Suspects and Those Accused of the Commission of a Crime" (adopted by the State Duma on June 21, 1995 and enacted on July 20, 1995) is virtually unobserved. The places for detaining suspects before trials (*Sledstveniy isolator*—SIZO—preliminary investigation wards, a kind of lock-up, and *Isolator vremennogo soderjianija*—IVS—isolation for provisional containment wards) have long been torture cells. The previous law provided 2.5 square meters of living space per suspect or accused. The new law stipulates an increase to 4 square meters by 1998. But the actual living space in big cities (Moscow, St. Petersburg, etc.) remains at .5 square meters. On hot summer days, inmates die from heat and strokes. In Moscow's *Matrosskaya tishina* (Sailor Quiet) SIZO during the autumn of 1995, relatives of prisoners had to stand in line for 10–14 days to get to the counter where they could hand over parcels containing foodstuff or clothes for prisoners. The time allowed for meeting with prisoners is not observed—only one instead of three hours is permitted. Prisoners are not provided with necessary medical aid. Tubercular patients share the same chamber with others, contrary to legal norms. According to information from the Novgorod region's Office of the Public Prosecutor, eight prisoners in a penal colony died of tuberculosis during nine months in 1994. For nine months of 1995 the figure was 19 dead. Almost all of them had been delivered from the Moscow SIZO with consumptive diseases and in extremely grave condition. After the delivery to the colony to serve their sentences, these convicts lived for 5 to 10 days before they died. Food for convicts is by no means satisfactory. It is insufficient in quantity and awful in quality. According to information from the Novgorod region's Office of the Public Prosecutor, 38 percent of prisoners suffered from weight deficiency. This is typical of all places of imprisonment across the country (Abramkin, 1996, p. 141; also 1993).

Many of the recent legislative and other normative acts are directly in contradiction to the Constitution. For example, the Decree of the President of Russia of June

14, 1994 (No. 1326), "On Urgent Measures for the Protection of the Population from Banditry and Other Manifestations of Organized Crime," is inconsistent with Articles 4, 10, 17, 19, 22, 34, 50 and 55 of the Constitution. The Decree provides for 30 days imprisonment of a suspect before laying a charge (under Arts. 90 and 122 of the Code of the Criminal Process the period is three days, or in some cases up to a maximum of ten days), the initiation of an investigation before a criminal case is instituted (contrary to Article 109 of the Code of the Criminal Process), and searches by third parties. Detention is warranted by a public prosecutor and is not to be reviewed in court (a direct violation of Art. 22 of the Constitution). The resolution of the State Duma, "On Guarantees of Constitutional Rights and Freedoms of Citizens by Combating Crime," accuses this Decree of "disorganizing the system of justice" (Russian Federation, 1994). Yet the Duma also makes laws which contain unconstitutional and anti-democratic provisions, such as the acts "On Detective and Operational Activity in the Russian Federation," "On Combating Organized Crime," or "On Bodies of Federal Service of Security in the Russian Federation."

Media: Manipulation and Cooperation

The totalitarian regime in the former USSR which lasted for seventy years constructed an enormous repressive mechanism, comprising the Militia, the *Komitet Gosudarstvenoj Besopasnosty* (the KGB—the Committee for State Security) and the penitentiary (GULAG) system. Mass fear of this mechanism should be taken into account when assessing the following information.

Until recently, the Militia, the State Security bodies and all the other institutions of the totalitarian state were beyond criticism. (The year 1985, the beginning of *perestroika*, marks the conventional benchmark, although the face of contemporary Russia took shape a little later, in 1987–1990.) The mass media, as the faithful ideological servant of the state, backed the indisputable authority of the *Vserossijskaja Chrezvychainaja Komissija* (VChK—the National Extraordinary Commission), the *Narodniy Kommissariat Vnutrennich Del* (the NKVD—the People's Commissariat for Internal Affairs) and the *Ministerstvo Vnutrennich Del* (the MVD—the Ministry for Internal Affairs) as the "staunch champions of the working people." Public opinion was of no concern. No sociological research was carried out to determine it. It was conditioned by the mass media, beginning with children's magazines and newspapers. For example, eleven issues of the 1938 edition of *Murzilka* (a monthly journal for small children) carried articles about the feats of Soviet State Security men. The issue numbered six is quite remarkable. It contains a piece about "Uncle Kolya" (Nikolai Ezhov, a chief of the NKVD whose bloody reign is immortalized in the people's memory as *Ezhovshina*), a paragraph entitled "Long Live the Soviet Intelligence Service," a story called "The Story of a Gunsmith," and a letter from a schoolboy who aspired to serve in the Militia. *Ezhovshina* refers to the period 1934–38 when many persons were killed by the NKVD, then headed by Ezhov, in order to eliminate all those who did not support Stalin outright. Many were killed without reason. Later on the NKVD was renamed KGB. The "Story of a gunsmith" is an account of spying by capitalist countries.

Following 1985, with the advent of *glasnost*, the situation became more complex. The mass media were able to criticize the police and to publish information about the growing crime rate and the social phenomena of drug abuse, prostitution, homosexuality and suicide. The first reaction of a public used to having such information smothered under top secret protection was one of shock. However, public fear about crime, drug addiction, prostitution and homosexuality is so great that a critical attitude toward the Militia goes hand in hand with the belief that if the forces of law and order were to step up the fight against crime that not only would crime be beaten but economic problems would also be solved.

In contemporary Russia, the relations between the police and mass media are not easy to define. On the one hand, the mass media carry information about crime growth and many unsolved, serious crimes (for example, the assassinations of Alexander Men, a progressive orthodox priest who spearheaded the democratic movement; Vladislav Listyev, a well-known television journalist; and other journalists, businessmen, and bankers). The media discuss abuses of power, corruption in the police and tortures in the course of investigations and in penitentiary institutions. All this discussion creates a negative image of militiamen. (The derogatory slang term for the Militia now is "ment"—which is impossible to translate but describes a person held in low regard.)

On the other hand, the image of a militiaman has not yet lost its former attractiveness: a crime fighter, protector of citizens and a brave boy who risks his health and life for the sake of justice. The mass media still present this image in television and radio programs, as well as in the written press.

Not infrequently, the mass media play into the hands of the existing state power or the opposition. In conditions of long-lasting social and political instability, any information becomes politically tinged. It is said in public discussion that if the crime rate grows quickly, the authorities will use that fact to ask for more powers, more financial resources and more emergency measures. It is also argued by others, though, that the authorities are good for nothing and that their resignation is required together with the victory of the opposition. Others will argue that the authorities know how to handle the situation, are in a position to cope with difficult problems and the population need not worry. The information may also be interpreted as a lie offered by the authorities, as one more attempt to deceive the population, and that such a government ought to resign to be replaced by a more trustworthy opposition.

Yet the publication of police statistics and news about the MVD, the *Federalnaja Sluzba Besopasnosti* (the FSB—the Federal Service of Security), the Office of the Public Prosecutor General and others is a significant democratic achievement. And though much information is still a departmental secret, the mass media are gaining access to information. They carry urgent information about criminals who are wanted and about victims in need of help.

CONCLUSION

Compared with their Western colleagues, the Russian Militia are significantly worse off with regard to equipment, transport, buildings, etc.. Russian officers' salaries are far smaller than those of their Western counterparts, even though they are still a good deal better than those of other workers such as doctors, teachers and the staff of scientific and cultural establishments. There is a shortage of staff, although a far greater problem lies in the shortage of professionalism. Professional qualifications among employees of the Militia are very low as many older professionals have moved into the private security sector or established detective agencies. Employees are inefficient and lack the motivation to tackle well-organized crime, including the drug trade, or to provide security for the population. There is also widespread corruption.

The strong dependence of the Ministry of Internal Affairs and the local Organs for Internal Affairs on the political games of the power structures results in the Militia becoming a toy in the hands of political adventurists. One of the most recent examples of this phenomenon is the non-registration and inaccurate recording of crimes since the end of 1993–1994. This is happening under the direction of the Ministry of Internal Affairs and allows the Ministry and politicians to speak of a "stabilization" and control of crime. It is obvious that these falsehoods encourage criminals and make for an even more vulnerable population. It is worth mentioning that a similar practice of massive cover-up of reported crimes was carried out in the USSR until 1983.

In today's Russia, there exist the legal conditions and the relative freedom of speech and the press to challenge economic, social and political programs. Yet, at the same time, under the existing social conditions, there is no real possibility that the police can meet the challenges of policing in a democratic fashion. Moreover, the tendency to resurrect an authoritarian regime remains strong. It is the author's opinion that Russians are in need of a change of mentality; they need to appreciate the value of the social and economic conditions necessary to support a democracy. But this is and will be a very long, difficult and agonizing process.

NOTES

[1] The acronym GULAG (*Glavnoe Upravlenie Lagerei*) stands for the main administration of penitentiary camps in the Soviet Union. The terrible Soviet penitentiary system is described in the famous book by Solzhenitsyn, *The Gulag archipelago*. The mix of totalitarian control and utopian vision is captured in the slogan on the gate of the Solovki camp, one of the first in the GULAG system—"Happiness For Everyone Through Violence."

REFERENCES

(Russian titles translated by the author)

Abramkin, V. (Ed.). (1993). *Prison reform in the former totalitarian states*. Moscow: Center for Prison Reform.

——. (1996). *In search of a solution: Crime, criminal policy and prison facilities in the former Soviet Union.*. Moscow: Human Rights.

Afanasiev, V., Gilinskiy, Y., & Golbert, V. (1995). Social changes and crime in St. Petersburg. In Evald, U., (Ed.). *Social transformation and crime in metropolics of the former eastern block countries. Findings of a multi-city pilot study* (pp. 162–181). Bonn: Forisom Verlag Godesberg.

Amnesty International newsletter (1993, September). Russian Federation: Survey of new legislation. London: Authory.

——. (1995, January-June). CIS: List of facts. London: Authory.

Aromaa, K. & Lehti. M. (1995). *The security of Finnish companies in St. Petersburg*. Helsinki: Yliopistopaino.

Borisov, A., Dugin, A., & Malygin, A. (1995) *Police and militia in Russia: Pages of history* (original in Russian). Moscow: Nauka.

Chervakov, V., Shapiro, V., & Sheregi, F. (1991). *International conflicts and problems of refugees* (original in Russian). Moscow: Institute of Sociology (Issues I and II).

Cohen, S. (1980). *Folk devils and moral panics*. London: Basil Blackwell.

Conditions of Crime in Russia in 1995 (original in Russian). (1996). Moscow: MVD.

Conquest, R. (1968). *The Soviet police system*. New York: Praeger.

Crime and delinquency in the USSR: Statistical review 1989 (original in Russian). (1990). Moscow: Juridical Literature.

Crime and delinquency: Statistical review 1992 (original in Russian). (1993). Moscow: MVD.

Crime and delinquency: Statistical review 1994 (original in Russian). (1995). Moscow: MVD.

Dikselius, M. & Konstantinov, A. (1995). *Russia's criminal world* (original in Russian). St. Petersburg: Bibliopolis.

Dolgova, A. (Ed.). (1994). *Change of crime in Russia* (original in Russian). Moscow: Criminological Association.

Dolgova, A., & Djakov, S. (Eds.). (1989). *Organized crime* (original in Russian). Moscow: Juridical Literature.
———. (1993). *Organized crime—2* (original in Russian). Moscow: Juridical Literature.
———. (1996). *Organized crime—3* (original in Russian). Moscow: Juridical Literature.
Egorshin, V. (1994). *Militia. Police. Sheriffdom* (original in Russian). St. Petersburg: Economics and Culture.
Everyone's Newspaper (Obshaya Gazeta) (1995), No.50, No. 51.
Gilinskiy, Y. (1994), Shadow economics and organized crime. In Sluziciy, E. (Ed.). *Youth: Figures, facts, opinions* (original in Russian) (pp. 77–84), No. 2., St. Petersburg: Petropolis.
———. (Ed.). (1996). *Organized crime in Russia* (original in Russian). St. Petersburg: St. Petersburg Branch of the Institute of Sociology of the Russian Academy of Science.
Gilinskiy, Y., Luneev, V., & Michailovskaja, I. (1995). *Crime, militia, population* (original in Russian). Paper presented to the State Duma, Moscow.
Human rights abuses in the armies of the countries of the Commonwealth of Independent States (1992). Frankfurt: Internationale Gesellschaft für Menschenrechte.
Iljin, V. (1994). The Nature of racialism. *Boundary*, No. 5, 189–204.
Juviler, P. (1976). *Revolutionary law and order*. New York: Free Press.
Izwestia.(1995, March). No. 11.
Lacqueur, E. (1994). *Black hundred: The rise of the extreme right in Russia*. Washington, DC: Problems of Eastern Europe.
Luneev C. (1995). Criminality in international conflicts (original in Russian). *Sociological Studies*, No. 4, 103–11; No. 7, 99–109.
Makarevich, L. (1995). Russian businessmen are disappointed by the President's message (original in Russian). *Financial Times*, No. 2.
Milukov, S. (Ed.). (1995). *Criminology* (original in Russian). St. Petersburg: Higher School of the MVD.
MittleEuropäische Polizeiakademie. (1995), *Lehrbrief*, Vienna, No. 1, No. 2.
Moscow News (in Russian). (1995). No. 83.
On the Militia for Civil Safety. (1993). In *Complete collection of the statements of the President and Government of the Russian Federation* (original in Russian). Moscow: Government of the Russian Federation, No.7, Art. 562.
On the Militia: RSFSR Law. (1991). In *Communications register of the People's Deputies of the RSFSR and the Supreme Soviet of the RSFSR.* (original in Russian). Moscow: The Supreme Soviet of the RSFSR, No. 16, Art. 503.
Ovchinski, V. (Ed.). (1996). *Basis to combat organized crime* (original in Russian). Moscow: Intra-M.
The population of Russia (original in Russian). (1993). Moscow: Goskomstat.
Russian Federation. (1994). *Complete collection of the legislation of the Russian Federation 1994* (original in Russian). Moscow: Government of the Russian Federation, No. 12, Art. 1369.
Russian statistical annual 1994 (original in Russian). (1995), Moscow: Goskomstat.
Shelley, L. I. (1996). *Policing Soviet society: The evolution of state control*. New York: Routledge.
Shkolnikov, V., Mesle, F., & Vallin, G. (1995). *Life expectancy and mortality of the population in Russia in 1970–1993: Analysis and forecast* (original in Russian). Moscow: Health and Environment Foundation.
State of crime in Russia in 1995 (original in Russian). (1997). Moscow: MVD.
Vardanian, R. (1995). The Influence of economic reforms on the process of international migration (original in Russian). *Sociological Studies*, No.12, 58–70.
Vitovskaya, G. (1993). *The forced migration: Problems and prospects* (original in Russian). Moscow: Institute of Sociology, Issue III.
The white book of Russia: Observations and recommendations in the field of human rights(1994). Frankfurt: Internationale Gesellschaft für Menschenrechte.
Zaslavskaja, T. (Ed.). (1995). *Where is Russia going? Alternatives for social development. Volume II* (original in Russian). Moscow: Intercenter.
Zaslavskaja, T. & Arutjunjan, L. (Eds.). (1994). *Where is Russia going? Alternatives for social development, Volume I* (original in Russian). Moscow: Intercenter.
Zemskov, V. (1990). The compulsory migrants. (The unpublished documents of NCVD-MVD, USSR) (original in Russian). *Sociological Studies*, No. 11, 3–17.

8

Challenges Facing Democratic Policing in South Africa

JEFFREY LEVER and ELRENA van der SPUY

INTRODUCTION

As befits its name, South Africa is the southernmost country on the African continent. Its area is 471,446 square miles. According to the census of 1996, the population is approximately 38 million, divided into four main groupings: southern Bantu speakers (29 million) who are conventionally referred to as Africans, white descendants of settlers (about 5 million), coloreds or people of mixed racial background (3 million), and people of Indian descent (1 million). This division of the population into four "racial" groups was a product of diverse migrant streams and of a white-dominated colonial history. Entrenched by the *apartheid* system since 1949, this nomenclature is by no means accepted happily by all South African citizens. But South African history and present-day society can hardly be understood without references to these overriding social divisions that are part of the everyday consciousness of most South Africans.

Settlement by people of European origin (at first Dutch, German and French speakers, who coalesced into the Afrikaner group), and later by Britons, began in 1652. These European settlers overcame indigenous resistance over a period of 250 years. Twentieth-century political history has been dominated, first, by competition between the two white groupings (British and Afrikaner) and by the attempt of the leading white political parties to secure white supremacy. Prolonged electoral strife between political parties representing the ethnically mobilized Afrikaner block and the predominantly English-speaking remainder of the white population culminated in the victory of the Afrikaner Nationalist Party in 1948. The party had come to power with a twofold manifesto: to entrench Afrikaner control of the white-dominated state, and to segregate the non-white population geographically, socially, and economically. The latter goals were to be achieved by a policy of racial separation and subordination that became locally and internationally notorious as the *apartheid* system.

The second political trend was continued resistance, both violent and passive, by the non-white population to their exclusion from political and economic participation and success through the policy of *apartheid*. The policy evoked bitter opposition from the non-white population, which was able by the 1980s to count on the sympathetic attention of the leading Western powers. A combination of grassroots revolt, international pressure and the reconceptualization of Afrikaner self-interest in the dawning post-Cold War period led to the Nationalist Party leadership's decision in February 1990 to seek a negotiated compromise with the leading forces in the black population. Here the dominant organization was the African National Congress, led by the moral authority of Nelson Mandela, finally released from nearly three decades of imprisonment in that year.

Recent South African political history is now conventionally divided into two periods. First, there is the period of liberalization following President F. W. de Klerk's speech of February 2 1990, which announced, inter alia, the tolerance of major extra-parliamentary organizations, most notably the African National Congress. This period, in which the white-elected National Party government remained in power, lasted until May 10, 1994. Second, there is the period of democratization, beginning in May 1994, when the African National Congress-dominated Government of National Unity under the new President Nelson Mandela was inaugurated following the general election of April 1994. Since the transformation of policing in South Africa has spanned both periods, the following discussion will necessarily make reference to each of them. The main emphasis however will be placed on the second period in which a new regime based on majority rule and intent on the consolidation of a new democratic order has been in power.

THE CONCEPT OF DEMOCRATIC POLICING

Reforming the South African police to reflect the emerging democratic order has been a key priority during both the liberalization and democratization periods. Under F.W. de Klerk (with qualification, due to the activities of an alleged "Third Force" intent on fomenting political violence, aided by elements of the South African Police), the most significant step was the curtailment of police powers by the shelving of the country's draconian security legislation. The parties to these ongoing political negotiations were agreed upon the ideals of a professional and constitutionally constrained police. The second element of democratic policing was the adoption of the rhetoric of community policing—a transparent and not altogether successful attempt to bestow popular legitimacy on a highly unpopular body. The third element comprises the ongoing process to deracialize the police personnel. The South African police consisted of a roughly fifty-fifty division between black and white personnel for several decades before the 1990s. But whites continued to dominate the senior ranks and their numbers failed to conform to the consensus that the public institutions of the new state should move toward reflecting the overall demographic composition of the total population.

ORGANIZATIONAL CHALLENGES

Democratization of the Public Police

To superimpose a system of constitutional rule on the foundations of colonial order required a radical reconceptualization of governance itself. Since April 1994, the Government of National Unity has undertaken a major policy review of the state machinery as a whole. As a consequence each Ministry of State has become the object of organizational transformation. In this situation of major system overhaul, the reform of the police institution is one part of a much broader program of restructuring governance. In turn, democratization of the public police constitutes one phase in the transformation of policing more broadly. Coming to grips with the democratization of policing requires more than just a focus on the remaking of the South African Police. In the context of a deeply divided polity, non-state forms of policing had long assumed an importance which gave social ordering in South Africa a distinctive cast. Any discussion of the democratization of policing in the country must also pay attention to this fact.

The debate on democratization of the police in South Africa has been a heated one, understandably so given the repressive legacy of an institution responsible for the maintenance of *apartheid* rule. Although the issue of reform of the South African Police was placed on the agenda in the early period of liberalization, a concerted program of action awaited the formation of a restyled Ministry of Safety and Security. The policy parameters, which were intended to guide the transformation of the police, were derived from three principal sources: the *Interim Constitution* of 1993; the *Reconstruction and development program* (1994), which served as the ANC's election manifesto; and various administrative guidelines for transformation of the civil service, such as the *Public service act of 1994* and the *White paper on the transformation of the public service* (November 1995; and updated September 18, 1997). The rhetoric of these documents indicated that the new government expected a radical shift in political mission, cultural orientation and operational practices of the public police. Like other state departments, the police were required to develop a new political mission, set service standards, define outputs and create performance indicators as well as monitoring and evaluation mechanisms. It was also expected that the police would develop "partnerships" with the private sector and with NGOs and CBOs. By 1995 official acceptance of the notion of community participation in practices of social ordering weaved its way through debates on policing (as it did in justice and correctional services).

After April 1994, the first of a series of interlocking tasks faced the new political elite of the renamed police administration. This was the rationalization and amalgamation of the eleven officially separate police administrations created by virtue of the *apartheid* program. The unification of the former homeland and national forces into one renamed central structure (the South African Police Service) consisting of some 145,000 members was a time-consuming effort. Beyond the technical processes of rationalization and amalgamation lay "the more fundamental priority of transforming the Police Service to improve the quality and accessibility thereof to all the people of South Africa" (SAPS, *Status Report: Transformation of the South African Police Service*, 1996, p. 11).

Energies had to be focused along a wide and troublesome front. Police policy-makers pursued a six-fold approach to the transformation of a notably recalcitrant organization. On the institution building front, the decentralization of communication structures, the development of information management, and the delegation of powers to provincial, area and local levels has been pursued as a part of institution building. Second, in terms of the goal of transformation of service delivery, progress has been claimed in redistributing policing resources, institutionalizing community policing, restructuring the detective service, developing a crime information management center, and reshaping parts of the public order division and other specialist units such as protection services and border control. Third, in the area of human resource development, new regulations for labor relations have been implemented, social and psychological services for police restructured, basic training revamped and new leadership programs developed. Fourth, with regards to the promotion of a service ethos, demilitarization and the deployment of other strategies to affect a culture change has become official policy. Fifth, the goal of democratization has led to the creation of mechanisms of civilian oversight and accountability. Finally, various initiatives have been devised to embrace equal opportunity principles and to render the organization more representative in both racial and gender terms.

These have been the items on the shopping list for police reform. Their mere declaration, of course, should not be construed as actual achievements. The organizational realignment has had to contend with the most intractable of practical problems: lack of finances, insufficient infrastructure, underdeveloped operational capabilities, pockets of resistance, corruption within police ranks, labor unrest and low morale. These have combined to render the project of institutional reform extremely difficult.

The Discourse of Community Policing

Since the period of liberalization, a common policing discourse has arisen in South Africa in which the concept of community policing has stood central. Prior to 1994, the idea of community policing had been seized upon—some might say hijacked—by the existing leadership of the South African Police as the *leitmotif* of its seriousness regarding reform. A trial run in pre-independence Namibia with changing the public image of the police, then still ultimately under South African control, was no doubt influential. Provisions requiring the adherence to community policing were incorporated into the Interim Constitution agreed to by the major political parties in December 1993.

With the change of government in 1994, the doctrine of community policing continued to provide the broad philosophical base for the intended democratization of the SAP (now renamed the South African Police Service, SAPS) by the new police elite (*Green paper*, 1994; *South African Police Service Act*, 1995). The doctrine has proved particularly instrumental in diffusing the kinds of policing values associated with the emerging international policing regime (Marenin, 1998). As such it has contributed to an increasing shift of local policing rhetoric toward the democratic western norm.

In the evolution of operational guidelines for community policing the SAPS *Policy framework and guidelines on community policing* (1997) deserves particular mention. Written as a manual for the police, the framework provides a definition of the concept, lists the definitive characteristics of community policing, the general principles on which it is based and considers various strategies for implementing it at station level. It defines community policing as

> a philosophy that guides police management styles and operational strategies and emphasizes the establishment of police-community partnerships and a problem-solving approach responsive to the needs of the community (p. 1).

A key institutional mechanism for entrenching community policing and for democratizing policing at the local level has been that of Community Police Forums (CPFs). CPFs are intended to eradicate the historical legacy of popular hostility towards the SAPS by institutionalizing formal liaison structures between the police and the public. As such they constitute the main legal expression of the doctrine of community policing. A great deal of attention has gone into devising operational guidelines in terms of which of these structures can be put to work at the level of police stations. However, the achievements on this front are very uneven. At a recent inter-provincial meeting on community policing, an inventory was drawn up about the performance of community police forums. The picture which emerged from this attempt at national stock-taking is a bleak one. In each and every report from the nine provinces, CPFs were said to contend with problems ranging from lack of trust between police and communities, racial animosity and political rivalry within communities, personal vendettas between forum members and inadequate infrastructure, to a general lack of interest among the public (SAPA *Report*, Policing Training, October 21, 1997). The institutionalizing of community police forums and boards at station, area and provincial levels thus continues to be bedeviled by a range of problems.

At the same time, it is clear that police managers are now more aware of both the potential and the limitations of community policing as a means toward police reform rather than an end in itself. The political romanticism of earlier years has given way to a more pragmatic appreciation that community policing has no single meaning and is not endowed with intrinsic democratic effect. Growing incidences of vigilantism has forced a measure of introspective assessment. Clarification of the parameters within which "active citizenry" needs to be pursued is now demanded. In turn, this has contributed to a realization that the challenges facing the democratization of policing in South Africa reach beyond the level of the state. These challenges require a critical look at non-state forms of policing.

Democratization of Non-state Forms of Policing

The legacy of authoritarian rule extended its reach into the non-state terrain of policing. Living under conditions of siege, the *apartheid* state actively encouraged some forms of non-state policing in the black community, such as the anti-ANC *witdoeke*.[1] Part of the historical baggage, as Seegers (1996, pp. 311–12) has pointed out, is that the South African state's monopoly of force has long been broken. In the context of political reform, the regulation of non-state forms of policing and the inter-relationship between state and non-state actors had to be tackled. Recent discussions provide proof of the perceived urgency to extend the debate on democratization of policing structures beyond a narrow focus on the South African Police Service. A debate on how to deal with the remnants of youth-based paramilitary formations, usually aligned to the ANC, such as Self-Defense Units (SDUs), has raised many creative suggestions but few viable solutions. There have been attempts in some parts of the country to incorporate SDUs into the public police as a type of auxiliary. Others, however, are skeptical of a strategy, which, under the slogan of demilitarization, resorts to essentially military means to solve a social problem. The spread of anti-crime committees in residential areas plagued by high rates of violent crime and gang-related activities has posed new questions concerning the regulation of policing energies at the grass-roots level. Similar issues face the regulation of neighborhood watches, which in some of parts of the Western Cape, for example, have grown to a sprawling urban-based private policing phenomenon. Finally, there is the private security industry, the personnel of which now outnumber those of the SAPS by two to one. All of these spheres may overwhelm the initiatives which the state has tried to harness within Community Police Forums.

At no time has the question concerning effective regulation of policing capacity—whether situated at the level of the state, civil society or the market—been of more critical importance. If the future of policing indeed lies at the dual or plural level (Brogden & Shearing, 1993; Shearing 1996), then the need for democratizing the rules of the game is urgent. Instructive lessons have been learned. When put to the test in 1997, vague and utopian definitions of community involvement in policing held the potential to erode rather than sustain the capacity for effective governance. It has become critical both to define more clearly the terms of the partnership between state and non-state actors and to create the institutional capacity which can enforce adherence to the agreed-upon rules. These are matters which require skillful resolution if strategies in pursuit of safety and security are to serve the spirit of constitutionalism. In their absence the democratization of policing is bound to suffer a premature death.

OPERATIONAL CHALLENGES

The Crime Situation in South Africa

Very high rates of serious crime present one of the most critical challenges to the new democratic regime in South Africa. The incidence of homicide, armed robbery, rape, assault and vehicle hijacking, not to mention widespread property theft, amounts to a crisis of the social order in which democracy is equated with a weak state. In such a situation the state police enjoy little credibility as guarantors of the social fabric. Those who can afford it turn to private security. South Africa is coming to resemble the "crime-wrecked states of Latin America"—if indeed it is not outperforming them on this score already (Shaw, 1996). Of course, South Africa has always been a violent society in comparative terms—its legacy of colonialism, industrialization and urbanization under conditions of coercive inequality saw to that. On the one side, a colonial-type state criminalized the very movement of people, their consumption of liquor and their resistance to oppressive conditions. On the other, a subjugated population often viewed criminal activity against the dominant section of the population as legitimate, or in Fanonesque fashion vented its frustration within its own ranks in violent ways. The turn to militant resistance from the 1960s onward merely added a new dimension of "political violence" to the crime-ridden strife of South African society. Such a perspective is altogether indispensable in assessing the current crisis.

The inadequacy of official crime statistics makes an accurate assessment of the problem impossible. But certain trends are relatively clear. Existing high levels of serious crime began to escalate from the mid-1980s onward. The picture is complicated by the intensification of political clashes from that time which were also recorded in the crime statistics. A further escalation in serious crimes took place from 1990 onward. "Recorded levels of almost all crime increased for the period 1990 to 1994. Most crimes increased phenomenally during this period: assault increased by 18 per cent, rape by 42 per cent, robbery by 40 per cent, vehicle theft by 34 per cent and burglary by 20 per cent" (Louw & Shaw, 1997, p. 4) In 1994, when political violence had greatly subsided, a new high of some 26,000 cases of murder were reported, a proportional incidence about four times greater than that in the United States, for example.

The latest statistics from the Crime Information Management Centre (CIMC) of the SAPS suggest that this veritable crime wave may have peaked by 1996. The CIMC issues quarterly reports on the incidence of the twenty most serious crimes. The most recent report (CIMC, *Quarterly Report*, 3, September 1997) claims that eleven of these crime categories have shown a decreasing incidence in 1997, and all nine of the rest have stabilized at previous levels. These categories cover some 74% of all reported crime. If largely correct about the broad trends, these figures provide at least a glimmer of optimism in a generally bleak picture.

As is clear from all three CIMC Quarterly Reports for 1997, the reliability of their crime statistics is very much open to question. Perhaps the major weakness is the "lack of a strong positive crime information management culture at ground level (that is at stations and units)" (*Quarterly Report*, 1, 1997, p. 4). In other words, it is debatable that the primary data gatherers and inputters—the police on the ground—do their job at all adequately. (Research on this issue seems urgently required.) The inputted data are still very crude. The CIMC headquarters in Pretoria is struggling to finalize and then, more formidably, gain grass-roots compliance with a criminological crime code which will begin to describe the nature of each reported offense in proper detail for central analysis (*Quarterly Report*, 3, 1997, p. 7).

Presentation of data is further debilitated by the nature of census figures, which do not make synchronization of police jurisdictions and census tracts possible. Thus, at present, data are only broken down to the provincial level. A further complication arises from the reliability of census figures themselves widely agreed to be dubious at best. The latest census conducted in 1996 came up with a total population of some thirty-eight million people, much below the previous "guesstimates" of around forty-one million. The statistical result is that working with the latest preliminary census results to calculate the incidence of crime increases crime rates. The figures from the stations in the former homeland system are also agreed to be of questionable value. Finally, underreporting of crime is considered to be quite prevalent, either from skepticism as to police effectiveness, remaining hostility to the institution or for more South African reasons, such as unwillingness to report murders committed during witch hunts (*Quarterly Report*, 2, 1997, p. 8).

The CIMC places its faith on the overall trends which these admittedly unreliable data appear to demonstrate, while conceding that the totals themselves may be quite wide of the mark. Under the circumstances, it may be wise to admit that we simply do not know the real proportions of crime in the country, apart from the familiar fact that they are extremely high. Research by the CIMC and others is, however, of use in underlining what is to be expected: that the crime situation is variable over the country; that Gauteng province (which includes Johannesburg and its overcrowded urban sprawl) suffers disproportionately from serious crime; and that the picture painted in the media which stresses so-called high-profile crimes (against the rich, white or prominent) fails to convey the full facts.

A recent report by Louw and Shaw (1997) is valuable in disaggregating some of the overall crime statistics and testing the most plausible explanations. The point which these researchers are at pains to stress is that it is the country's poor "who arguably suffer the brunt of high crime levels" (p. 1). While "the wealthy are victims of property crime … the poor are the victims of violent crime as well as being victims of property crime" (p. 6). Thus African males, who make up the bulk of the nation's poorer population, are twenty times more likely to be victims of homicide than whites. The consequences of this crime for the poor are likely to be far more devastating, depriving families of wage earners and overburdening household resources. Black women are at extreme risk of rape. Louw and Shaw's report also looks at one of the CIMC's own counter-intuitive findings, namely that the homicide and assault rate in the low-density and rural Northern Cape Province is the highest in the country. The report's findings, based on qualitative research in the Kimberley area, pinpoints the strong link between patterns of alcohol use and violent crime, in the process demonstrating that most homicides are "assaults that went too far" between acquaintances at weekends. In an equivalent study of the Eastern Cape Province (including the former Transkei), Louw and Shaw sketch the variability inherent in South African crime patterns. Here and elsewhere, where petty stock theft is rife and police often altogether absent, distinctive policing strategies are indicated.

The work by these researchers is a useful corrective to media depictions of the current crime wave. Nevertheless, their findings do not contradict what one would expect from even a cursory knowledge of the South African scene. Social dislocation, substance abuse and poverty form a tangle of pathology which will be extremely difficult to eradicate. In parts of the nation, this particular syndrome is tied up with the activities of gangs, especially in the colored communities of the Western Cape and elsewhere, although it is by no means exclusively restricted to them (Pinnock & Douglas-Hamilton, 1997). The gang issue is a longstanding one in South Africa, which has recently been exacerbated by the activities of private groupings such as People Against Gangsters and Drugs which have spiraled into vigilante organizations, themselves ready to resort to violence.

To the activities of long-established gangs and new vigilante groupings must be added the category of "organized crime" to which the SAPS today devotes considerable resources. The illegal immigrant issue referred to elsewhere in this discussion is heavily influenced by current perceptions that international "crime syndicates" are now for the first time deeply involved inside South Africa. The CIMC devotes a considerable part of one of its reports to the issue (*Quarterly Report*, 2, 1997). Russian, Chinese and Nigerian networks are singled out for particular attention, being held responsible for a major share in drug, firearm, gold, diamond and endangered species smuggling, as well as sophisticated fraud techniques. A specialist arm of the SAPS, the Organized Crime Unit, has recently been formed to counter this new trend, working in conjunction with existing specialized units concerned with gold, diamond and endangered species smuggling.

The current crime situation in South Africa thus presents complex domestic and international aspects with which an understaffed and underskilled police service must struggle to cope. Government itself is particularly sensitive to media reports about the so-called high profile crimes, as President Nelson Mandela's *Political report to the 50th national conference of the African National Congress* in December 1997 made eminently clear. The spate of killings of white farmers in the last months of that year comprised only a handful of the thousands of brutal murders in South Africa, but these are the ones which capture an international audience, with grave consequences for foreign investment, tourism and general international confidence in the South African regime. (See for example the British *Sunday Times*, December 21, 1997, on this issue.) The South African media are hardly alone in their highlighting of crimes against the more privileged white sector of the population (for example, see also the *New York Times*, September 9, 1997, "Where even the police get robbed"). Domestically, the high rates of crime are driving up the emigration of the more skilled, swelling the ranks of the private security industry and intensifying the social distance between white and black, rich and poor as the former seek shelter in the enclave suburbs sprouting in the major cities (Shaw, 1996). The country's crime problem, which puts the fashionable social constructionist debates about crime-related moral panics in Western democracies (O'Malley, 1996) in a rather radical chic light, remains a matter of life and death for both the population and the South African democracy.

Migration

Patterns of migration to South Africa have reflected its racialized past. With some halts due to Afrikaner nationalist opposition, successive governments encouraged the permanent immigration of northern Europeans. On the other hand, hundreds of thousands of black workers, mainly from Mozambique and Lesotho, entered the country each year to work on the mines and farms. Such workers were viewed as strictly temporary and were compelled to return home on expiration of their contracts. However, due to the attraction of an industrializing economy, much illegal migration from neighboring states into South Africa took place. This migrant stream has been accentuated since the 1980s, with tens of thousands of Mozambicans fleeing a war-torn country into the eastern Transvaal. At present, South Africa thus shelters a large population of undocumented immigrants-cum-refugees from African states. With the easing of a racist entry control after 1990, thousands of newcomers from Asian and other nations have also settled. Always a recipient of considerable numbers of immigrants of diverse origin (including indentured laborers from the Indian subcontinent in the second half of the nineteenth century whose descendants now number nearly a million people), South Africa now faces a more complex and varied immigration problem. The consequences for policing are direct and difficult.

Up to 1985 the South African Police were the instrument of the white-dominated state's attempts to control the movement of its own African population in accordance with the policies of segregation and *apartheid*. A sort of "internal immigration" control function of great coercive power was thus allotted to the police, causing widespread hostility toward them. Control of external immigrants went hand in hand with this function, and thousands of non-South African migrants were deported each year to nearby countries. This latter situation continues. The new democratic government faces considerable popular pressure to act against undocumented immigrants who are believed to compete for inadequate work opportunities and to contribute to the high rate of criminality.

In response to the new situation, immigration policy is under review. Throughout the twentieth century, the South African state tightened up immigration law in order to control the entry and facilitate the removal of immigrants viewed as undesirable. Since 1937 the cornerstone of this policy has been the *Aliens Control Act*, which has undergone successive amendments up to 1995 (Peberdy, 1996). Sweeping powers were awarded to officials to remove immigrants without correct documentation and without recourse to the courts. The general trend of this policy was directed against persons of non-European origin, both African and Asian, and in earlier years against would-be Jewish immigrants.

In 1986, as the South African state extended its reluctant reform of racially discriminatory policies, the overtly racist clause of the *Aliens Control Act* was removed (Peberdy, 1996). The coercive, non-rights based character of immigration law remained, however, and has become the focus of recent policy review. Academic work commissioned as part of the policy reform process has almost universally condemned the current legal position regarding undocumented or illegal immigrants (Friedman, 1996; Reitzes, 1997). A Green Paper drafted as the basis for policy change has called for a new approach based on the acceptance of a "rights-based enforcement" policy of immigration control, and the creation of new administrative structures to implement such a policy (*Draft green paper on international migration*, 1997). But the liberalization of immigration law and practice has yet to gain the agreement of the politicians whose main concern appears to be the growing grassroots opposition to the presence of the so-called illegals.

Thus, both existing law and popular pressure demand that the South African Police Service act against undocumented migrants. Working in concert with the South African National Defence Force, the SAPS conducts regular sweeps in the major urban areas which result in the arrest of hundreds of people—the majority of whom turn out to be "illegal immigrants" who are detained and deported without recourse to judicial review. Many such migrants from neighboring states promptly return to South Africa in search of work, which they cannot find in their own societies. The consequence is that everyday police practice has come to be characterized by considerable abuse. Undocumented migrants allege that they cannot expect police protection and that they are the objects of police corruption and of criminal activity by South Africans who view them as soft targets. Although the popular perception is that the illegals are disproportionately involved in the current crime wave, research has not borne out this contention.

South Africa's long, relatively open borders, and its many uncontrolled small airfields make effective immigration control practically impossible. The actual extent of the undocumented migrant population is unknown and probably unknowable, although the figures of up to eight million which have been bandied about are probably extreme over-estimates (Reitzes, 1997). For the South African Police Service, the situation is complicated by the upsurge in drug-trafficking and arms-smuggling. These activities are by their nature often cross-border matters, demanding some form of police monitoring and control. The SAPS has responded by establishing a new border control and policing unit of 1,600 officers (Ministry for Social Welfare, *Framework for a national drug master plan*, 1997). The immigration issue, as researchers have pointed out, has become conflated with criminality, as international "crime syndicates" are viewed as partially responsible for the proliferation of drugs and arms, in addition to the popular perception of the immigrant role in the rise in street crime (Friedman, 1996; Reitzes, 1997). The large-scale presence of undocumented immigrants and a creeping tide of grass-roots xenophobia thus add to the intractability of the problems facing the South African Police Service.

RESPONSES TO CHALLENGES

Legislation

The overall political structures and accumulating security legislation bestowed sweeping powers of search, arrest and detention on the South African Police under the old order. Especially notable in this respect were the *Suppression of Communism Act* of 1950, the *General Laws Amendment Act* of 1962, the *Internal Security Act* of 1976 and the *Criminal Procedures Act* of 1977 (Dugard, 1978). By means of these laws, the police were able to take action against political opponents of the regime in an almost untrammeled fashion and to use their firearms with impunity. After 1962, the police were empowered to detain persons incommunicado for long periods during interrogation. Taken together with other racially motivated and coercive legislation, such as the various pass laws, these legal measures gained South Africa a well-deserved notoriety as a police state (Brewer, 1994; van der Spuy, 1989). Physical abuse, intimidation and torture by police officials became routine. Recent evidence before the Truth and Reconciliation Commission has only served to underline what were well-known facts in the lives of South Africans prior to 1990.

Events since 1990 have radically changed this situation, on paper at least. During the period of negotiations between 1990 and 1994 a number of agreements put the existing draconian provisions of the security legislation largely in abeyance. The adoption of an *Interim Constitution* in 1993, a new *South African Police Service Act* in 1995 and the final Constitution in 1996 created a legal order under which the police are restrained by very progressive-sounding norms of accountability and legality. The inclusion of the Bill of Rights in the 1996 Constitution is intended to lay the foundations of a *Rechtstaat* influenced by the most up-to-date constitutional thinking. Section 35 of the Bill of Rights entrenches the rights of accused persons to access to courts within 48 hours, the right to remain silent, not to be compelled to make incriminating confessions and various other provisions which bring South Africa into line with liberal constitutionalism—and in the process remove from the police some of the most potent weapons they had been accustomed to exercise. Other provisions of the Bill of Rights with direct relevance for previous police practice include Section 12, "Freedom and security of the person" and Section 14, "Privacy." In terms of fundamental legal rights, then, the wheel, on the face of it, has turned full circle from the old South Africa.

The *South African Police Service Act*, No. 68 of 1995, replacing the original Act of 1912, contains several elements which anticipated the final constitutional order. First, this Act mandates the responsible Minister to establish a "Secretariat for Safety and Security." It allows, but does not oblige, provincial police ministers to do the same. The principle of civilian oversight of the police services is introduced by means of this body. As laid down in Section 3 (1) of the Act, the new Secretariat has wide-ranging functions of policy formulation, review, research and evaluation to aid the Minister. The Secretariat is also to "promote democratic accountability and transparency in the [Police] Service" Section 3 (2)(c). The Secretariat is now functioning under the direction of Azur Cachalia, formerly a major figure in the ranks of the ANC and well versed in police matters. Positioned between the Minister and the operational head of the SAPS, the Secretariat is necessarily the Minister's creature. The potential for friction between the new body and the serving heads of the SAPS, which seems eminently possible, has by all accounts indeed been a feature of the new police regime.

Second, the Act of 1995 takes over from the *Interim Constitution* of 1993 the requirement that Community Police Forums (CPFs) and Boards be established at station, district and provincial levels (Section 7). The Act stipulates that CPFs shall be "broadly representative of the local community," but lays down no precise procedures for their composition (Section 19 (1)). Designed to ensure transparency, accountability, communication, police-community cooperation, partnership, service improvement and joint problem solving (Section 18 (1)(a)(b)(c)(d)(e)(f)), CPFs in practice have run into many problems. As the Member of the Executive Committee for policing the province of Gauteng noted recently, "the central core function of the CPFs was to monitor police, but we now have cases where they control the police stations" (SAPA *Report*, December 10, 1997). In contrast, there have been complaints of local station commanders dominating their CPFs or of a general local indifference to involvement.

Finally, the 1995 *Police Service Act* provides for the establishment of an additional civilian oversight body in the form of an Independent Complaints Directorate (Sections 50–54). A long history of police failure to investigate public complaints against its members lies behind this new provision. The Directorate is to "function independently from the Service" (Section 50 (2)) and, according to the 1996 Constitution, must "investigate any alleged misconduct of, or offense committed by, any member of the police service in the province" concerned (*Constitution of the Republic of South Africa*, 1996, Section 206 (6)). All deaths in police custody must be reported to the Complaints Directorate. If it so decides, it may refer cases to the relevant Attorney-General for possible prosecution. It is as yet too early to assess the workings of the Directorate, but with 700 policemen awaiting trial in the province of Gauteng alone in December of 1997 for offenses ranging from rape to armed robbery, its task may prove onerous indeed, if it can cope at all with its statutory mandate.

The provisions with regard to civilian oversight and accountability aside, the 1995 Act is notable for its confirmation of the long-standing centralized and national character of the South African Police. In just about every significant respect, the head of the service, the National Commissioner, has overriding powers for the way in which policing is to be structured and priorities assigned. The only obligation regarding the structure of the SAPS which the Act lays upon the National Commissioner is to establish a National Public Order Policing Unit, but this in fact has long existed in various guises. Otherwise the National Commissioner has the final authority to decide upon the size and distribution of the service, its division into various functional arms, the training establishment scattered among various police colleges, and the appointment of provincial police commissioners.

Proponents of a truly decentralized police system adaptable to varying local needs and shielded against central political interference thus met with failure in the policy debates which preceded the finalization of legislation (Shaw, 1996). The Act allows for the creation of municipal and metropolitan police services (one such has long existed in Durban) but they are to be subordinate to the SAPS itself (Section 64).

The impact of the new constitutional order and the 1995 Act on actual police operations is difficult to assess. Incorporated into the latter Act is the stricture that members of the SAPS "may use only the minimum force which is reasonable in the circumstances" (Section 13 (3)(b)). This is a striking departure from twentieth-century police practice in South Africa. But a closer scrutiny of both the 1996 Constitution and the 1995 Act (also as amended in 1997, Act No. 41 of 1997) suggests that discretionary use of coercive police power may lurk in the loopholes. The Constitution and the *State of Emergency Act*, No. 86, of 1995 allow for the use of sweeping powers which would abrogate the very liberal new rights should the President and National Assembly consider that conditions such as "general insurrection or disorder" (Act 86, 1995, Preamble) warrant it. Of more import for everyday police operations, the *Police Service Act* of 1995 provides for quite extensive powers of search and arrest which *prima facie* seem incompatible with the broad thrust of the Bill of Rights. Acting under written authorization of the Provincial Commissioner, who may delegate these powers right down the ranks, police may "without warrant search any person, premises or vehicle ..." in pursuit of specified objectives such as suspected trafficking in illegal immigrants, during roadblocks and at other times (Section 13 (6)(7)(8) of *South Africa Police Service Act*, 1995). These powers are now in quite wide use at a time when most of the public seems more concerned about the high crime rates than their constitutional rights. But whether they would survive the judicial review of the Constitutional Court is an interesting question which has yet to be answered.

In sum, new legislation which includes the basic law of the land now enjoins upon the SAPS high standards of constitutionalism and respect for the citizenry, which differ sharply from the previous era. It does so, however, within the framework of a centralized police service whose provincial arms seem little more than administrative agents of the national headquarters. The details of the most relevant laws suggest that the police may act more decisively and with less constraint than the large print of the Constitution declares. Public opinion for the moment is more likely to back a law-and-order police crackdown than insist upon strict adherence to constitutional injunctions. In this sense the new legal order is perhaps more of a pliable mold than the straightjacket it appears to be at first sight.

International Cooperation

The opportunities for cooperation and exchange between the South African Police Service and other police agencies have increased dramatically in recent years. This situation stands in marked contrast to that prevailing under *apartheid* when contact with the international policing brotherhood, at least at the official level, was effectively curtailed. In the 1980s, some attempt was made at cultivating alliances ostensibly for purposes of "fighting a mutual enemy, Communism" with intelligence agencies in Argentina and Chile (Dienaar, 1988, p. 621). Such cooperative ventures however, were the exception. During the long era of racial rule the South African Police, as the coercive arm of a pariah state, remained subject to political isolation.

Under conditions of political liberalization and democratic transition, contacts between the police in South Africa and other police agencies have become a regular feature. A major symbol of the end of isolation for the South African Police was its readmittance to Interpol in 1993. The SAPS now has its own Interpol office in Pretoria, which is already handling a high volume of work under serious resource constraints with regard to personnel and technology (SAPS, *Scenario 1997–2001*, Annexure A).

Aside from its operational insertion into a global policing community via institutions such as Interpol, the transformation of the South African Police's international status stemming from democratization has occurred in two major ways: assistance from overseas sources eager to improve police capacity in the country and regional cooperation with the police agencies of southern Africa. In this respect the SAPS displays a paradoxical character as a recipient of aid internationally on the one hand while taking the lead in policing regionally on the other. It is clear that for the main Western nations concerned, the cooperative networks currently being cemented offer a crucial mechanism through which organizational reform of the police both inside South Africa and in neighboring states may be brought about.

International Assistance

Large-scale assistance from the democratic world (predominantly Western Europe and a sprinkling from other Commonwealth nations) has become a central component of international efforts to promote and sustain "good governance" in South Africa and other countries experiencing political and economic changes (Marenin, 1998). Originally driven by a need to transform an *apartheid* institution into a legitimate organ of state social control, Western efforts have more recently been sustained by the emergence of crime as one of the most central challenges to future stability in South Africa. The scale of international assistance has increased significantly after 1994. Various bilateral and multilateral agreements exist which bind institutions involved in policing to a large number of donor countries and intergovernmental institutions. An inventory of donor-assisted projects for 1997 indicates that the areas earmarked for donor assistance traverse a very wide terrain. There are currently 53 projects being conducted in the following areas: public order policing, community policing, protection services, high risk services, detective services, equity, training of trainers, management training, station commissioners training, management capacity building, corruption, human rights, serious violent crimes and assistance to members of the Executive Councils at the provincial level.[2]

Foreign assistance in the area of police training deserves particular mention. Since 1994 these training initiatives have come to embrace different levels of the police organization, *inter alia*, top management, middle-ranking officers, specialist divisions, trainers, as well as foot soldiers. Valuable lessons seemed to have been learned concerning the conditions under which optimal utilization of foreign training assistance can take place. An ambitious project by a Multi-National Implementation Team in 1995 and 1996 has led to the restructuring of basic police training and the development of a field training program for recruits (*Training Evaluation Group*, 1996).

Donor approaches have varied from direct engagement with the SAPS to support for NGOs operating in the policing field. By 1997 the *ad hoc* approach to police assistance characteristic of earlier years was giving way to more administrative formalism. Bureaucratic rules guiding the process of tendering, administering and monitoring of donor-assisted projects have been formally promulgated. There is also a trend toward favoring multilateral agreements which bind the recipient institution into contractual obligations with a collectivity of donors. Furthermore, the current preference is for assisting larger, more consolidated programs which embrace the transformation vision for the criminal justice system as inscribed in the

National Crime Prevention Strategy (1996). As a consequence, donor assistance regarding policing is now operationalized as part of a broader program towards "cooperative governance" in the criminal justice system as a whole.

Donor assistance to the SAPS has contributed to an increasing shift, particularly at the rhetorical level, of local policing ideas toward the internationally dominant Western norm. This drift toward universalism in policing discourse has been augmented by the wide support that the operational philosophy of community policing enjoys and the ease with which the managerialist ethos has been incorporated into local policing debates. The internationalization of South African policing cultures has thus become a given. The years to come are bound to witness an increasing intermeshing of local and global police developments and of further expansion of the opportunities for coalition building between local policing elites and their international counterparts.

However, it cannot be assumed that the process of internationalization is a simple or internally consistent transfer of good things. Theories and practices imported from abroad may not prove effective when they are put to the test under vastly different policing conditions. The benefits of donor assistance are rendered more uncertain by the problems pertaining to local conditions, ineffective management of donor assisted projects and by inter-institutional rivalry unleashed in a competitive struggle for donor resources within the police institution. Within the larger picture of policing in South Africa, international aid may in the long run make scarcely more than a dent in the endemic problems of police management.

Regional Police Cooperation

The South African Police now stand at the center of a widening regional web of cooperation in southern and central Africa—again a marked change from the days when South African police agents were hostile elements compared to the rest of Africa. Much of the current impetus for cooperation in the region comes from newly permeable borders and the concomitant accessibility they offer to organized crime. The present thrust toward collegial interaction in pursuit of common regional goals has been given institutional form by the establishment of transnational policing structures, such as the South African Regional Police Chiefs Co-ordinating Organization (SARPCCO).

Formed in August 1995, SARPCCO draws representation at the level of police chiefs from eleven countries in the region. The central objective of SARPCCO is to "promote, strengthen and perpetuate cooperation and foster joint strategies for the management of all forms of cross-border and related crimes with regional implications" (*SARPCCO, Constitution*, Article 3.1). In pursuit of this objective there is a commitment to an exchange of information, joint management of criminal records and the creation of structures in support of combined operations on cross-border crimes. The latter include vehicle theft, illicit firearms and drug trafficking, illegal immigrants, corruption within law enforcement agencies, fraudulent travel documents and armed robberies. However, despite good intentions, the consolidation of policing energies in the region is bound to encounter a range of obstacles.

The Permanent Co-ordinating Committee of Heads of Investigation Services—a key structure within SARPCCO—has been propagating the need for joint operations in order both to "build trust and confidence between our operational officers" and to galvanize policing resources around "priority crimes" (SARPCCO, 1997, p. 3.) This move toward joint operational deployment around priority crimes at the regional level dovetails with the joint deployment of police and military units in "high density operations" in "hot spots" at the domestic level. By late 1997, joint operations between law enforcement agencies inside the country and in the Southern Africa region seem destined to become a regular feature of the policing landscape.

However, the realization of this collaborative capacity remains dependent on a number of factors. Current obstacles include court delays in disposing of cases, non-harmonized laws, lack of extradition treaties and inadequate manpower. Thus, while there is considerable commitment among key players in the region to enhanced cooperation and the exchange of intelligence and standardized operational procedures, the practicalities involved are likely to be difficult to resolve.

Regional policing structures such as SARPCCO are increasingly viewed as critical conduits for institutionalizing democratic policing at a time when human rights abuses on the part of police and security forces remain a common feature of security conduct in the region. Bodies such as Amnesty International (1997) now look toward regional structures as an avenue for institutionalizing a culture of respect for human rights in post-colonial Africa. Both the global and regional policing scenes provide new opportunities to a number of players for setting an agenda for policing reform which is compatible with the ideal of professional and democratic policing. Clearly the question as to who sets the agenda and who steers the process in aid of global and regional police cooperation is an important one. By all accounts, it is the South Africa Police which seem poised to assume a dominant role in the region. Its own vision of the future of policing is being shaped in the context of international police aid in which the influence of western governments and police agencies looms large, and it is this vision which it is expected in turn to pass on to its partners in the region. It remains to be seen whether the emerging regional approach to crime and security can combine the specific interests of the West with the particular needs and realities of governance of criminal justice in Southern Africa.

Professionalism

Police professionalism and police legitimacy, although conceptually distinct, are intimately connected in practice. Weitzer (1995, p. 83) defines police legitimacy as the "acceptance of the moral authority of a police force and its right to enforce laws and issue commands," and professionalism as "organizational autonomy from external forces and fidelity to rational-legal norms such as impartiality and political neutrality." In the context of radical reform, advances at the level of police professionalism remain dependent on the capacity to create the structural conditions within which police legitimacy can grow. This conflation makes it difficult, in the South African case, to separate out issues of police professionalism from the more politically entangled debates on police legitimacy.

The debate on police professionalism in the new South Africa has evolved through fits and starts. With the inception of the period of liberalization from 1990, official quarters began to invoke the concept of an "apolitical police professionalism" (Minister of Law and Order, *Hansard*, April 1990, col. 6971–6972). But the incongruence between political statements and policing realities on the ground gave an almost surrealist twist to official claims concerning impartial policing.

Nevertheless, the kind of anticipatory reform characteristic of the 1990–1994 period in South Africa led to some crucial organizational changes which promoted the notion of an apolitical professional police service. The most important of these involved the abolition of a powerful and notorious political police agency, the Security Branch. Its personnel were transferred to the National Criminal Investigation Division in April 1991 (*Hansard*, April 1991, col. 7542). Further, the restructuring of public order policing was undertaken through the formation of the

Internal Stability Division in March 1992, to replace the likewise notorious "riot squads" of the past (*Survey of Race Relations Annual Report*, 1992/3, p. 127). The creation of a Community Relations Division in December 1992 also reflected the growing importance being attached to community-police consultation and accountability.

A concerted program of action to steer the project toward legitimacy and professionalism had, however, to await the takeover by the new political elite. After April 1994, both the Bill of Rights in the Interim Constitution of 1993 and the doctrine of community policing provided the overarching framework within which the political values and operational rules of professional policing came to be defined. The effect on policing discourse was evident in the Draft Policy Document issued by the new Minister of Safety and Security, Sidney Mafumadi, in 1994. Noting the need for a "fundamental reassessment of the nature of policing in South Africa," the document included in a section on "A Professional Police Service" the comment that "the Police Service should offer a respectable, professional career … . In order to achieve true professionalism, the police profession must be demilitarised. The system of discipline should be informed by professional, rather than military values" (*Draft Policy Document*, 1994, p. 18).

Mirroring experiences in other parts of the world, training and education in South Africa were regarded as crucial vehicles through which police professionalism at a more general level could be advanced. Subsequent to the rationalization and amalgamation of the diverse number of police administrations inherited from the *apartheid* order into a unitary system, uniform standards were applied throughout the police institution. Basic police training in particular came to be viewed as an important area for inculcating new apolitical orientations congruent with democratic notions of police professionalism. Dismantling the paramilitary edifice of the organization at the entry gates was viewed as a symbolically important step towards demilitarizing the institution. In January 1995, the foreign-aided Basic Training Pilot Program commenced with 1,760 students. The explicit aim of the program was to produce "community police officers" equipped to police a democratic polity with due regard for human rights. The Basic Training Pilot Program came to be informed—if often in piecemeal and uneven fashion—by the broader philosophy of community policing, the ethics of human rights and the managerial and operational principles associated with police professionalism.

Despite various constraints, the pilot program succeeded in bringing the content of basic police training in line with the internationally accepted practice of probationary training. Under guidance of the Multi-National Implementation Team (MIT), the pilot project had some measure of success in imparting a new concept of police professionalism. But more importantly, the Basic Training Pilot Program confirmed that the search for police professionalism would require large-scale investment on a broad front. Only in doing so could one begin to affect the cultural orientation of an institution shaped by racially repressive politics. The prospects for such a major investment of resources, which ideally would include a massive intake of new recruits undergoing the new-style basic training as well as extensive retraining of existing personnel, are, however, slim. After the first batch of new recruits in 1995, financial constraints led to a moratorium on new recruiting for the SAPS. Induction of only the second group of new recruits occurred during 1998.

Within the past two years, a certain ideological trade-off between a professional, efficient police versus a democratic, community-oriented one has been surfacing. In both the discussions which led to the issuing of the *National Crime Prevention Strategy* in 1996 and current debates on a new *Policing White Paper* (1997), a de-emphasis of the notion of community policing occurred. In its place, there is a shift toward the development of performance-based indicators to measure police effectiveness and efficiency. In the emerging policing discourse, the emphasis is on making "war on crime." The *National Crime Prevention Strategy* (1996) sets out the need for cooperative governance in crime prevention around a number of strategic priorities. This drift toward prioritization of crime problems enjoys "national priority status" and is further operationalized in key policing documents such as the latest SAPS *Policing Priorities for 1997/8* (1997).

Also symptomatic of this more technocratic bent within the SAPS was the appointment in August 1997 of Meyer Kahn, the head of one of South Africa's largest business conglomerates, as a seconded two-year chief executive officer of the SAPS. Kahn's appointment had been solicited personally by Deputy-President Thabo Mbeki. Given the rigid and constitutionally defined command structure of the SAPS, Meyer's position is anomalous to say the least. It signalled that the country's leadership lacked confidence in the abilities of George Fivaz, the SAPS National Commissioner, to shake up the organization. Fivaz, a white Afrikaans-speaker whose lack of past involvement in security matters furnished him with a suitably clean image for the new era, was also a sop to the white-dominated force. The SAPS is now under the control of a black Communist (the Minister), an Indian radical (the Director of the Secretariat), an Afrikaner (the National Commissioner), and a reputably foul-mouthed businessman. The combination makes sense in South African terms, and Meyer's reputation as a hard-nosed no-nonsense type indicates the seriousness with which the new governing class is approaching the overhaul of the country's police.

A further consequence of deepening technocratization is a steady process of organizational specialization within the police as well as in other components of the justice system. New specialist divisions have been created on a wide front to fight priority crimes in the terrain of drugs, illegal weapons, economic crimes, internal corruption and sexual violence. This thrust towards specialization is bound to affect prevailing notions of the police professional. At one level it seems destined to detract from the social prestige which in the early days of community policing was attached to a generalist definition of the term. Specialist divisions may tend to attract resources at the expense of the street-level uniformed police. Policing managers now seem inclined to project the creation of "pockets of excellence" as a strategic response to the needs of the time.

An important spin-off of this growth in specialization has been an appreciation of the crucial importance of police intelligence. Enhancing the intelligence capacity of the SAPS in turn requires investment in both infrastructural and other (human) resources. Technical assistance from abroad is increasingly being directed towards capacity building of specialist divisions equipped with high technology. The modernization of technology and the creation of operational capacity are also considered key components toward building a professional police institution adept at operating in a global environment. In this respect, the private sector organization Business Against Crime (BAC) has become a key player in technology transfers to the SAPS.

Mirroring the changes in criminal justice discourse elsewhere in the western world, policing debates in South Africa have also become more concerned with the policing needs of certain categories of victims—of child abuse, domestic violence and violent sexual crime in particular. This emerging victim-centered approach to policing (as in the criminal justice system as a whole) has given new meaning to the concept of police professionalism. "Sensitivity training" around the particular policing needs of such groups is now a popular catchword. In recent years, non-governmental organizations active in the field have become closely involved in lobbying for the kinds of changes required to develop professional expertise in the terrain of victim-centered policing. There are moves afoot to train police and other justice service personnel in victim empowerment support. Official doctrine has now incorporated the concept of a generic victim-support curriculum (SAPS, *NCPS Victim Empowerment Program*, 1997).

The higher echelons of the SAPS are awash with the very latest ideas of the modern professional police officer in all possible guises. But unfortunately there is reason to be skeptical of how effectively the new discourse is trickling down into the daily practice of a demoralized and undertrained rank-and-file. There is a certain tension between notions of professionalism and the more populistic versions of the democratic policing ideology. It is perhaps significant that the trend is toward the former, under the cover of a continuing rhetoric about a democratic service. Carnage on the streets, widespread suspicions of corruption and high levels of incompetence certainly underline the need for a professionalist ethic. Extreme stress on the democratic nature of the new SAPS is more likely to lead to rank-and-file cynicism than confidence in the overall direction of police management. The worst case, but altogether possible, scenario is, of course, one in which neither professional nor democratic policing come to characterize realities on the ground.

CONCLUSION

The South African case illustrates the centrality of policing in the transition to the hoped-for consolidation of a new democratic regime. Widespread violence and consequent mistrust of the state police and other armed forces had been one of the key issues during the pre-1994 negotiations between the ANC and the National Party government. Lack of social order in South Africa's plural society after formal democratization signifies that policing functions remain a high priority for the new regime.

However valid, these abstract considerations do not go to the heart of the policing dilemmas in South Africa today. Under the specific conditions of South African society, the agenda of police democratization presents a more than usually formidable challenge whose outcome cannot readily be predicted. The issues which concern more stable societies with regard to the proper balance between police effectiveness and police accountability often seem trifling in comparison. At a time when many fear that the social fabric is tearing apart—or rather continuing to tear apart—Hobbes may be seen as preferable to Locke or Rousseau when it comes to political philosophy. Yet Hobbes in the past had a white face, and the visage of the new South Africa must necessarily be multicolored and not as pale in its upper echelons as it presently is.

The democratization of the South African police and its indissoluble concomitant of Africanization are driven as much by demography and power politics as by ideology. That the latter is important is not in question: there is simply no other credible alternative policing philosophy on offer to that presented in the new global discourse. That it coincides with the interests of the new political elite is convenient indeed. But its design reflects the post-modernity of the West rather than the pre-modernity of most of the globe. Its transplantation therefore hangs in the balance.

These considerations suggest that a global policing discourse is not a wholly benevolent development. To the extent that its normative rhetoric is mistaken for palpable realities it may serve to mask the real challenges facing police agencies in less blessed corners of the post-colonial world. In this, as in many other intellectual areas, the indigenization of policing to meet local needs should look seriously at a counter-discourse which begins at its own center.

NOTES

[1] *Witdoeke* (literal translation "white bands") refers to large groups of vigilantes—identified by a piece of white cloth tied to their bodies—who battled with ANC-aligned "comrades" in the squatter areas in the Western Cape in 1986–87. Collusion between state security forces and the *witdoeke* in these battles resulted in many deaths as well as the large scale destruction of squatter camps in the Crossroads and KTC areas (CIIR, 1988).

[2] The tally is taken from a print-out received from the SAPS Head Office of current foreign assisted projects (Personal communications, Director Du Plessis, SAPS, December 1997).

REFERENCES

African National Congress. (1994). *Reconstruction and development program*.
——. (1997). *Political report to the 50th national conference*.
Amnesty International. (1997, April). *Southern Africa: Policing and human rights in the Southern African Development Community*. London: Report by the International Secretariat.
Brewer, J. (1994). *Black and blue: Policing in South Africa*. Oxford: Clarendon Press.
Brogden, M., & Shearing, C. (1993). *Policing for a new South Africa*. London: Routledge.
CIIR. Catholic Institute of International Relations. (1988). *Not everyone is afraid: The changing face of policing in South Africa*. London: Author.
Dienaar, M. (1988). *The history of the South African police 1913–1988*. Silverton: Promedia Publications.
Dugard, J. (1978). *Human rights and the South African legal order*. Princeton, NJ: Princeton University Press.
Friedman, S. (1996). *Migration policy, human rights and the Constitution*. Pretoria: Paper prepared for the Task Team on Immigration Green.
Louw, A., & Shaw, M. (1997). *Stolen opportunities: The impact of crime on South Africa's poor*. Unpublished manuscript.
Marenin, O. (1998). The goal of democracy in international police assistance programs. *Policing: An International Journal of Managment and Policy*, 21, 2, 159–167.
O'Malley, P. (1996). Post social criminologies: Some implications of current political trends for criminological theory and practice. *Current Issues in Criminal Justice*, 8, 26–39.
Peberdy, S. (1996). *A brief history of South African immigration policy*. Paper prepared for the Task Team on Immigration Green Paper.
Pinnock, D., & Douglas-Hamilton D., (1997). *Gangs, rituals and rites of passage*. Cape Town: African Sun Press, with The Institute of Criminology, University of Cape Town.
Reitzes, M. (1997). *Undocumented migration: Dimensions and dilemmas*. Pretoria: Paper prepared for the Task Team on Immigration Green Paper.
SARPCCO. (1997, July). *Constitution and Report by The Permanent Co-ordinating Committee of Heads of Investigation Services*. SARPCCO 2nd Annual General Meeting, Cape Town, South Africa.
Shaw, M. (1996). *South Africa: Crime in transition*. Unpublished report.
Seegers, A. (1996). *The military in the making of modern South Africa*. London: Tauris Academic Books.
Shearing, C. (1996). Reinventing policing: Policing as governance. In O. Marenin (Ed.), *Policing change, changing police* (pp. 285–307). New York: Garland Publishing Co.
South African Press Association (SAPA). (1997). *Reports*.
South Africa, Government of. (1996). *Constitution of the Republic of South Africa*.
——. (1997). *Draft green paper on international migration*.
——. (1994). *Draft policy document*.
——. (Various). *Hansard*
——. (1993). *Interim Constitution*.
——. (1996). *National crime prevention strategy*.
——. (1997). *Policing white paper*.
——. (1994). *Public service act*.
——. (1995). *South African Police Service act*, no. 68.

——. (1997). *South African Police Service amendment act*, no. 41.
——. (1995). *State of emergency act*, no. 86.
——. (1992/93). *Survey of race relations. Annual report*.
——. (1994). *Reconstruction and development program*.
——. (November 1995 and September 1997). *White paper on the transformation of the public service*.
South Africa, Government of, Ministry for Social Welfare. (1997). *Framework for a national drug master plan*.
South Africa, South African Police Service. (1994). *Green paper*.
——. (1997). *NCPS victim empowerment program*.
——. (1997). *Policing priorities for 1997/8*.
——. (1997). *Policy framework and guidelines on community policing*.
——. (1997). *Scenario 1997–2010*.
——. (1996). *Status report: Transformation of the South African Police Service*.
——, Crime Information Management Centre (1/97, 2/97, 3/97). *Quarterly reports: The incidence of serious crime*.
Training Evaluation Group. (1996). *Report on basic training in the South African Police Service*, submitted to the Office of Development Assistance, Embassy of the United Kingdom, Pretoria, South Africa.
van der Spuy, E. (1989). Literature on the police in South Africa: An historical perspective. *Acta Juridica*, 262–291.
Weitzer, R. (1995). *Policing under fire*. Albany, NY: State University of New York Press.

9

Challenges of Policing Democracies: The Case of Austria

MAXIMILIAN EDELBACHER and GILBERT NORDEN

This chapter will examine the difficulties and challenges of democratic policing in Austria, using data from the security-executive forces and the results of previous sociological investigations. For that purpose, the organization of the security forces will be considered, the relationship between the security-executive forces and citizens will be described, and the development of security in the country will be analyzed. We will assess how much some measures that were recently taken have contributed to a better implementation of the tasks of policing, and suggest further measures and reforms as responses to the challenges of democratic policing.

GENERAL SITUATION AND CONTEXTS

With its 84,000 square kilometers (about 31,000 square miles), Austria is a relatively small country situated in the core of Europe. Today, there are about eight million people living (legally) in the country. About a fifth of them live in Vienna, Austria's capital. Since 1995, Austria has been a part of the European Community (EC).

Constitutionally, Austria is a federal democratic republic, comprised of a federal government and nine federal provinces (*Bundesländer*). At the federal level, the legislature (Federal Assembly) includes the *Nationalrat* (National Council) and the *Bundesrat* (Federal Council). The *Nationalrat* is directly elected by the people, in a proportional election system and has a term of office of four years. The *Bundesrat* represents the provinces according to their populations and members are elected by their respective *Landtag* (the provincial legislature). The most important parties are the Social Democratic Party (centrist/left), the People's Party (Christian democratic/conservative), the Free Party (right-wing), the Green Alternative and the Liberal Forum. Executive power is exercised by a popularly elected President and by the Federal Chancellor and Federal Ministers who are appointed by the Federal President.

On a per capita basis, Austria is the tenth-wealthiest country in the world. The prosperity per person is slightly above the European Community's average. The majority of Austrians are materially well off. The rate of unemployment is relatively low, and the unemployment rate among juveniles is the lowest in the EC (Bartunek, 1997, pp. 524ff.). In addition, the extent of social services is comparatively high in Austria. Due to the reasons stated above, and because cooperation between employers and employees is highly institutionalized (the so-called "Austrian Social Partnership") (Tàlos & Kittel, 1995), Austria is considered a country of economic as well as social peace (Hanisch, 1994, pp. 484ff.). By international comparisons, the readiness to go on strike is very low in Austria.

However, Austria has also been strongly influenced by the severe changes that have taken place in Europe within the last decade. The breakdown of the communist regimes in Eastern Europe can certainly be considered as the most important event. The iron curtain fell in the years 1989–1990 and, in consequence, the "New Eastern Countries" lifted their travel restrictions. As general conditions have been, and still are, very difficult there and since the Western countries have been, and still are, highly privileged, there has been a strong migration movement, a movement reinforced by the war in the former Yugoslavia. Before Austria could react with restrictive immigration policy measures, it experienced a strong wave of immigration (Reinprecht, 1995). Between 1988 and 1993, net immigration amounted to about 360,000 persons, thus more than doubling the number of foreigners living in Austria (Findl, 1997, pp. 827ff.). As a result, Austria has become a country with one of the highest proportion of foreigners. At present, in 1996, throughout Austria, about 9 percent of the population who live legally in the country are foreigners. In Vienna, the proportion of foreigners among the Viennese population has increased to 18 percent (Halàsz, 1997, p. 129). In addition, there are also a number of foreigners who are in Austria illegally. Estimates place their number at about 150,000 all across Austria. Thus, the Austrian security-executive forces which had been spared the trouble of policing a multi-ethnic society for a long period of time, now have to deal with such problems as well.

Opening the borders to the East has also led to an increase in transnational illegal activities. This increase, however, must also be examined within the context of the four freedoms which are found within the EC and which also apply to Austria: freedom of movement for services, goods, and capital, and freedom of residence. As a result of the Schengen Treaty, there no longer exist border controls among most countries of the European Community. However, at the same time, there has been an increase in controls at the external common border of the Community. Consequently, the extent of crime among the different member states may become more or less equal over time. Further, it cannot be questioned that interior security and judicial cooperation, which has been proclaimed the "third pillar" of the Community (according to the Maastricht Treaty), has to be realized. (The "first pillar" refers to economic intra-community cooperation and the "second pillar" stands for a common foreign and security policy). Thus, the question regarding a "New Police" in a "New Europe" arises (Busch, 1995: Morie, Murck & Schulte, 1992).

THE SECURITY-EXECUTIVE FORCES IN AUSTRIA

History

Austria is one of the countries which is characterized by the "continental police-system." This system differs from the "Anglo-Saxon system" in its legitimization (that is, by central governments rather than by local authorities), its structure (a centralized instead of a decentralized organization), and in its functions (the fight against crime is merely one task among others) (Shelley, 1996, p. 3). Historically, the current security-executive forces are an outgrowth of the "police welfare state" of the "Benevolent Despotism" of the Hapsburg Empire (Girtler, 1981, p. 14). In that era, "administration" mainly meant "police administration." Policing was thought of as the execution of laws for the maintenance and the promotion of the general welfare (Davy & Davy, 1991, p. 13). Police power was practically unrestricted then. At the end of the eighteenth and in the early nineteenth century, the concept of an "absolutistic welfare police" was replaced by the concept of a "security-police," with warding off dangers as its most important and unique task. However, this concept was perverted by the Metternich[1] regime of neo-absolutism, with its strong police reign (Beidelt, 1896–1898, pp. 117ff; Pelinka, Haller & König, 1990, p. 8).

Consequently, in the 1848 Revolution, "the main criticism regarding the absolute state was focussed primarily on the police force, whose aim it was to maintain a solidified political system" (Girtler, 1980, p. 23). In the late aftermath of the revolution, with the rise of political liberalism, important legal steps were taken in order to limit police power; these included the law regarding the protection of personal freedom (1862), the law regarding the protection of domestic rights (1862), and the basic state law regarding the general rights of citizens (1867).

The structures of the security-executive forces go back to those days as well and have proven to be remarkably stable. The "gendarmerie" (also referred to as *"Landessicherheitswache"*—"security-guards of the country"), which operates in rural areas, was established in 1849; the "police," which are found in the cities, was established in 1869 (as the "k.k. [kaiserlich.königlich] *Wiener Sicherheitswache*"—the "Imperial and Royal Viennese Security-Guards"); and lastly, the criminal police was established in 1870 (as the *"Institut der k.k. Polizeiagenten in Wien"*—the "Institution of Imperial and Royal Police-Agents in Vienna") (Bögl & Seyrl, 1992, pp. 23, 26; Gebhardt, 1997; Hesztera, 1994; Hirschfeld, 1989; Jäger, 1990; Keplinger & Stierschneider, 1994, p. 51; Miedl, 1988; Zima & Hochenbichler, 1969). This organizational structure of the security-executive forces was taken over, in nearly unchanged form, by the "First Republic" after the Empire had collapsed in 1918. As a matter of fact, the security forces represented the only authority capable of providing order in the post-war era; they made a large contribution to the transition from the old state authority to the new one. "The police-organization that had its roots in the previous century thus became the legally protected force of the new state" (Girtler, 1980, p. 33). During the years between World War I and World War II, the Austrian police were famous all over the world. Austrian police officers were sent, for example, to Chicago to educate officers of the Chicago Police. Interpol was founded in Vienna in 1923 and its headquarters were in Vienna in those days.

The organization of the security-executive forces of the Second Republic (after 1945) adopted the organization of the security-executive forces of the period between the wars, thus carrying over organizational features of the security-executive forces of the Monarchy as well. It still has "hardly changed" (Girtler, 1981, p. 14). Even in the present day, some features of police bureaucracy can be encountered that go back to the time before democratization in Austria took place (Pelinka, Haller, & König, 1990, p. 9). It is no wonder, therefore, that some police officers said in the course of an internal survey that the police bureaucracy still reminded them strongly of the Metternich era (Röglin, 1989), especially the explicit police hierarchy as well as the strong centralization of the organization.

Organization and Structure

Police power is exclusively assigned to the federal government in Austria. The Federal Minister of the Interior is responsible for the security-executive forces. After him or her, the General Director of Public Security (*Generaldirektor für Öffentliche*

Sicherheit) is the head of all security forces. In each federal province of Austria, the Director of Security (*Sicherheitsdirektor*) is the authority in security matters. He or she is appointed by the Federal Minister of the Interior, but must be acceptable to the Head of the provincial government (*Landeshauptman*).

The security-executive forces include the federal police, the federal gendarmerie, the state police, the criminal police—Interpol, special units of the federal ministry, the border gendarmerie, and customs. The federal police protect fourteen cities in Austria with a total of about 2.8 million inhabitants. The police in each of the cities is headed by a Chief; the President of the Viennese police is at the same time Director of Security, because Vienna is the capital and a federal province as well. In 1995, the federal police were 13,431 officers strong (10,844 security-guards and 2,587 non-uniformed detectives).

Approximately 5.2 million people living in rural areas are protected by the federal gendarmerie. The gendarmerie are under the Central Command (*Zentralkommando—GZK*), which is headed by a general. There are eight provincial commands (*Landeskommandos—LGK*). Next in the hierarchy are the Regional Commands (*Bezirkskommandos—GAK*). The federal gendarmerie employed 12,697 officers in 1995.

The state police services the state and protects its functionaries and visitors against attacks. The criminal police—Interpol, for example, conducts anti-drug investigations, offers criminalistic services, compiles statistics and maintains international liaisons. Special units provide protective services against suspected terrorist and special security threats.

Due to the centralization of decision making and a strong hierarchical organization, demands are passed from higher to lower levels of the security-executive forces and result in the over-regulation of police actions. The fact that, due to instructions that were done away with only in 1993, the gendarmes on duty were only allowed to watch Austrian TV (ORF 1 and ORF 2) and no foreign TV program, even if they agreed to pay for them out of their own pockets, may serve as a distinctive example.[2]

All relevant surveys of recent years have come to the conclusion that there exists, among police officers, a high degree of dissatisfaction concerning the organization. Many police officers think that hierarchical authority is too predominant; they criticize the usage of a "command-language" (that is mostly regarded as useless); and, finally, the relationships with their bosses are often judged to be problematic (e.g., Meggeneder, 1995, p. 84; Wolf & Korunka, 1993, p. 119). Their superiors are particularly reproached for being inflexible and for passing on "useless" plans from politicians, without checking if these tasks are realistic or what they might mean for the police officer on the beat. There is a distinct contrast between "Street Cops" and "Management Cops," to the degree that one can speak of "two cultures of policing," as Reuss-Ianni (1983) has done in her study of the New York police. Additionally, there are tensions in the management of the Austrian security-executive forces as a result of the simple fact that too many jurists and lawyers have been made senior managers in the past.

Apart from the organization and the problems of supervision, a distinct bloatedness of the police-bureaucracy also goes back to the imperial tradition (Heindl, 1991). Compared to the world standard, which is one police officer to 700 citizens, and to some Western countries, with up to 500 citizens per police officer, Austria is more heavily policed. In Austria, the police/gendarmerie to civilian ratio is about 1 to 300. According to a survey conducted in the year 1992, 54 per cent of all Austrians have a police or gendarmerie station in their immediate residential area, that is, within a 15-minute walking distance of their households (Eichwalder, 1993). Today (in 1997), this percentage may be a little bit less because there has been some amalgamation of stations and reductions in posts in the recent past. There are about 1,300 police and gendarmerie stations (offices) in Austria today.

As Austria is such a well-policed society (Das, 1994), private security services are, comparatively speaking, of minor importance. Still, they have been manifesting themselves in the course of the last few years due to the partial privatization of some public security services as, for example, the security controls at the conferences of the "Organization for Security and Cooperation in Europe" in the Viennese Hofburg, or security controls at the airports and in the court room. As a consequence, the (public) security officer to private security guard ratio is 4 to 1 in Austria, (compared to 1 to 2.5 in the USA and 1.25 to 1 in Germany (Diederichs, 1992).

The Austrian security-executive forces are mainly organizations of men. Only about 5 per cent of the police officers are women (2 percent of the gendarmerie and 8 percent of the police).[3] Austria decided relatively late to make the security-executive services accessible to women. The gendarmerie have been employing women only since 1984, and the police have been hiring women only since 1990. Before those dates, women had been assigned only to work in the information service or as traffic wardens. Consequently, women can only rarely be found in senior management positions. However, the share of women in mid-level positions may increase, since a high proportion of women can be found among the applicants for police jobs.[4]

As far as the preference of the political parties is concerned at the moment, police officers more closely reflect the political views of society than is the case with gender distribution. It is striking how many police officers support J. Haider's (a right-wing politician) political views—about one-third. The support of his populist FPÖ (*Freiheitliche Partei Österreichs*—Freedom Party of Austria), which espouses and promotes a populist, reactionary, and nativist ideology, among the police reflects a protest against the politicization of the security-executive forces by the current political establishment (which is controlled by the Social Democratic Party and the People's Party).

PRACTICES OF DEMOCRATIC POLICING

The concept of democratic policing in Austria can be found in the specific language of law and police regulations as these have evolved over time, in concert with political changes as Austria moved from a monarchic, absolutist system toward its version of participatory, consociational democracy. Police regulations spell out the obligations of the police, in terms of work required, procedures to be followed and rights of the public to be protected.

Tasks and Relations to Citizens

The tasks of the security-executive forces are laid down (with a few exceptions, which are regulated in special laws, for example, traffic control) in the Security Police Act (SPA) of 1991. The law came into force on May 1, 1993. It consists of six parts: the structure of authority, the tasks of authority, the competence of the police, data protection provisions, penal provisions and legal protection provisions (Fuchs, Funk & Szymanski, 1993; Hauer and Keplinger, 1993; Noll, 1991).

Paragraph 1 of the SPA defines the contents of the regulations by general titles ("organization of the security-administration and the performance of the security-police"). According to the legal definition of paragraph 2 (2), the administration of security includes the security police; the passport and registration offices; the aliens branch; the control authorities when entering or leaving the federal territory; the control of weapons, ammunition, shooting and explosives as well as matters regarding the press, club organizations and social gatherings. According to paragraph 3, the tasks of the security police include the maintenance of public peace, order, security and the general duty of giving assistance. Related to these tasks and duties, the terms *warding off dangers*, *crime prevention* and *special protection* play an important role. Tasks include the following.

(1) "Special protection" refers to specific conditions or people, such as the protection of helpless people, the protection of national governmental institutions, the protection of foreign state representatives and the protection of public places;
(2) The prevention of probable and dangerous assaults against life, health, freedom, morals, property or environment. This refers to those cases for which there exist clear indications supporting the belief that a criminal offense is under preparation;
(3) The prevention of further dangerous assaults after one hazardous attack has already been executed. The operation of the security-executive forces is focused on "re-securing measures" after an executed and achieved hazardous attack; and
(4) Protection against an impending dangerous attack threatening life, health, freedom and property. If it is possible, security authorities have to inform the relevant population about the threat of a hazardous attack.

When fulfilling their tasks, the security-police can only interfere with the rights of persons if such a measure is provided for in the law, no other means are suitable to fulfill the task, and the relationship between cause and expected success is maintained. It is the success-oriented and careful police officer who serves as a model for the exercising of the authority and discretion of the police, or, put in different words, "sparrows must not be shot at with cannon balls and tanks must not be attacked with catapults" (N.N, 1993, p. 20).

These limitations on the means of intervention and powers of the security police, as well as the duties to ward off dangers (paragraph 21), to engage in prevention (paragraph 25)[5] instead of repression, to seek conflict mediation (paragraph 26), and to provide assistance in general (paragraph 19), are the fundamental objectives and principles of the Security Police Act. Underlying the Act is the concept of a "democratic" security-police, a concept which embraces a "balance of powers and duties" between the security-police and citizens, and sees the security-police as an integral part of democratic society and police officers as "security experts" who work in a people-oriented and effective way (Das, 1995, pp. 5ff; Jones, Newburn & Smith, 1996). As a whole, the regulations of the SPA are rather demanding for police officers (many find them too elaborate for practical police work), and, as a result (one can hypothesize), street personnel do not adhere to them strictly.[6]

More feasible and practical may be the following task priorities which have been formulated by a consultancy firm (TC, 1994, p. 55). These were developed as guidelines specially for the gendarmerie. (1) Assistance and conflict mediation as well as an intelligence presence (in order to ward off danger and to prevent it); (2) Roadworthiness; (3) To fight crime and to care for victims; and (4) Cooperation and networking with other institutions.[7] Similar priorities could also be set for the police.

The security-executive forces had already provided general assistance, mediated conflicts, and prevented crime (in varied, but informal ways) in the past even before they were obliged to do so by the SPA. Austrian citizens see the security-executive forces as fulfilling a social and assisting role. For the public, the police/gendarmerie are an important "unspecific remedy-agent" or "aid institution" (Hanak, 1986, p. 26), which can be called upon for many problems. Citizens see the police as a service organization which is competent to solve all kinds of disturbances, conflicts and abnormalities. This interpretation was derived from the analysis of emergency calls (compare Dreher & Feltes, 1996; Feltes, 1995). The most frequent reasons to dial the emergency number were requests for help or services (for the helpless, drunk people, etc.) and intervention in conflicts (neighborhood conflicts, disturbance of the peace, private quarrels, molesting, etc.), as well as traffic problems (accidents, etc.). "Real" crime, that is a police radio patrol response to an actual or suspected crime, accounted only for a small part of all the emergency calls (Hanak, 1991; TC, 1994, p. 33).

The Police at Work

In this respect, discussions about the fight against crime distract from the actual daily work of the police/gendarmerie, although crime is a special challenge of policing (as will be described below). But these findings also change the notion of the special tasks of police officers. The daily work of the security-executive forces mainly consists of assisting in emergency situations, keeping the public peace, and easing or mediating in conflict situations. The latter tasks must also be analyzed in context—in light of a probably expanding atrophy (or decline) in the ability of individuals to solve conflicts by themselves. This is why the security-executive forces are also involved in cases which would be considered of minor importance, as, for example, when the neighbor living one floor above waters his or her flowers and water drips on the balcony below or, as reported from Graz recently, if a four-year-old child plays in the sandbox of a house he or she doesn't live in. When analyzing conflicts, one also has to point out the Austrian habit of the quasi-institutionalized cooperation among the main societal groups which prevents conflicts among them, whereas the interests of individuals within these groups are often neglected. Thus, individuals are frustrated and instead of conflicts among groups, intense conflicts between individuals arise.

Such conflicts can also be observed in the traffic situation of the country. Aggression among drivers is a widespread phenomenon and leads (in combination with an over-estimation of one's driving abilities) to a bad traffic milieu. According to recent surveys and recent observations, 0.4 percent of all drivers cross when the traffic light is red, 12 percent do not use their indicators when turning and changing lanes, 45 percent do not use their safetybelts, and 62 percent of all drivers in Vienna speed. This behavior and, additionally, the generous attitude towards drinking and driving (66 percent of drivers admit having driven at least once while drunk) lead, compared to international statistics, to a high percentage of accidents. Traffic control is a difficult task for the security-executive forces to carry out, as citizens quite often consider being stopped for the sake of control an interference with their right of free movement and, hence, fines are met by a lack of understanding. However, the enforcement of traffic regulations (traffic control, the prevention of violations and the investigation of accidents), which is considered a very important if sensitive task, provides the most frequent contact between the security-executive forces and citizens (Institut für Konfliktforschung, 1990).

Citizens' Perceptions of Police Work

The results of a recent survey conducted in March and April 1997 regarding the satisfaction of citizens with such contacts in general (KF & P Consulting, 1997) have to be seen in this light as well. According to the survey, 69 percent of the Austrians who needed the help of the police/gendarmerie within the last twelve months were either "completely satisfied," "very satisfied" or "satisfied" with the service they received. "Customer satisfaction" with the security-executive forces is higher than with other local authorities, yet it is much lower than satisfaction with general practitioners (doctors), the staff of hospitals, pharmacies, chemist's shops, banks, insurance agents, post offices, telecom, public carriers, air lines, travel agencies, tourist industry, garages, filling stations, mail order houses, food shops, restaurants, boutiques, furniture shops, construction and do-it-yourself markets, electronic goods markets, and newspapers, with which, on the average, 90 percent of customers were satisfied. Further, the rate of complaints against the security-executive forces is comparatively high and, typically enough, the majority of these complaints refer to the problem of traffic. What is more, many of the complainants are not satisfied with the handling of their complaints, despite the fact that the Security Police Act was designed to facilitate the complaints procedure and provide a more "citizen-friendly" handling of the complaints.[8] In short, the comparatively poor "customer-orientation" of the security-executive forces, already observed in the course of past surveys concerning citizens' experiences with different institutions (Fessel + GfK, 1984b; IMAS, 1992), has continued.

When considering these survey results, it has to be kept in mind that assessments regarding contacts of customers/citizens with institutions do not only depend on such factors as the actual process of the contact, but are influenced, as well, by the status of the person, that is whether the contact is in the role of a customer or in the role of a person who must submit to sovereign action. Hence, it is not surprising that the security-executive forces achieve poorer results in such surveys. In any case, the toughness of the police apparatus can be distinctly noticed. Many police officers are aware of these problems, yet they are forced by their superiors and the official regulations to issue a fine, even though warnings or instructions, etc. might be more appropriate. This is why a vast majority of police officers experience conflicts between official regulations and the notion of community policing.

Based on their experiences during contact situations, many police officers think they do not get enough recognition from the public. According to an internal survey in the year 1993, about half of all police officers share the opinion that their profession is perceived negatively by the population. This impression would not have been true even of the Viennese police in the early 1970s, which had somewhat of a crisis in those days (Fischer-Kowalski, et al., 1982). Today, this impression is certainly wrong. Existing studies regarding public acceptance of the police show that although Austrians are critical about police officers and their general status or behavior on the one hand (e.g., Pichler, 1993, pp. 72ff; SWS, 1990), they appreciate their "hard work," especially when fighting crime, and thus respect officers on the other hand (e.g., IMAS, 1989, 1993). This respect is especially noticeable in rural areas. The institution "security-executive forces" is seen as an undeniable powerful institution which represents the state (e.g., Market 1992; Österreichisches Gallup-Institut, 1985). It is quite popular (e.g., IFES, 1982, pp. 5ff.; Plasser & Ulram, 1991, p. 138; Rosenmayr, 1980, p. 161) and citizens trust it. In comparison with other societal institutions, the police/gendarmerie achieve a very positive ranking within the hierarchy of confidence (see Table 9.1).

It is obvious that, being an organization that services all citizens, the security-executive forces also require more public confidence than a particularistically ori-

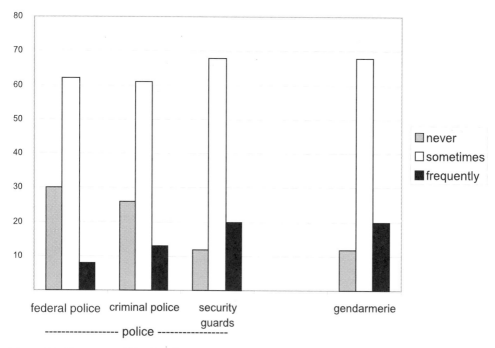

Figure 9.1. Frequency of Conflicts Between "Offical Regulations" and "Citizen-Friendliness" within the "Police" and "Gendarmerie" (source: Meggeneder 1995, 78)

ented organization (for example, the church, trade unions, etc.). Yet, the security-executive forces also enjoy a higher degree of confidence than those political institutions in a narrower sense (government, parliament, etc.) which regulate the democratic conversion of the wishes of the population, and also serve all. In addition, the security-executive forces are also considered more trustworthy than the judicial branch, which has been affected by diverse scandals and affairs during the last ten years, beginning with the judicial escape of U. Proksch[9] and ending with the recent judgment of an extremely right-wing judge. At any rate, this special confidence in the security-executive forces provides the basis of cooperation with citizens in the fight against crime and in routine police work.

OPERATIONAL CHALLENGES

Crime and Security

Normal Crime

According to estimates and official data, every fifth Austrian in the history of the Second Republic was at some point sentenced according to the criminal law. Examining changes in the number of reported criminal acts, it can be seen that these increased moderately in the 1960s and early 1970s but then, in spite of a certain "decriminalization" due to a substantial reform of the criminal law in 1974[10] (Pilgram, 1995, p. 492), increased greatly after that period (Hanak & Pilgram, 1991, p. 18). Between 1975 and 1994, reported crime increased about 70 percent; taking the increase in population into consideration, the increase in crime (registered criminal

Table 9.1. The rank of the "police"/"gendarmerie" within the hierarchy of institutions, according to the degree of confidence expressed by the population (by survey results)

Source	Time of survey	Rank
Plasser/Ulram (1991, p. 138)	1989	1. before Courts of Justice, Offices/Boards, Army, Government, Parliament, Media, Political Parties
Fessel+GfK (1991)	1991	1. before Courts of Justice, Offices/Boards, Church(es), Army, Government, Parliament, Trade Union, Media, Political Parties
Market (1992)	1992	3. after Physicians and Banks, but before Craftsmen, Parsons, Judges, Lawyers and 20 other occupations, professions and branches of the economy
SWS (1995a)	May 1995	1. before Trade Union, Army, Austrian Broadcasting Corporation, Workers' Chamber, Economic Chamber, Parliament, Newspapers, Government, Catholic Church, Political Parties
SWS (1995b)	June 1995	1. before Workers' Chamber, Economic Chamber, Trade Union, Parliament, Government, Army, Austrian Broadcasting Corporation, Newspapers, Catholic Church, Political Parties
SWS (1995c)	August 1995	1. before Austrian National Bank, Parliament, Workers' Chamber, Government, Army, Economic Chamber, Austrian Broadcasting Corporation, Trade Union, Newspapers, Catholic Church, Political Parties
SWS (1995d)	November 1995	1. before Austrian National Bank, Workers' Chamber, Economic Chamber, Trade Union, Austrian Broadcasting Corporation, Newspapers, Government
SWS (1996)	May 1996	1. equal with Austrian National Bank, before Workers' Chamber, Economic Chamber, Trade Union, Austrian Broadcasting Corporation, Newspapers, Government
Fessel + GfK (1996)	1996	1. before Courts of Justice, Boards/Offices, President, Army, Trade Union, Parliament, Church(es), Government, Media (Newspapers, TV), Political Parties

actions/100,000 inhabitants) was still about 60 percent. Within the past few years, the number of reported criminal acts has decreased slowly.

The increase in reported criminality is affected by: 1) a rapid increase of urbanization (an increase of the population in urban zones, where the criminal rate is higher) and the development of urban-like conditions (with the corresponding opportunities for criminals) in many rural communities because of the expansion of tourism; 2) a change in attitudes (which increases the chances that certain crimes, such as sexual abuse of women or children, are reported more frequently than they used to be, and, 3) an increase in economical activities as a result of economic growth, the spread of insurance coverage (which leads to more filings), and the growth of special opportunities (apartments are more often unattended because households are smaller and more women go out to work; more and more people have a second residence, which is mostly unattended; there are more and more cars of interest to thieves, etc.).

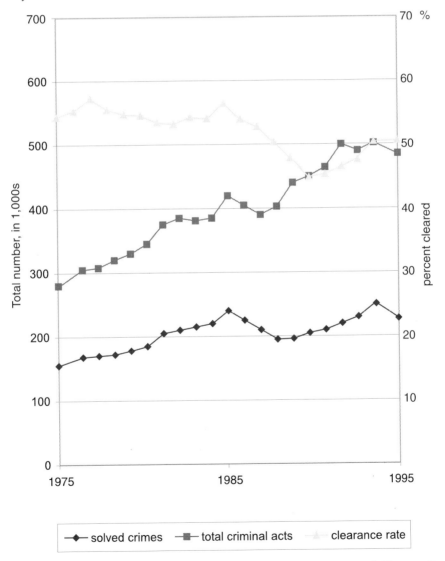

Figure 9.2. Number of Registered Criminal Acts, Number of Solved Crimes, and Clearance Rate in Austria, 1975–1995 (Source: Oesterreichisches Statistisches Zentralamt, 1996: 405)

In addition, there is also a stronger integration of Austria in international economic conditions and thus increasing economic exchange processes, as well as an increased international mobility in the course of which more crime has been imported (see Figure 9.3). The extent of foreigners' share of crime (measured by convictions) tripled in Austria between 1985 and 1992. Since then it has been decreasing slightly. Today, the share of foreigners in crime is four times the share of crime (convictions) by youths, which has decreased mainly due to a reform of the criminal law for youths in 1988 which introduced the possibility of "Out-of-Court Settlements of Offenses." The number of criminal charges brought against youths has remained constant, however (Pelikan & Pilgram, 1993, pp. 271ff.).

The increase of criminality among foreigners cannot be linked to foreign workers (whose crime rate is lower than the Austrian average) but is connected to other foreigners, that is tourists (who often come for a day), passengers, private guests or illegal foreigners. When calculating the extent of criminality of the whole foreign population (which includes foreign workers, the other foreigners who live here legally, the daily average number of tourists who stay overnight, the estimated average number of tourists who stay only during the day, private visitors, passengers and foreigners illegally in Austria) and if the younger age structure and the higher proportion of men among foreigners is taken into account, one finds, statistically speaking, the same level of crime as with native people (Pilgram, 1993; Pilgram, Hanak, & Morawetz, 1992). The security-executive forces, however, have acted (and frequently still act) under the impression that foreigners are more responsible for crimes than the native population. Therefore, foreigners were (and still are) over-policed in Austria.

As quite a few criminal foreigners only visit Austria in order to commit criminal actions and then leave the country, the category "crime-tourism" has become recognized (Dearing, 1993). Due to the increase of this so-called "crime-tourism," the chances of clearing crimes are obviously reduced. The clearance rate, which had already sunk in the long run from 80 percent in 1958 to 57 percent in 1985, sank within a short time to a low of 44 percent in the year 1990 (Hanak & Pilgram, 1991, p. 20). Since then it has increased somewhat (see Figure 9.2). In 1996, 51 percent of reported criminal acts were cleared. It cannot be judged here, as Lenz & Mason (1992) have argued, whether these numbers are real or "polished" statistics. The same inability also holds true for the contrasting argument that many criminal acts, probably committed by right extremist criminals (e.g., a series of letter bomb attacks directed against persons who support foreigners and immigrant issues), have not yet been cleared because right extremist elements within the security-executive forces cooperate with these criminals.

When analyzing the structure of reported criminality, one can see that the great majority of criminal acts are offenses against someone else's property or wealth (in 1996, about 67 percent of all reported criminal acts). But material losses which result from such crimes are increasingly mitigated by insurance coverage (Hanak & Pilgram, 1991, pp. 202ff.). At the time, when the borders to the East were opened, criminal acts against personal property and wealth increased especially in the Eastern part of Austria and, above all, in Vienna. They have stayed at the same high level or have only slightly decreased within the past few years. Information concerning the development of criminal acts against property and wealth in the capital Vienna contained in the official statistics has been re-enforced by relevant surveys among the population and victims. In 1980, about every fourth Viennese interviewed said that he or she had been robbed or become the victim of a burglary (IFES, 1980); in 1993, property crime had touched every second Viennese (Nielsen, 1993); two years later, in 1995, about 10 percent of the Viennese who were interviewed stated that they "had been the victim of an offense against property within

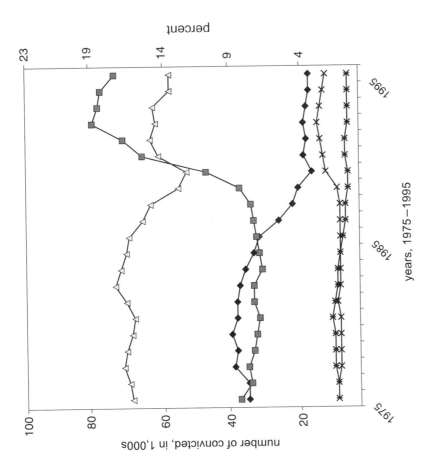

Figure 9.3. Total convicted, as well as share of foreigners and youth among convicted, in Austria 1975–1995. (Source: Oesterreichisches Statistisches Zentralamt, 1996: 405)

the last year" (IFES, 1995). Along with the increase in crime, there has also been an increase in security precautions among the population (e.g., precautionary measures against burglary, such as safety locks and doors) (Market, 1996 Österreichisches Gallup-Institut, 1994).

As the result of crime developments, Vienna has lost the special rank as one of the most secure towns in Austria that it had had until the mid-1980s, due to its relatively low level of crime for a city of its size (Hanak & Pilgram, 1993). Now its crime rate (per 100,000 inhabitants) is only slightly lower than the crime rate of Innsbruck and Salzburg, the towns with the highest crime rates in Austria. Yet, internationally speaking, Vienna, as well as Austria as a whole, is still very safe (International Crime Victimization Survey, 1996). Compared to German cities, Vienna, though it is a large city, ranks together with the relatively small German cities of Solingen, Wuppertal and Pforzheim near the bottom of criminal statistics—while Frankfurt is in a leading position. The figures for car theft in Vienna are lowest among the European cities. The drug problem, and with it the problem of drug-related crime, have not yet reached the same dimensions as in some Western European cities and those of the USA, although it has increased in comparison to the 1980s. Youth gangs, a problem in the early 1990s when about twelve gangs made trouble in Vienna, have almost completely disappeared due to a specifically formed police task force (which has been dissolved recently) and because of changes in youth culture (greater individualization, etc.). Finally, with five murder cases per 100,000 inhabitants annually, Vienna ranks among the safest capitals in the world.

In all of Austria there are about 180 murder cases a year; 65 percent to 80 percent of these are "relationship murders;" most of them are cleared. This, and the fact that the majority of violent incidents (physical injuries) observed by police radio dispatched patrols can be attributed to domestic abuse (violence in the family), indicates that international professional killers from different criminal gangs have (so far) only operated on a small scale in Austria. Yet, according to the estimates of experts and the security-executive forces, organized crime is on the increase in this country.

Organized Crime

Organized crime is characterized by work-sharing and the conscious and intentional co-operation of several people for a given period of time in order to commit criminal actions and to gain large profits, frequently by exploiting the modern infrastructure (Ettmayer, 1990, p. 88; Lesjak, 1995). This form of crime has existed in Vienna for a long time. However, it used to be called the "Vienna underworld." This label also meant that this form of crime was still in the hands of Austrians or Viennese. The dangers from organized crime, which today is mainly controlled by foreigners, has changed as well. Organized crime is often characterized by an enormous amount of brutality, which, in the "floodlight-society" (a society in which only extremes are noticed) (Allerbeck, 1982, p. 82) quite understandably leads to "public excitement about a generally increasing violence in our society which, however, cannot be supported by objective data" (Karazman-Morawetz & Steinert, 1995, p. 93). Yet today, organized crime has such power that it can have an impact on whole state structures (KSÖ, 1993, pp. 8ff.; Edelbacher, 1993).

At the moment, it is estimated that about 25 percent of all crime in Austria is committed by organized groups. According to Kemper (1996), Austria has even become the European Community headquarters for espionage and organized crime. Lenz and Mason (1992) see the development in this country in a similar, dramatically exaggerated way and claim that Austrian politicians are responsible for it. The

Austrian population does not see the problem in such a drastic manner. It is true that half of the Austrians believe "in the presence of the Mafia in their own country," but 37 percent think "one cannot talk about this as a great danger" (IMAS, 1995, p. 2). It may well be that the development of organized crime has been "overdramatized" by police officers. Attempts in Germany to find exact statistics on the development of organized crime show only a slight increase, or even a decrease, during the last few years (Pütter, 1997, p. 24).

It is, obviously, extremely difficult to provide an empirically confirmed account of organized crime since one of its traits, and one which makes it successful, is to operate in the "dark" (Mayerhofer & Jehle, 1996). One can also refer to organized crime as a "clandestine industry." The following, though, cannot be questioned. Its main areas of activities are the smuggling of people, radioactive substances and cultural goods; arms and drugs trafficking; the illegal selling of stolen cars; the extortion of protection fees; international financial fraud (involving, for example, checks, credit cards, and money laundering); economic crimes; computer-related crimes; illegal technology transfers; business espionage; the forgery of documents; and environmental crime. The "Russian" or "Eastern Mafia" (as the organized crime groups from the former Soviet Union are generally referred to), as well as the Italian Mafia, use Austria primarily as a "recreation zone" and as an "investment country." In addition, the Chinese Mafia (which causes a special problem for the security-executive forces) has established subsidiaries in Vienna and throughout Austria. Additionally, syndicates of former Yugoslavians (who are especially active in burglaries, car thefts, and dealings in drugs and arms), who partly operate from Bratislava (Slovak Republic), and Turks (who mainly work in drug trafficking) are active in Austria. Austrians are represented in the classical domains of organized crime, such as procuring and gambling. But Austria is not a country from which transnational crime originates. Yet organized crime constitutes a special challenge for the security-executive forces.

Feelings of Safety among the Population

In general, people feel safe in Austria these days. According to a survey done in 1996 (Market, 1996), more than 80 percent do. Comparatively speaking, Austrians even feel safer these days than do the citizens of other member countries of the European Community, with the exception of Scandinavians. The Viennese consider their city to be safer than other major cities in Europe (e.g., Munich, Paris, London, Rome) (Österreichisches Gallup-Institut, 1993), and Austrians consider only Switzerland as being safer than, or equally as safe, as their own country (Market, 1996; Österreichisches Gallup-Institut, 1994). This perceived security has even become a source of their national pride. In surveys, Austrians take most pride in "Austria being a tourist country," but this trait is followed immediately by "internal security, little criminality," which is even mentioned before "social security," "Vienna being a Congress city," etc. (Sandgruber, 1995, p. 533). For quite a number of tourists public security is a (further) reason for spending their holidays in this country; this reason is, therefore, also especially important for Austria's position as a tourist country; to a large extent, tourists feel even safer here than they do in their respective countries of origin (Dr. Haerpfner, 1989). Furthermore, perceived security is also important for Austria's position as an industrial country, since a considerable number of international firms settle here for this reason (Aiginger & Peneder, 1997).[11]

Compared with the situation in the US, this impression may be true. However, feelings of safety among the Austrian population vary to a certain extent. For example, the Viennese feel safer in autumn than they do in spring. That is because they usually spend their summer holidays in less safe (southern) countries and their experiences there influence the assessment of the situation here. What is more, feelings of safety are also dependent on political propaganda. Politicians use data regarding the criminal situation as "political instruments"[12] in order to achieve their aims (Pilgram, 1990). Consequently, there exists during times of elections a certain feeling of insecurity among some sections of the population, and the call for "a more efficient fight against crime" is then high on the list of problems which Austrians want to have dealt with by the government. Within a few weeks after the elections, this wish loses its salience and high ranking and is then placed in the middle range of concerns again.

Additionally, feelings of safety also depend on the place of residence. In general, inhabitants of smaller villages or towns feel safer than inhabitants of the big cities. About half of the citizens feel "safe about crime" and a quarter feel "rather safe" in Vienna (Diess, 1995b). However, there are differences among districts and residential areas. In Döbling, an upper middle/upper class district, 100 percent of the inhabitants "feel rather safe in their immediate residential area," whereas in Favoriten, a traditional working class district, only 25 percent feel so (OGM, 1995b). In the housing estate "Am Schöpfwerk," located in a "bad" residential area in Meidling, only 16 percent feel safe.[13]. Generally speaking, in residential areas with a comparatively high proportion of unemployed juveniles, feelings of safety among the community are lower than elsewhere. Further, in accord with international studies, it can be shown that when native residents perceive that too many foreigners reside in their neighborhood, their, expectations regarding their territorial, cultural and psychological security needs are difficult to meet.

Feelings of safety also depend on age and sex. Younger people as well as men feel safer. Elderly women, who, objectively speaking, are the least endangered and least likely to become victims of a crime, feel the least safe.

The Need to Maintain Order in Public Space

One of the reasons for feelings of insecurity is an exaggerated media coverage of crime. The security-executive forces can hardly do anything about this. Informing journalists regarding criminal cases requires a tightrope walk between official secrets, data protection and the right of the public to be informed. The security-executive forces are also often confronted with an "American style of journalism." Reporters collect information much more "illegally" and sometimes even "brutally" than they did earlier. They frequently do not care whether or not officials have finished their investigations, something could be destroyed by their reports, fear is evoked among the population, or people lose their reputation. During the last few years, journalists have not hesitated to manipulate fear by associating crime with particular groups of foreigners in their reports. Perner (1992) has called these journalists "desk perpetrators."

Real criminal experiences, however, only explain a small part of feelings of insecurity among the population. What mainly makes people feel insecure is noise from and conflict with their neighbors; carelessness and hindrances in public space (parking on pavements, the blocking of entrances, etc.); unusual uses of public space (loudly dressed youth; alcoholics and drug addicts; tramps that linger in parks, railway stations or in squares; foreigners who repair their cars in the street, etc.); being molested in public space (unwanted contacts in parks or in the streets, such as being panhandled, badgered or addressed; women being molested by men; being frightened by big dogs, etc.); and so on (Karazman-Morawetz, 1996, p. 22).

Public fear and feelings of insecurity result more often from different aspects of disorder within the community than from criminal acts. The generally widespread feeling of safety in Austria has to be seen in the light of the relatively secure public order which now exists. Yet police officers only have few or weak formal measures for the maintenance of public order; for example, they have the authority to command "problematic people" (tramps, etc.) to leave certain public places. Mainly, though, they have to use informal measures, such as the application of communicative and argumentative skills, to achieve a kind of de-escalation in each case and thus meet the challenge of maintaining order in public spaces.

Migration

Since the opening of the eastern boundaries, Austria is situated on the crossroads of two migration routes and movements. People come from the East and head to the West or they come from the South and migrate to the North. As a "Schengen country" Austria has to protect 1,460 kilometers of the external border of the European Community: 1,259 km. of eastern border (Czech Republic, Slovakia, Hungary, Slovenia) and 201 km. of western border (Switzerland, Liechtenstein). This is a difficult task for the border-gendarmerie. In order to better protect the border with Hungary, the border-gendarmerie is assisted by the Austrian Army. Both organizations are confronted with an (again) increasing number of immigrants, refugees and asylum-seekers trying to cross the border illegally, the majority of whom are being smuggled by Mafia organizations. Dealing with these illegal entrants—among whom are many people from Romania and the former Yugoslavia—is a challenge to the norms of democratic policing, because of the need for compassionate treatment demanded by international norms and the prejudice of the public regarding immigrants in a country that perceives itself as a "non-immigration-country" (Faßmann & Münz, 1992). According to a critical assessment of the performance of police officers by Team Consult in 1989–1990, however, the security-executive forces have mastered this challenge, but not quite as well as they have confronted the challenge of crime in general (Kogler, 1991, pp. 14f.; TC, 1990).

PROFESSIONAL CHALLENGES

The Challenge of Humanity: Abuse of Power and Human Rights

Despite, or because of, many constitutional guarantees, there are complaints made against police violence and human rights violations by the security-executive forces in Austria. Between 180 to 350 complaints each year are made about the police and about 30 to 90 cases a year about the gendarmerie. During the last few years there have been, again and again, spectacularly presented media reports about "police-attacks."[14] The Austrian ombudsman, the anti-torture committee of the European Council and of the United Nations, and the human rights organization Amnesty International have seriously criticized the security-executive forces for reported cases indicating serious violations of rights. These charges were investigated, however, by the Austrian police themselves and not—as often demanded—by a group of independent persons. Most charges have been determined to be unfounded.

As a matter of fact, the problem of "inadmissible violence while being on duty" seems to have existed to a much higher extent about one or two decades ago. In the course of a survey held in the town of Salzburg in 1978, 3 percent of the people interviewed said they had been "abused" by a police officer. This figure seems rather high, although the survey was conducted at a time of many mass demonstrations and in handling them police officers were under stress and thus prone to use violence; it might also be the case that some young men among those interviewed just tried to make themselves more interesting by making such statements. Further, 40 percent of the people then interviewed said that "the police sometimes use corporal punishment even if they do not react in self-defense" (IfG, 1978). In a survey conducted throughout Austria in 1984, about one-third of the people interviewed stated that such incidences "happen more often than one can hear or read about" (Fessel + GfK, 1984a, question 14). In a survey among male youths in Vienna in 1986, 71 percent of those questioned stated that "the police should be strictly forbidden to beat criminals during interrogations;" 10 percent, however, pleaded for "the right of the police" to do so (Schulz, 1993, p. 223). The "Awareness Report of the Austrian Chamber of Lawyer's Day, 1988" noted that "the impression cannot be denied that in some parts of the Viennese Police, especially in the Viennese Security Office, abuse and maltreatment are inherent in the system and are perceived as tacitly accepted."[15] And in a survey in the whole of Austria in 1990, 48 percent of those questioned thought that "violent acts by the police occur" at least "occasionally" (OGM, 1990).

Police brutality occurs in the context of a strict, hierarchical organization. Hierarchy results in aggressions being built up and passed down the chain of command until they finally arrive at the bottom of the organizational pyramid where they are "stored" by the officers on duty. It can, therefore, be suggested that these aggressions are then expressed against the person encountered and "treated in the course of official duty." Typically enough, a recently published book concerning this topic is called *Whipping Boys*, a title which is meant to suggest that "many" officers are "whipped" from above and "some" then "pass on the whipping to the ones below" (Anschober, 1995, p. 10).

Corruption

Police corruption has not yet become a big problem in Austria. Only seven cases were reported in 1994. This is a further reason why the confidence in the security-executive forces is so high in Austria. According to public opinion surveys, some groups, especially politicians, city executives and mayors, executives of insurance companies, builders and architects, and managers and entrepreneurs are perceived as prone to corruption; whereas corruption—comparatively speaking—is not expected of policemen and gendarmes (IMAS, 1986; Österreichisches Gallup-Institut, 1988). However, since the pay of police personnel is not especially high and most police officers work overtime (about 5 percent of the police officers in Vienna even have [officially] an additional occupation),[16] an increase in police corruption is generally feared because of the corruptive influence of organized crime.

MEASURES TO BETTER RESPOND TO CHALLENGES AND NEW TASKS

Enhancing Technology

Of the measures that have already been taken, those concerning "organized crime" have to be pointed out, especially the setting up of a special task group of police officers to fight organized crime (in 1993), taking legislative measures (since 1994) such as new and newly defined offenses regarding membership in a criminal organization and money laundering,[17] as well as procedural provisions for the protection of witnesses, in particular the possibility to grant anonymity to witnesses in danger. These have proven to be useful measures. At present, the security-executive forces are working on a special "data bank on organized crime" ("OK-Datei") which will facilitate criminal investigations. In autumn 1997, the "connecting the dots method" (the combining and use of data, for example, among the army, the state police and the fiscal authorities, in order to make the search more efficient) was introduced, and in 1998 "bugging operations" (electronic surveillance and bugging of people and apartments who are suspected or have links to suspects) will be possible under certain, legally defined circumstances, provided that a judge approves in each case. As both methods involve fundamental interventions into the basic rights of privacy and freedom, and as there were worries that citizens could become "glass people," the political debate on this topic has been rather contentious. In the end, supporters of these measures succeeded in this debate.[18]

International Cooperation

In order to fight organized crime, but also due to the consequences of Schengen (for example, Austrian police officers can work now in Germany in order to track criminal elements who have operated in Austria), international cooperation in which Austria has been involved for some time already (for example, setting up the Middle European Police Academy was an Austrian initiative) will have to be (even further) expanded. This will lead to an "Europeanization" of national police-systems (Shelley & Vigh, 1995). In the course of this process, the Austrian security-executive forces will change as well. Generally, they will have to be more flexible in the future; they will have to introduce a different managerial ideology (more delegation of responsibility); and they will have to approach their tasks more professionally and at the same time in a more humane way.

Administrative Reforms

A "dismantling of the hierarchy" will be an essential prerequisite (Edelbacher, 1995, p. 17), and is already necessary due to the rising educational level of police officers (Kogler, 1991, pp. 7, 112). Today, many of the recruits to the police and gendarmerie schools are highschool graduates. Twenty to thirty percent of the recruits drop out of police schools; however, those who stay will be a factor for change in the organization. The educational system of the security-executive forces will have to be reformed also. In no other public organization in Austria, except in the Army, can such a high degree of internal education effort be found as in the security-executive forces. As diverse concepts developed in the last few years suggest, besides the internal police education system there should also be more education in common with other professional groups (for example, in pedagogical academies and academies for social welfare work) in subjects which do not directly relate to police work. The wide gap between the security-executive forces and universities which exists now should be closed. A further reason for expanding educational opportunities and requirements is that the police in a democracy should have working relationships with all segments and institutions of society.

The dismantling of the hierarchy would also facilitate the demand for a better and more humane approach to dealing with citizens. Worth mentioning among steps which should be expanded in order to improve relations with citizens, are Community Policing Projects (Dölling & Feltes, 1993; Fehervary, 1996), such as: Criminal Advisory Services (founded in 1974;[19] at present there are 143 such information centers in the whole of Austria); contact officers (introduced in 1977; in Vienna, for example, there are 280 such officers today); youth contact officers (since 1984); contact officers especially for foreign populations (since 1996); "security places" and "security committees," in which police officers, politicians, citizens, firemen, and representatives of rescue parties, associations and housing construction cooperatives cooperate with each other (since 1991); "meet your cops" programs (since 1994); "mobile police stations" (since 1995); etc.[20]

Furthermore, by introducing a so-called "complaints management system," such as can be found in some larger firms, the security-executive forces could become more citizen-friendly. Complaints should be seen as chances to get to know something about the needs of the population, and should be handled correspondingly. Positive effects in this direction (developing closer relationships with the people) could also result from the promotion of (more) women into management positions.

In order to better address the foreign community in Austria, it might be advisable to recruit and employ people from this section of the population; at this time they are not allowed to join the security-executive forces because they are not Austrian citizens. The employment of foreigners is especially advisable because of the fact that immigration will certainly continue, even if to a reduced extent (IFES, IWS & KDZ, 1990). There are about 150,000 Hungarians, Poles, Czechs and Slovaks who would like to settle down in Austria and work there. Generally, it should be brought home to police officers that foreigners are no more prone to criminality than are native Austrians and that immigrants are unfairly targeted by policing policies.

Another step must be improvements in the working conditions (equipment, infrastructure, etc.) of police officers. This would also help reduce stress levels, which are rather high at the moment (Meggeneder, 1995; Wolf & Korunka, 1993) and would certainly enable police officers to meet the challenges of democratic policing (even) better than they are doing now.

CONCLUSION

Democratic policing depends to a large extent on the democratic organization of the police. According to Pelinka, Haller & König (1990, p. 74), the degree of democratic organization of an institution can be seen in the extent to which it is open and transparent, especially to critics, and controlled from outside. In this sense, the Austrian security-executive forces surely have democratic deficits, especially concerning the control of police deviance. Furthermore, it sometimes seems that the organization of the security-executive forces itself, with its strong hierarchical order, prevents democratically inclined police officers from coming closer to citizens and, thus, to act more democratically. Nevertheless, police officers have done good police work so far, as one can see in the fact that Austria is a safe country—except for roadworthiness, perhaps. It will be Austria's great challenge to maintain or improve this safety in this part of the "New Europe."

NOTES

[1] Metternich was Austrian State Chancellor from 1821–1848.
[2] Motz-art, *Salzburger Nachrichten*, April 24, 1993. These instructions have been removed.

3   In comparison, about 37 percent of all civil servants in the federal administration in Austria are women.
4   See also, *Frauenansturm auf die Innsbrucker Polizei* (Women's siege of the Innsbruck police), *Die Presse*, November 25, 1996, p. 11.
5   The wording is this: paragraph 25 (1): To prevent dangerous assaults on the life, health and the property of the people, it is incumbent on the security authorities to encourage the willingness and the ability of the individual to obtain knowledge of a threat to his[her] objects of legal protection and, correspondingly, to prevent assaults. (2): Moreover, it is incumbent on the security authorities to encourage plans that serve the prevention of dangerous assaults on life, health and the property of the people.
6   There is, however, no sociological research on this kind of problem in Austria, with the exception of one academic thesis (Artner, 1993).
7   For example, to assign duty shifts in such a way that most of the police officers work at that time when they are most urgently needed (in the evening) and not during the daytime when everything is relatively peaceful.
8   A complaints commission, established externally to the police organization in each of the federal provinces in Austria, investigates and decides on complaints. The members of this so-called "Independent Administration Senate" are nominated by the government of each federal province.
9   A famous swindler and confidence man who could, due to his contacts with politicians, escape the jurisdiction of the courts—which thereby revealed themselves to be politically influenced—for some time.
10 The criminal code was changed from an offense-related law (*Tatstrafrecht*, act-related criminal law), according to which a criminal is charged for the crime he or she committed regardless of his or her personality, background or state of mind, to an offender-oriented law (*Täterstrafrecht*, actor related criminal law), which means that the personality, the background and the state of mind of a criminal is taken into account when he or she is sentenced. This ongoing liberalization has led to a more thoughtful application of prison sentences and punishments, and the average duration of imprisonment has been drastically shortened.
11 For example, an American working in Vienna described his experiences as follows:

> When performances are over around ten, throngs of all ages, with the normal numbers of single, elderly and female, leave the theaters and the opera and routinely make their way home, or elsewhere, on mass transit or foot. Unlike in the U.S., they don't necessarily scurry into cabs bound either for home or for expensive watering holes or remain far removed from contact with surrounding streets and neighborhoods. In Vienna, fear does not accompany people out at nights, as it never comes along with me, nor have I ever heard it mentioned as a topic of conversation. True, newspapers report burglaries and domestic assaults, even political violence occasionally, but the individual's sense of safety is quite secure. Whether people stay out for a short time, or as I often did, for a good part of the night, they walk or ride in tranquility, their thoughts free to roam far from the haunting fears of others.... If you can't sleep would you feel free to venture out of doors in any American large city or even suburb? If duty or pleasure keeps you from home, do you travel happily in your own city? Recent optimistic figures about the large decline in American crime rates are long overdue in terms of making the country more livable, but they are so new and as yet so undecipherable in terms of the factors that may have led to this that no one seems convinced the leap ahead, or back into the way things used to be, will last.... But in Vienna, no one seems afraid for personal safety, for violent robbery or rape. If you own good jewelry you need not deny yourself the pleasure of wearing it. And a fur openly worn attracts no attention or hostility either. The individual seems quite self-sufficiently safe, any individual.... (Bernheim, 1997, p. 15).

12 For an example see, *Mißbrauchte Kriminalität* (Abused criminality), *Die Presse*, September 28, 1996, p. 13.
13 See *Meidlinger Bezirksjournal*, May 1993, p. 7.
14 The latest report can be found in *Profil*, November 11, 1996, pp. 46–51.
15 Cited in *Wiener Zeitung*, November 22, 1988, p. 4.
16 *Kurier*, August 11, 1993, p. 15; August 12, 1993, p. 15.

[17] Austrian banks are now obliged to report suspicious transactions to the police.
[18] According to surveys done in 1995, more than two thirds of Austrians supported the authority of the police to use "bugging operations" and, depending on how the questions were worded, about half to about seven out of ten respondents supported the "connecting the dots method" (Diess, 1995a; OGM, 1995a).
[19] Their establishment was stimulated by Crime Prevention Projects in Great Britain.
[20] Neighborhood watch programs, however, are not possible in Austria, because they would remind the population of the "spy-on-each-other-system" during the Third Reich.

REFERENCES

Aiginger, K., & Peneder, M. (1997). *Qualität und Defizite des Industriestandortes Österreich* [The quality and deficits in the position of Austrian industry]. Wien: Österreichisches Institut für Witschaftsforschung (Austrian Institute of Economic Research).

Allerbeck, K. R. (1982). *Ein Generationenkonflikt* [A generational conflict]. In H. Janig, P. Hexel, K. Luger, & B. Rathmayer (Eds.), *Jugendprobleme—Jugendprotest* [Youth problems—youth protest]. Stuttgart. (Cited from Schulz, 1993, p. 199).

Anschober, R. (1995). *Die Prügelknaben der Nation. Chronik einer Schande* [Whipping boys for the nation. A chronicle of shame]. Wien: Edition Va Bene.

Artner, D. (1993). *Die Funktion von Rechtsnormen im Polizeialltag. Die Anwendung von Rechtsnormen durch Sicherheitswachebeamte* [The function of legal standards in the working day of the police. The application of legal standards by security officials]. Wien: Institut für Soziologie, Grund- und Integrativwissenschaftliche Fakultät der Universität Wien, Diplomarbeit (Institute of Sociology, Faculty of Basic and Integrative Sciences of the University of Vienna, Thesis).

Bartunek, E. (1997). *Die Arbeitskräfteerhebung der EU—Internationale Daten für 1995* [The growth in the labor force of the European Union—International data for 1995]. *Statistische Nachrichten* (Statistical News), 52, 7, 520–540.

Beidelt, I. (1896–1898). *Geschichte der Österreichischen Staatsverwaltung 1740–1848. Band II* [History of Austrian public administration. Volume II]. Innsbruck: Wagner.

Bernheim, M. (1997, February 14, 15). Self-sufficiently safe. *Wiener Zeitung EXTRA*.

Bögl, G. & Seyrl, H. (1992). *Die Wiener Polizei. Im Spiegel der Zeiten. Eine Chronik in Bildern.* [The police of Vienna. In the mirror of their times. A chronicle in pictures]. Wien: Edition S (Verlag Österreichische Staatsdruckerei).

Busch, H. (1995). *Grenzenlose Polizei. Neue Grenzen und polizeiliche Zusammenarbeit in Europa* [Police without borders. New borders and police cooperation in Europe.] Münster: Westfälisches Dampfboot.

Das, D. K. (1994). Can police work with people: A view from Austria. *The Police Journal, 68*, 334–346.

———. (1995, May). Challenges of Policing Democracies: Executive Summary. *Proceedings of the Second International Police Executive Symposium*. Onati, Spain.

Davy, B. & Davy, U. (1991). *Gezähmte Polizeigewalt?* [Tamed police violence?]. Wien: Service Verlag.

Dearing, A. (1993). *Der sogenannte Kriminaltourismus. Der Ladendiebstahl und die Öffnung der Ostgrenzen* [The so-called criminal tourism. Shoplifting and the opening of the eastern borders]. In A. Pilgram (Ed.), *Grenzöffnung, Migration, Kriminalität. Jahrbuch für Rechts- und Kriminalsoziolgie* [Opening of borders, migration, criminality. Yearbook of the sociology of law and criminal sociology] (pp. 181–194). Baden-Baden: Nomos Verlagsgesellschaft.

Diederichs, O. (1992). *Redaktionelle Vorbemerkung. "Sicherheit ist Kapital." Private Sicherheitsdienste in Zahlen und Fakten* [Editorial preface. "Security is capital." Private security services in numbers and facts]. *Bürgerrechte & Polizei /CILIP* (Civil Rights and the Police/CILIP) 43, 4f, 24–31.

Diess. (1995a). *Umfrage "Lauschangriff und Rasterfahndung"* [Inquiry "Eavesdropping and electronic data base searches"]. Salzburg/Wien: Author.

———. (1995b). *Umfrage "Sicherheit in Wien"* [Inquiry "Safety in Vienna"]. Salzburg/Wien: Author.

Dölling, D. and Feltes, T. (1993). *Community policing. Comparative aspects of community oriented police work*. Holzkirchen: Felix Verlag.

Dr. Haerpfner-Datenanalyse-Datenpräsentation. (1989). *Sicherheit und Fremdenverkehr in Österreich. Öffentliche Sicherheit und persönliche Sicherheit als Faktor bei ausländischen Gästen in Österreich. Studie im Auftrag des Kuratorium Sicheres Österreich*. [Safety and tourism in Austria. Public and personal safety as a factor for foreign guests in Austria. A study commissioned by the Kuratorium Safe Austria]. Wien: Author.

Dreher, G. & Feltes, T. (1996). *Notrufe und Funkstreifenwageneinsätze bei der Polizei* [Emergency calls and radio patrol car responses by the police]. Holzkirchen: Felix Verlag.

Edelbacher, M. (1993, October). *Das Tor zum Osten*. [The door to the East]. *Der Kriminalbeamte* (The Detective), 8–13.

——. (1995, September). *Kriselt's in der Polizei?* [Is there a crisis in the police?]. *Der Kriminalbeamte* (The Detective), 16–18.

Eichwalder, R. (1993). *Erreichbarkeit von bestimmten Einrichtungen. Ergebnisse des Mikrozensus September 1992* [The attainability of certain institutions. Results of the micro-census of September 1992]. *Statistische Nachrichten* (Statistical News), 10/1993, 853–859.

Ettmayer, W. (1990). *Muß es immer mehr Verbrechen geben? Sicherheit und Kriminalität als neue Herausforderung* [Must there be more and more crime? Safety and criminality as a new challenge]. Wien: Politische Akademie, Schriftenreihe Standpunkte, Band 24 (Political Academy, Opinion Series, Volume 24).

Faßmann, H. & Münz, R. (Eds.). (1992). *Einwanderungsland Österreich? Gastarbeiter—Flüchtlinge—Immigranten* [Immigrationland Austria? Guest workers—refugees—immigrants]. Wien: Bundesministerium für Unterricht und Kunst, Österreichische Akademie der Wissenschaften, Institut für Demographie (Federal Ministry of Education and Art, Austrian Academy of Sciences, Institute of Demography).

Fehervary, J. (1996). *Community policing—nun auch in Wien* [Community policing—now also in Vienna]. In W. Hammerschick (Ed.). *Die sichere Stadt—Prävention und kommunale Sicherheitspolitik* [The safe city. Prevention and the politics of community security] (pp. 149–167). Baden-Baden: Nomos Verlags-Gesellschaft.

Feltes, T. (1995). *Notrufe und Funkstreifeneinsätze als Meßinstrument polizeilichen Alltagshandelns* [Emergency calls and radio responses as a measure of day to day police work]. *Die Polizei* (The Police), *86*, 157–188.

Fessel + GfK. (1984a). *Umfrage "Image der Polizei"* [Inquiry "Image of the Police"]. Wien: Author.

——. (1984b). *Umfrage "Bürokratie." Im Auftrag der Bundeskammer der gewerblichen Wirtschaft* [Inquiry "Bureaucracy." Survey commissioned by the Economic Chamber]. Wien: Author.

——. (1991). *Umfrage "Institutionsvertrauen"* [Inquiry "Trust in Institutions"]. Wien: Author.

——. (1996). *Umfrage "Politische Indikatoren"* [Inquiry "Political Indicators"]. Wien: Author.

Findl, P. (1997). *Demographische Lage im Jahre 1996*. [The demographic situation in the year 1996]. *Statistiche Nachrichten* (Statistical News), *52*, 812–832.

Fischer-Kowalski, M., Höllbacher, M., Köckeis, G., Leitner, F., Schäfer, I., & Steinert, H. (1982). *Polizei und Öffentlichkeit. Endbericht. Eine Untersuchung der Wiener Sicherheitswache und ihres Verhältnisses zur Bevölkerung 1972* [Police and transparency. Final report. An investigation of the security forces of Vienna and their relations to the public]. Wien: Institute für Höhere Studien (Institute of Higher Studies).

Fuchs, H., Funk, B.-C., & Szymanski, W. (1993). *Sicherheitspolizeigesetz—SPG* (Security Police Act—SPG). Wien: Manz.

Gebhardt, H. (1997). *Die Gendarmerie in der Steiermark von 1850 bis heute* [The gendarmerie in Styria from 1850 until today]. Graz: Leykam.

Girtler, R. (1980). *Polizei-Alltag. Strategien, Ziele, und Strukturen polizeilichen Handelns* [The working day of the police. The strategies, goals, and structures of police actions]. Opladen: Westdeutscher Verlag.

——. (1981). *Überlegungen zur Polizeibürokratie aus der Sicht der Soziologie* [Reflections on police bureaucracy from the perspective of sociology]. *Zeitschrift für Verwaltung* (Journal of Administration), *1*, 13–19.

Halàsz, L. (1997). *Städtvergleich: Berlin–Wien* [Comparison of the cities Berlin and Vienna]. *Statische Mitteilungen Stadt Wien* (Statistical Information City Vienna), *97*, 2, 3–27.

Hanak, G. (1986). *Polizei und Konfliktverarbeitung im Alltag* [The police and the processing of everyday conflict]. *Bürgerrechte & Polizei/CILIP* (Civil Rights and the Police/CILIP), 25, 26–39.

———. (1991). *Polizeinotruf. Intervention über Aufforderung. Ergebnisse einer empirischen Untersuchung zum Polizeinotruf in Wien* [Emergency calls to the police. Intervention by request. Results of an empirical investigation of emergency calls to the police in Vienna]. Holzkirchen: Felix Verlag.

Hanak, G. & Pilgram, A. (1991). *Der andere Sicherheitsbericht. Ergänzungen zum Bericht der Bundesregierung* [The other security report. Supplement to the report of the federal government]. Wien: Verlag für Gesellschaftskritik.

Hanak, G. & Pilgram, A. (1993). Ziffern zur Aufklärung—Kriminalitätsentwicklung in Wien unter dem Einfluß von Ostgrenzenöffnung und Migration [Data for an explanation—the development of criminality in Vienna under the influence of the opening of the eastern borders and migration]. In *Beiträge zur Stadtforschung, Stadtentwicklung und Stadtgestaltung, Band 47* [Contributions to city research, city development and city design, volume 47] (pp. 55–78). Wien: Magistrat der Stadt Wien, MA 18—Stadtentwicklung und Stadtplanung (City Council of Vienna, MA 18—city development and city planning).

Hanisch, E. (1994). *Der lange Schatten des Staates. Österreichische Gesellschaftsgeschichte im 20. Jahrhundert* [The long shadow of the state. Austrian social history in the 20th century]. Wien: Ueberreuter.

Hauer, A. & Keplinger, R. (1993). *Handbuch zum Sicherheitspolizeigesetz* [Handbook for the Security Police Act]. Eisenstadt: Prugg Verlag.

Heindl, W. (1991). *Gehorsame Rebellen. Bürokratie und Beamte in Österreich 1780 bis 1848. Studien zu Politik und Verwaltung 36* [Obedient rebels. Bureaucracy and officialdom in Austria from 1780 to 1848. Studies in politics and administration 36]. Wien/Köln/Graz: Böhlau.

Hesztera, G. (1994). *Die Geschichte der Gendarmerie. Die Österreichische Bundesgendarmerie feiert ihr 145-jähriges Bestehen* [The history of the gendarmerie. The Austrian federal gendarmerie celebrates its 145th anniversary]. *Öffentliche Sicherheit* (Public Security) 7–8, 29–32.

Hirschfeld, A. (Ed.). (1989). *Die Gendarmerie in Österreich, 1849–1989* [The Gendarmerie in Austria, 1849–1989]. Wien: Bundesministerium für Inneres (Federal Ministry of the Interior).

IFES (Institut für empirische Sozialforschung. Institute for empirical social research). (1980). *Subjektives und objektives Sicherheitsempfinden in Wien* [Subjective and objective security in Vienna]. Wien: Author.

———. (1982). *Sicherheitsempfinden der Österreicher* [Feelings of safety of Austrians]. Wien: Author.

———. (1995). *Leben in Wien. Basisstudie* [Life in Vienna. Basic study]. Wien: Author.

IFES/IWS/KDZ (Institut für empirische Sozialforschung/Institut für Wirtschafts- und Sozialforschung/Kommunalwissenschaftliches Dokumentationszentrum. Institute for empirical social research/Institute for industrial and social research/Documentation center for community research) (Ed.). (1990). *Wien 2010. Entwicklungstendenzen bei wachsender Bevölkerung und offenen Grenzen* [Vienna 2010. Developmental tendencies as a result of growing population and open borders]. Im Auftrag des Magistrats der Stadt Wien, MA 18—Stadtentwicklung und Stadtplanung (Study commissioned by the Magistracy of the City of Vienna, MA 18—city development and city planning). Wien: Author.

IfG (Institut für Grundlagenforschung. Institute for basic research). (1978). *Umfrage "Polizei"* [Inquiry "Police"]. Salzburg: Author. (Cited in *Wiener Zeitung*, December 3, 1978, p. 6).

IMAS (Institut für Markt- und Sozialanalysen. Institute for market and social analysis). (1986). *Umfrage "Korruption in Österreich"* [Inquiry "Corruption in Austria"]. Linz: Author.

———. (1989). *Umfrage "Polizei als Hüterin der Sicherheit"* [Inquiry "The police as guardian of security"]. Linz: Author.

———. (1992). *Umfrage "Kundenfreundlichkeit"* [Inquiry "Customer friendliness"]. Linz: Author.

———. (1993). *Umfrage "Ansehen von Berufen"* [Inquiry "The prestige of occupations"]. Linz: Author.

——. (1995). *Umfrage "Kriminalität aus der Sicht der Deutschen und Österreicher"* [Inquiry "Criminality as seen by Germans and Austrians"]. Linz: Author.
Institut für Konfliktforschung (Institute for the study of conflict) (1990). *Quantitative Studie "Das Image der Sicherheitsexekutive"* [Quantitative study "The Image of security executive forces"]. Im Auftrag der Bundesministerium für Inneres (Commissioned by the Federal Ministry of the Interior). Wien: Author.
International Crime Victimization Survey (1996). *Öffentliche Sicherheit* (Public Security), 7–8, 97, 2.
Jäger, F. (1990). *Das große Buch der Polizei und Gendarmerie in Österreich* [The great book of the police and gendarmerie in Austria]. Graz: Weishaupt.
Jones, T., Newburn, T., & Smith, D. J. (1996). Policing and the idea of democracy. *The British Journal of Criminology 26*, 182–198.
Karazmann-Morawetz, I. (1996). *Was macht Stadtbewohner unsicher? Unsicherheitserfahrungen in zwei Wiener Stadtvierteln und ihre strukturellen Hintergründe* [What makes city dwellers insecure? Experiences with insecurity in two quarters of Vienna and their structural background]. In W. Hammerschick (Ed.). *Die sichere Stadt. Prävention und kommunale Sicherheitspolitik* [The safe city. Prevention and the politics of community security] (pp. 17–38). Baden-Baden: Nomos Verlags-Gesellschaft.
Karazman-Morawetz, I. & Steinert, H. (1995). *Schulische und außerschulische Gewalterfahrungen Jugendlicher im Generationenvergleich. Ergebnisse einer Repräsentativumfrage bei Jugendlichen, Erwachsenen, und Lehrpersonen in Österreich* [Cross-generational experiences with violence by youth in and out of school. Results of a representative inquiry among youth, adults, and educational personnel]. Wien: Institut für Rechts- und Kriminalsoziologie (Institute of Sociology of Law and Criminal Sociology).
Kemper, E. (1996). *Verrat an Österreich* [The betrayal of Austria]. Wien: Zeitschriften-Buch.
Keplinger, R. & Stierschneider, C. (1994). Die Geschichte der Linzer Polizei [The history of the police of Linz]. *Öffentliche Sicherheit* (Public Security), 1–2, 46–57.
KF & P Consulting (Kreutzer Fischer & Partner) (1997). *Österreichischer Kundenbarometer* [Austrian customer barometer]. Wien: Author.
Kogler, R. (1991). *Kritische Würdigung polizeilicher Arbeit und Reformansätze unter besonderer Berücksichtigung der Anwendung von Marketingwissen für sicherheitsbehördliche Dienste* [A critical assessment of police work and reform initiatives, with particular reference to the use of marketing knowledge for security services]. Wien: Wirtschaftsuniversität Wien, Dissertation (University of Economics Vienna, Master's Thesis).
KSÖ (Kuratorium Sicheres Österreich. Kuratorium Safe Austria). (Ed.). (1993, October 6–8). *Forum Sicheres Österreich 1993. Gesellschaft und Kriminalität—Versuch einer volkswirtschaftlichen Rechnung. Bericht über die gleichnamige Veranstaltung in Maria Alm, Salzburg,* [Forum Safe Austria 1993. Society and criminality—an attempt at an economic accounting.] Report on the same-titled conference at Maria Alm, Salzburg. Wien: Author.
Lenz, T. & Mason, K. (Pseudonym) (1992). *Die schutzlose Gesellschaft. Die Ohnmacht der Polizei im Kampf gegen das internationale Verbrechen am Fall Österreich* [The unprotected society. The impotence of the police in the fight against international crime. The case of Austria]. München: Universitas.
Lesjak, K. (1995). Phänomen "OK" [Phenomenon "Organized Criminality"]. *Öffentliche Sicherheit* (Public Security), 95, 8–12.
Market. (1992). *Umfrage "Vertrauenswürdigkeit von Branchen"* [Inquiry "Trustworthiness of sectors of the economy"]. Linz: Author.
——. (1996). *Umfrage "Sicherheitsbewußtsein"* [Inquiry "Awareness of security"]. Im Auftrag des Kriminalpolizeilichen Beratungsdienstes (KBD) (Commissioned by the Criminal Advisory Service). Linz: Author.
Mayerhofer, C. & Jehle, J.-M. (Eds.). (1996). *Organisierte Kriminalität. Lagebilder und Erscheinungsformen. Bekämpfung und rechtliche Bewältigung* [Organized crime. Stock-taking and forms. The fight against and legal control]. Neue Kriminologische Schriftenreihe Band 103 [New criminological series volume 103]. Heidelberg: Kriminalistik Verlag.
Meggeneder, O. (1995). *Abara Kadabara—is a Kibara a Habara? Zur Arbeits- und Berufssituation von Polizistinnen* [Abara kadabara—is a police officer a friend? On the work and occupational situation of male and female police officers]. Linz: Universitätsverlag Tauner.

Miedl, W. (Ed.). (1988). *Die Grazer Polizei* [The police of Graz]. Feldkirchen: Informations- und Verlagsgesellschaft.

Morie, R., Murck, M., & Schulte, R. (Eds.). (1992). *Auf dem Weg zu einer europäischen Polizei* [On the road to a European police]. Stuttgart: Richard Boorberg Verlag.

N.N. (1993). Sicherheitsspezialisten [Security specialists]. *Öffentliche Sicherheit* (Public Security), 5, 20ff.

Nielsen (1993). *Umfrage "Diebstähle"* [Inquiry "Thefts"]. Wien: Nielsen.

Noll, A. J. (Ed.). (1991). *Sicherheitspolizeigesetz (SPG). Textausgabe samt Materialien* [Security Police Act. Text plus materials]. Wien/New York: Springer-Verlag.

Österreichisches Gallup-Institut (1985). *Das Bild der Justiz* [The image of justice]. Im Auftrag des Bundesministerium für Justiz (Commissioned by the Federal Ministry of Justice). Wien: Author.

——. (1988). *Umfrage "Korruption in Österreich"* [Inquiry "Corruption in Austria"]. Im Auftrag der Zeitschrift *trend* (Commissioned by the magazine *trend*). Wien: Author.

——. (1993). *Umfrage "Sicherheit in Wien"* [Inquiry "Security in Vienna"]. Im Auftrag. der Tageszeitung *Kurier* (Commissioned by the newspaper *Kurier*). Wien: Author.

——. (1994). *Sicherheitsbewußtsein in Österreich. Bekanntheitsgrad des Kriminalpolizeilichen Beratungsdienstes* [Awareness of security in Austria. Familiarity with the advisory service of the criminal police]. Im Auftrag des Bundesministerium für Inneres (Commissioned by the Federal Ministry of the Interior). Wien: Author.

Österreichisches Statistisches Zentralamt (Austrian Central Statistical Bureau). (Ed.). (1996). *Statistisches Jahrbuch für die Republik Österreich 1996* (Statistical yearbook for the Republic of Austria). Wien: Author.

OGM (Österreichische Gesellschaft für Marketing. Austrian society for marketing). (1990). *Umfrage "Polizei"* [Inquiry "police"]. Wien: Author.

——. (1995a). *"Bombenangst und Kompetenzen für Polizei"* ["Fear of bombs and the competency of the police"]. Wien: Author.

——. (1995b). *Umfrage "Sicherheit in Wien"* [Inquiry "Security in Vienna"]. Wien: Author.

Pelikan, C. & Pilgram, A. (1993). *Läßt sich "Erziehung ohne Zwang" im öffentlichen Bereich verwirklichen? Zum Jugendwohlfahrtsgesetz 1989 und zum Jugendgerichtsgesetz 1988* [Can "education without compulsion" be implemented in the public arena? On the youth welfare law of 1989 and the juvenile court law of 1988]. In Bundesministerium für Umwelt, Jugend, und Familie (Federal Ministry of Environment, Youth, and Family) (Ed.). *2. Bericht zur Lage der Jugend* [Second report on the situation of youth] (pp. 264–276). Wien: Bundesministerium für Umwelt, Jugend und Familie.

Pelinka, A., Haller, B., & König, I. (1990). *Die Aussensicht der Polizei. Band 1. Qualitative Studie* [The external image of the police. Volume 1. Qualitative study]. Im Auftrag des Bundesministerium für Inneres [Commissioned by the Federal Ministry of the Interior]. Wien: Institut für Konfliktforschung.

Perner, R.A. (1992). *Menschenjagd. Vom Recht auf Strafverfolgung* (Manhunt. On the right to prosecution). Wien: Donauverlag.

Pichler, J. W. (1993). *Rechtsakzeptanz. Eine empirische Untersuchung zur Rechtskultur aus dem Blickwinkel der Ideen, Werte und Gesinnungen. Dargestellt am Beispiel einer österreichischen Demoskopie* [Acceptance of law. An empirical investigation of legal culture from the perspectives of ideas, values and predispositions. Depicted through the example of an Austrian survey] Wien/Köln/Weimar: Böhlau Verlag.

Pilgram, A. (1990). *Zur Sicherheitsinformation in Österreich. Wie das polizeiliche Definitionsmonopol über die "innere Sicherheit" hergestellt wird* [On security information in Austria. How the police monopoly on defining "domestic security" is accomplished]. *Kriminalsoziologische Bibliografie* (Criminal-Sociological Bibliography), 17, 3–36.

——. (1993). Fremdenkriminalität [The criminality of strangers]. *Kontraste* 5, 17ff.

——. (1995). Die Zweite Republik in der Kriminalpolitik [Criminal policy in the second republic]. In R. Sieder, H. Steinert, & E. Tàlos, (Eds.). *Österreich 1945–1955* (pp. 485–496). Wien: Verlag für Gesellschaftskritik.

Pilgram, A., Hanak, G., & Morawetz, I. (1992). *Probleme der Sicherheits- und Kriminalpolitik in Wien in Zusammenhang mit der Ostgrenzenöffnung* [Problems of security and crime policy in Vienna, in relation to the opening of the eastern border]. Im Auftrag des Magistrats der Stadt Wien, MA 18—Stadtentwicklung und Stadtplanung [Commissioned by the

Magistracy of the City of Vienna, MA 18—city development and city planning]. Unveröffentlichter Forschungsbericht. Unpublished research report.
Plasser, F. & Ulram, P. A. (1991). *Staatsbürger oder Untertanen? Politische Kultur Deutschlands, Österreichs und der Schweiz im Vergleich* [Citizens or subjects? A comparison of the political cultures of Germany, Austria, and Switzerland]. Frankfurt/aM./Bern/New York/Paris: Peter Lang.
Pütter, N. (1997). *Organisierte Kriminalität in amtlichen Zahlen. Über die Aussagekraft der Lagebilder* [Organized crime in official figures. On the evidentiary power of basic information]. *Bürgerrechte & Polizei/CILIP* (Civil Rights and the Police/CILIP), 56, 15–25.
Reinprecht, C. (1995). *Österreich und der Umbruch in Osteuropa* (Austria and changes in Eastern Europe). In R. Sieder, H. Steinert, & E. Tàlos (Eds.). *Österreich 1945–1955,* (pp. 341–353). Wien: Verlag für Gesellschaftskritik.
Reuss-Ianni, E. (1983). *Two cultures of policing. Street cops and management cops.* New Brunswick/London: Transaction Books.
Röglin, H.-C. (1989). *Zum Selbstverständnis der Wiener Polizei* [On the self-image of the police]. Studie im Auftrag der Freunde der Wiener Polizei [Study commissioned by the friends of the Viennese police]. Unpublished paper, cited in *Die Presse,* November 15, 1989, p. 18. Vienna.
Rosenmayr, L. (Ed.). (1980). *Politische Beteiligung und Wertwandel in Österreich. Einstellungen zu Politik und Demokratieverständnis im internationalen Vergleich* [Political participation and value change in Austria. An international comparison of attitudes toward politics and the comprehension of democracy]. München/Wien: Verlag R. Oldenbourg/Verlag für Geschichte und Politik.
Sandgruber, R. (1995). *Ökonomie und Politik. Österreichische Wirtschaftsgeschichte vom Mittelalter bis zur Gegenwart* [Economy and politics. The economic history of Austria from the middle ages to the present]. Wien: Ueberreuter.
Schulz, W. (1993). *Städtische Jugendliche und ihre gesellschaftliche Integration. Ergebnisse einer empirischen Untersuchung* [Urban youth and their social integration. Results of an empirical investigation.] In Bundesministerium für Umwelt, Jugend, und Familie (Federal Ministry of Environment, Youth, and Family) (Ed). *2. Bericht zur Lage der Jugend* [Second report on the situation of youth] (pp. 199–227). Wien: Bundesministerium für Umwelt, Jugend, und Familie.
Shelley, L. I. (1996). *Policing Soviet society. The evolution of state control.* London: Routledge.
Shelley, L., & Vigh, J. (Eds.). (1995). *Social changes, crime and the police.* Chur: Harwood Academic Publishers.
SWS (Sozialwissenschaftliche Studiengesellschaft. Social scientific research society) (1990). *Das Sozialprestige der Facharbeiter* (The social prestige of skilled workers). Wien: Author.
———. (1995a-d, 1996). *Umfragen zu verschiedenen Themen* (Inquiries into various themes). Wien: Author.
Tàlos, E. & Kittel, B. (1995). *Sozialpartnerschaft. Zur Konstituierung einer Grundsäule der Zweiten Republik* [Social partnership. On the creation of a pillar of the Second Republic]. In R. Sieder, H. Steinert, & E. Tàlos (Eds.). *Österreich 1945–1955* (pp. 107–121). Wien: Verlag für Gesellschaftskritik.
TC (Team Consult AG). (1990). *Grobanalyse "Verwaltungsmanagement BMI"* [Rough analysis "Administration-management in the Federal Ministry of the Interior"]. Zürich: Author.
TC (Team Consult Austria). (1994). *Organisationsentwicklung bei der Bundesgendarmerie. Analyse der Strukturreformen bei der Gendarmerie unter besonderer Berücksichtigung der neuen Dienstzeitregelung* [The organizational development of the federal gendarmerie. An analysis of structural reforms in the gendarmerie, with special consideration of the new conditions of service regulations]. Wien: Author.
Wolf, C. & Korunka, C. n.d. (1993). *Belastung und Beanspruchung der österreichischen Exekutive. Forschungsbericht* [Workload of and demands on the Austrian executive branch. Research report]. Wien: Abteilung Arbeitsmedizin, Universitätsklinik Innere Medizin IV, Institut für Psychologie, Universität Wien.
Zima, H. & Hochenbichler, E. (1969). *100 Jahre Wiener Sicherheitswache. 1869–1969* [100 years of the Viennese security guards. 1969–1969]. Wien: Bundespolizeidirektion Wien (Directorate of the Federal Police Vienna)

10

The Challenges of Policing Democracy: The British Experience

GREGORY J. DURSTON

INTRODUCTION

England, a smallish island (228,356 square miles) lying off the northern coast of the European continent, is home to about 58.8 million people (in 1996). The official political designation, the United Kingdom of Britain and Northern Ireland, includes the areas of Britain, Wales, Scotland and Northern Ireland. The political and legal systems (parliamentary democracy and common law) which originated in English history can be found in many countries of the world, the result of successful colonization. Current politics is dominated, in iterative alterations, by two parties which arose during the successive enfranchisement of social groupings into the political life of the country, with occasional third party longings arising in response to specific domestic and international events. A constitutional monarchy represents and symbolizes an overarching organizational framework, a sense of national identity, and allows for the mitigation of political conflicts.

England enjoys a high standard of living and, partly because of its colonial history and continuing ties, is strongly integrated into the global economy. Yet there has been some reluctance to join into membership activities of the European Union as freely as some other European countries have done. The British people see themselves as somewhat aloof from Continental political and cultural practices and prefer the stability of their own traditions and norms.

THE CONCEPT OF DEMOCRATIC POLICING IN BRITAIN

The notion of "democratic" policing in Britain has been linked to a number of separate, but closely associated, concepts, none of which holds a monopoly of the field. However, a "blanket" description that encompasses most aspects of the topic could be said to be contained in the concept of police legitimacy. In Britain, historically, a ready public compliance in their own policing was a distinctive feature of the country's law enforcement system. Legitimacy has "popularly been known as 'policing by consent'." However, public consent was (and is) premised on the police being perceived as having a legitimate authority, that is, "the power he or she wields is recognized as legitimate by those who are subjected to it" (Waddington, 1991, p. 3). Nevertheless, consent is not irrevocable, and legitimacy can be lost by the police. Some commentators assert this has happened, to some extent, in recent decades in Britain. According to this analysis, although all police everywhere in the world are ultimately a coercive force (the State having a monopoly of the use of legal force), in Britain, for a long period, coercion could be heavily concealed. Unlike the American situation in which police officers overtly carried the tools of their authority: firearms, long night sticks, MACE gas and handcuffs even in the quieter towns, in Britain this was unnecessary because the police were a widely respected and supported institution.

In turn, legitimacy is premised on diverse factors that together might be considered to cover the theoretical ambit of democratic policing. These factors include (in varying degrees of importance): police accountability (one of the most important aspects), police legality, the use of minimum force in dealing with problems and conflicts (and an attendant lack of overt weaponry), evenhandedness in the use of police powers and operations, and political non-partisanship.

Legitimation

Some critiques of recent British policing have focused on the need for a revival and extension of these themes to restore a perceived loss of respect in the police as an institution. For example, it has been asserted that there is an urgent need for "protecting its [police] independence and extending accountability; giving more substance to its much vaunted neutrality" and for encouraging the police themselves to view formal police action as a final resort, only to be employed where informal and other means of conflict resolution have failed. This approach has the explicit aim of producing a force where welfare paradigms, backed up by a legal and "constitutional corral" for those who hold the office of constable, would be more salient in comparison to a policing ideology based on deterrence and a "simplistic 'law 'n order'" approach (Uglow, 1988, p. 148).

Historically, the initial replacement of the diverse policing agencies of the eighteenth century, such as part-time parish constables and night watchmen, with the "new" police forces, from their inception in London in 1829 and their gradual extension to other cities and counties later in that century, was, initially, fiercely opposed by many different social groups. These were concerned with the potential for government abuse of such a force for political repression, the loss of historic liberties, the financial expense involved, etc.. As Robert Roberts (1973, p. 100) observed of his Edwardian childhood in the northern slum of Salford, almost noone spoke in fond regard of the police and the poor in general "looked upon him with fear and dislike." Roberts felt that when it was said that the public held their bobby in esteem this actually applied only to the middle and upper classes; "these sentiments were never shared by the undermass." Gradually, however, most of these groups were won over, with working class opposition to the new force in some areas lasting longest. There appears to have been a growing acceptance of the institution, even among the lower social groups, a process that peaked in the 1940s and 1950s. As a result, for a period in the mid-twentieth century, it can be argued, the police in Britain enjoyed a level of public support that was unprecedented before or since. (It should also be noted that there was probably a large degree of ignorance of the reality of much policing on the part of a large section of the wider public at this time.)

It has been argued, with considerable plausibility, that the police gained a measure of legitimation after their foundation in 1829 by the successful employment of a number of different strategies, many of which are still important to democratic policing. There was, from the beginning, a marked contrast between the Metropolitan Police model which developed in England and that for the Royal Irish Constabulary found useful in the conquest and control of colonies. Such constabularies were bodies which were often (though not always) paramilitary in appearance, training, deployment and weaponry. A major difference between the two models was the adoption of a strategy of minimal force by the Metropolitan Police. (Their only routine weapon in Britain, the short truncheon, was even carried concealed until 1863.) Although violence by the police was, to an extent, a constant feature of Victorian and Edwardian England, "there is no doubt that the British police developed a tradition of containing industrial disputes and political demonstrations with minimum force when contrasted with the experience of other countries" (Reiner, 1992, p. 65). The police eschewed a military approach to such situations, that is one based on defeating an enemy, in favor of a policy of containment. Force was usually used with restraint (Waddington, 1991, p. 8). This was crucial to the development of police legitimacy as any blurring of the distinction between the military and the police potentially "represents a threat to the public" (Waddington, 1996, p. 22).

Additionally, the development of a tradition of non-partisanship, along with the avoidance of direct political control or involvement, reduced the widespread initial fear that the police would be used as a tool of central government. There was also an early and complete acceptance by the police of accountability to the rule of law, and the practice of strictly disciplining abuses of power by police officers. In a more subtle way, there developed an identification of the police with the people. From the beginning, constables were appointed from the working classes and their officers promoted from the ranks, Sir Robert Peel specifically rejecting any proposal of recruiting an officer class. Combined together, these tactics meant that "from a widely hated and feared institution, the police had come to be regarded as the embodiment of impersonal, rule-bound authority, enforcing democratically enacted legislation on behalf of the broad mass of society rather any partisan interest" (Reiner, 1992, p. 73).

Accountability and Control

Very closely linked in Britain to the notion of legitimacy, and thus democratic policing, have been the concepts of police control and accountability, seen as important elements to help promote a view of the police as a necessary body that deserves support. Accountability is at the root of good policing, especially "in a democracy [where] it is expected that those who wield public power must be fully accountable for this" (Reiner & Spencer, 1993, p. 3). At the same time, to be able to fulfill their purpose, the police have considerable powers above those of the ordinary public, so much so that it has been observed that the police are in some ways an "anomaly in democratic society." They have "an awesome power" to disrupt freedom and intrude into people's privacy, yet at the same time "democracy is heavily dependant upon its police … to maintain the degree of order that makes a free society possible" (Goldstein, 1977, p. 1). The regulation of control and accountability attempts to reconcile this inherent tension.

Centralization and Local Control

Today, there are more than 125,000 police officers in England and Wales (with many more in Scotland), organized in forty-three separate forces.[1] For historical reasons there is a bifurcated system of organizational and operational police control in England. Control is shared among the central government in London and local governments. The basic structures, norms, ideologies and practices of the police are governed by Acts of Parliament.

The Home Secretary has direct responsibility for the two London forces, one of which, the Metropolitan Police, is easily the largest of the British forces. (The other, the City of London force, policing the commercial area, is very small.) The provincial forces have their own police authorities and are not under the direct supervision and responsibility of the Home Secretary.

In the years after the Second World War, a progressive reduction occured in the number of small provincial forces by means of amalgamation, starting with the abolition of forty-five separate borough forces (Cornish & Clark, 1989, p. 619). The most significant change in police structure came in 1964, when the Police Act further reduced the number of forces in England and Wales from 122 to the present forty-three. The Royal Commission on the Police, set up in 1960, whose report resulted in the 1964 Act, had decided narrowly against recommending a national force. One of the reasons for this decision was the potential threat to police accountability. Amalgamations produced a strange mixture of different size forces, ranging from distinct county forces (such as Kent) to much larger groupings (such as the West Midlands Force). In 1988 the Chief Constable of Sussex, Roger Birch, advocated that serious consideration be given to the idea of a *de facto* national police force, albeit one that was regionally differentiated. This argument received some support from Sir Peter Imbert, the then-Commissioner of the Metropolitan Police (effectively the most important British police officer). Additionally, Sir John Wheeler, the Chairman of the all-party parliamentary Home Affairs Committee, suggested that the present number of forces were "insufficiently business-like" and proposed that they should be merged to produce between five and ten regional forces (McLaughlin, 1992, p. 475). Although this has not occurred, the Home Secretary announced to the Association of Chief Police Officers' conference, in July of 1996, that he would be establishing a National Crime Squad for England and Wales to conduct investigations into organized criminals, such as major heroin suppliers, and to support local police forces. He further announced that he would expand the role of the National Criminal Intelligence Service's role in collating information on criminal activity (*Constabulary Journal*, July/August 1996, p. 1).

The structure of provincial police government that was established and consolidated by the 1964 Police Act has been characterized as "explanatory and co-operative." It was a tri-partite organization under which power was divided between a local police authority, the Chief Constable (senior officer) for the force involved, and the Home Secretary (and thus central government). The routine "direction and control" of provincial police forces was placed in the hands of their Chief Constables (Section 5 of the Act), but this officer had, in turn, obligations to report on this control to the other two partners, the local police authorities and the Home Secretary. Chief Constables had to provide an annual report, and could be asked by the local police authority for further reports on any specific matter concerning the policing of their area (Section 12 of the Act). Should he or she feel that such a request was not "needed for the discharge of the functions of the police authority," he or she could refer the matter to the Home Secretary for adjudication.

The members of local police authorities consisted of a mixture of locally elected councilors (a majority, being two-thirds of their number—effectively the democratic element in the system), with the remainder being appointed lay magistrates (Justices of the Peace). The powers of the police authorities are generally exercised subject to final Home Office approval. Thus, although they interview and appoint the Chief Constable (as well as the deputy and assistant Chief Constables), the short list that they consider for these offices first has to be approved by the Home Secretary, who can also veto the final selection if he or she feels that it is inappropriate (but that would be unusual given that he or she has approved the list of candidates in advance). As a result, the Home Secretary is in many ways the final power in the system, not least because in recent decades central government (rather than local authorities) has provided most of the funds for policing, and such payments are subject to the certification of the local force as efficient by the Home Office's Inspectorate of Constabulary. This power was highlighted in 1992 when the Derbyshire force became the first English force in the twentieth century to be officially described as inefficient (*Law and Society Gazette*, September 18, 1991).

Over the years since the 1964 Police Act, a combination of statutory changes (in particular the 1985 Local Government Act), caselaw from the courts, and developments in organizational practice appear to have significantly increased the control of policing by central government. However, despite this, relations between police authorities and their Chief Constables, and between Chief Constables and the Home Office, have usually been cooperative rather than conflictive (though this has not always been so, especially during a period in the mid-1980s). Normally, Chief Constables grant their authorities a degree of real influence, while in turn the authorities respect the professional judgment and experience of their senior officers. This system, combined with the principle of constabulary independence, has also protected chief officers from overt Home Office intervention in most of their operations, while in practice these chief officers have normally implemented circulars from the Home Office (even though these were nominally only advisory).

In recent years, it has been argued, central government has wielded increasing direct influence over the police. This has been accomplished in a number of ways, usually at the expense of the local police authority and (to a lesser extent) the Chief Constable. For example, the Home Office has more actively guided policy by issuing greater numbers of circulars to the Chief Constables than previously. The police inspectorate, which reports directly to the Home Office, has allegedly been evaluating the efficiency of forces more thoroughly than before (producing the above-mentioned Derbyshire result). In part, the reduction of the power of the police authorities appears to have been a result of the mid-1980s reaction by central government to the policies of the more radical Labour councils in some large urban areas. Some have asserted that this process has marginalized the authorities. According to Eric Caines, a member of the Sheehy inquiry into police pay and ranks, some Chief Constables even "hold police authorities in contempt" though this is probably not widespread (cited in *The Economist*, February 26, 1994).

Part of the rationale for establishing the existing system of accountability in regional forces was to protect them against abuse by politicians, whether national or local, by ensuring that no single body had total control over the police — a situation sometimes referred to as "the glory of the British system." However, some critics have not been so laudatory, and have argued that the 1964 structure was seriously flawed. According to this analysis, since the police are primarily a local service, dealing mainly with local crimes and emergencies, the dominant voice in their control should be local government, representing the population in the area most immediately concerned. One admittedly partisan commentator in the mid-1980s went so far as to state that "it is a fundamental principle of a democratic system of government that people who are elected, or appointed, to exercise power over others should be accountable for their actions. This principle is not observed in the current system of policing in Britain" (Spencer, 1985, p. 1). These commentators have argued the case that the police be fully accountable to democratically elected authorities, in a "subordinate and obedient" style and clearly subservient to local government.

However, one of the traditional barriers to this approach has been the doctrine of constabulary independence emanating from common law. This doctrine regards the individual constable (or police officer) as an independent office holder under the Crown, being primarily accountable to the law itself, who cannot be commanded by a governmental department (whether local or national) on how and when to exercise his or her powers. Applying this doctrine, the courts have traditionally refused to uphold challenges to the professional judgment of constables about methods of law enforcement (short of a total failure to carry out their duties). The Divisional Court's recent ruling in the case of *R v Chief Constable of Sussex ex parte International Trader's Ferry Ltd* ([1995] 4 All ER 364) reiterated this doctrine in refusing (as a matter of domestic law) to interfere with the decision of the Chief Constable of Sussex to refuse to allow the deployment of police officers to Shoreham Harbour in sufficient numbers to ensure that lorries carrying live animals for export could be loaded onto ferries bound for the continent, despite the presence of large numbers of animal rights demonstrators. The senior judge's decision restated the established view that a Chief Constable has a duty to uphold the law and to keep public order. If he (or she) abandons that duty, the courts can intervene. However, unless this dereliction of duty is obvious, the courts will be reluctant to intervene in a policy decision made by a senior officer in exercising his or her discretion.

Until recently, the debate over control has largely been between supporters of the consultative versus subordinate approaches. However, in the light of recent proposals for reform, such as the recommendations of the Sheehy Report and the 1993 White Paper, it has been suggested that there may be a move toward a third type of accountability, sometimes termed as "calculative and contractual" (Reiner & Spencer, 1993, p. 17). The background to such a potential change is that during the 1980s central government provided the British police forces with very generous increases in personnel and resources (almost 50% in real terms). However, as crime continued to increase inexorably, the central government began to focus on police efficiency amid concerns that they were not cost effective. For example, despite increased spending on the Metropolitan Police (an increase from 541 million pounds to 1,008 million between 1981 and 1988) offenses recorded by the force in that period rose by almost 27 percent and the crime clear-up rate fell from 21 to 17 percent. As Kenneth Baker (1993, p. 451), a former Home Secretary, admitted in his memoirs written in the early 1990s, there was concern that, "although we had spent more in real terms since 1979 there had still been a substantial rise in crime. 'Where is the value for money?' asked my colleagues."

As a result, during the second half of the 1980s, the government's wider concern with cost effectiveness and financial accountability (in areas such as the Health Service) also focused on the police. In itself this was not surprising, as by this period the policing budgets for the United Kingdom amounted to over five billion pounds a year. The Audit Commission and the House of Commons Home Affairs Committee (which monitor public expenditure) expressed concern about perceived inefficiencies in what they considered to be an out-of-date policing system (McLaughlin, 1992, p. 474).

This concern resulted in a major, if rather subtle, transformation in police organization and philosophy, producing a period of change the like of which had not been seen since the early 1960s (Reiner, 1995). These changes were manifest, for example, in the White Paper on Police Reform and the Sheehy Report on police rewards and responsibilities published in June 1993; these reports were followed by the 1994 Police and Magistrates Courts Act and the 1995 Home Office Report on the Police's Core and Ancillary Tasks.

The Sheehy Inquiry was chaired by Sir Patrick Sheehy, the former chairman of a multinational corporation (BAT). This Inquiry, which lasted for a year beginning in

May 1992, focused on enhancing crime-fighting capacity and "value for money". It recommended the abolition of perceived unnecessary police ranks, the widespread use of fixed-term appointments for police officers, and locally-determined pay scales. The rationale for the proposed reforms was declared bluntly in the 1993 White Paper: "The main job of the police is to catch criminals." Despite this, there has been doubt expressed at the value of deterrence by detection rather than prevention (via, for, example, overt patrols) or as to how performance can be measured in an occupation where discretion is so important. The Government's aim was to make the police better at crime fighting by making them a more centralized and businesslike organization (Reiner, 1995). Pursuant to these recommendations, in January 1993 Kenneth Clarke (the then-Home Secretary) created twenty-five centrally appointed police authorities, which would have replaced the elected police committees with substantially (though not wholly) non-elected magistrates, businessmen and councilors. However, this plan was heavily modified after concerted political opposition in the House of Lords. A majority of members of police authorities remained elected, leaving a modified version of the 1964 system, albeit with much greater central government control via the Home Secretary's power to set national performance targets and fixed-term appointments for senior officers (Leishman et al 1996, p. 16).

At another level, the Sheehy proposals were also bitterly, and largely successfully, opposed by the Police Federation (representing junior police ranks), the Superintendents' Organization, and the Association of Chief Police Officers (representing senior ranks). The chairman of the Police Federation described it as a situation in which "the balance sheet has become the bottom line of policing" (*LancCon*, 1995, p. 5). At one rally alone, 20,000 of the 125,000 officers in England and Wales were present (Leishman et al 1996, p. 13). Most of the more controversial proposals in the Report, such as fixed-term contracts for most police ranks, were subsequently abandoned by the new Home Secretary Michael Howard.

Complaints Procedures

Another crucial area of police accountability (and thus democratic policing) has been the investigation of complaints by the public against the police. Until 1976, all complaints against officers were dealt with internally by the forces concerned unless there was the possibility of prosecution for a criminal offense when reviewed by the Director of Public Prosecutions. Concerns about bias in internal reviews led, in 1976, to the establishment by parliament of The Police Complaints Board as an independent body to review the files on complaints gathered by the police themselves. The Board had the power to overrule any internal decision but not to bring disciplinary actions. Although unpopular with officers, the Board was reliant wholly on police investigations and was consequently criticized for having an establishment composition and bias. It has been stated by some commentators that, in the light of the available historical evidence, the very notion of "perfect self-regulation" by the police is utopian. This is partly because the tendency for police officers to socialize among themselves necessarily "encourages the development of an elite group ... distanced from the population it allegedly serves." The inevitable result of this, aside from the "implicit dangers to democracy," is to foster a group identity which is aimed at protecting "its own" at all costs (Hain, 1979, p. 11). It has also been asserted that the police need to be persuaded of the benefits to themselves (via increased public confidence) as well as to the public which the effective handling of complaints creates (Goldsmith, 1991, p. 56).

In part reflecting these ideas, the *Police and Criminal Evidence Act 1984* introduced an element of investigative independence in the form of a new body called the Police Complaints Authority. The PCA has the power to directly supervise investigations in some cases, including a duty to do so in all cases where a complaint has been made about police abuse causing death or serious injury (Reiner, 1994, p. 748). However, only the most serious 5 to 10 percent of investigations by the police are "supervised" by PCA members. Like its predecessor, it has faced considerable criticism, which has sometimes come from the police themselves against whom complaints are considered. In the early years of the PCA's existence, at one of its annual meetings the Police Federation even passed a vote of no confidence in the PCA. Its then-chairperson, Mike Bennett, accused the PCA of "being hell bent on destroying the civil liberties of police officers" (cited in *New Law Gazette*, August 4, 1989).

However, many outsiders criticize the PCA for being excessively ineffective, slow and inactive. Statistically, the Authority took action in only 14.4 percent of the 5,516 cases where investigations were finished in 1988. Of these only 62 complaints resulted in subsequent criminal charges and 17 cases were recommended for formal police disciplinary charges. The remainder led to recommendations for informal sanctions such as advice, counselling, or admonishment (HMSO, 1991). Of course, a fairly low rate of action is probably inevitable to an extent . Many of the complaints are made by young males involved in trouble with the police—youths who often have previous criminal records and no independent witnesses to support their allegations. In these circumstances, in which there is complete denial of the allegations by the police, it would often be almost impossible to prove them to the requisite standard. Despite this weakness, there have been suggestions that a completely independent body should be set up, as has happened in some other countries (though these bodies do not necessarily appear to have gained greater public support as a result).

Interestingly, nearly half of policemen surveyed who had been investigated as a result of a complaint saw advantages in greater external involvement in the complaints procedure, in boosting public confidence and speeding up the dismissal of obviously unfounded allegations. However, a small majority felt that outsiders, as a result of their lack of experience, could never have a proper understanding of the pressures and realities of policing or of the type of people who often made such allegations. Typical of these views were observations such as "if you've never been in a situation of fear, you won't be able to understand why a police officer acted that way. A police investigation is fairer from the officer's point of view because police understand each other" and "outsiders wouldn't be aware of the slippery eels they are dealing with" (HMSO, 1991, p. 70).

Local Consultation

Another aspect of democratic policing in Britain has been the recent emphasis on the need for local consultation (as opposed to control) with the people being policed. Following Lord Scarman's recommendations in the wake of the inner-city disturbances of 1981, which included an emphasis on the need to prevent the distancing of the police from the public they served and for greater sensitivity to the views of local communities, Section 106 of the Police and Criminal Evidence Act 1984 provided that "arrangements shall be made in each area for obtaining the views of people in that area about matters concerning the policing of the area and for obtaining their co-operation with the police in preventing crime in their area."

Details as to how this was to be implemented were subsequently provided by the Home Office in one of its regular circulars to the various police forces (No. 2/1985) which recommended the formation of formal Police Consultative Committees (PCCs). Despite initial resistance from some senior officers, these committees are now universal. In London they are based on the city's boroughs, and in the provinces on Police Subdivisions.

Additionally, a scheme of Lay Visitors Panels (LVPs) was set up on a nationwide basis. The LVPs are made up of local people who make frequent, random and unannounced visits to police stations to check that the standards set out by the *Police and Criminal Evidence Act 1984* for arrested and detained suspects (with regard to accommodation, access to legal advice, etc.) are upheld. In the Greater London area, for example, there are currently forty-one Police and Community Consultative Groups (PCCGs) as well as thirty-eight Lay Visitors Panels. Some of the administrative costs associated with supporting them are financed from the Metropolitan Police Fund. The PCCGs are made up of members of the local area, drawn from the local authority rolls, residents' associations, crime prevention panels, lay visitors schemes, schools and churches, etc. They hopefully promote a degree of local partnership in crime prevention initiatives as well as in community security and road safety.

Thus, in the ethnically mixed and very poor London borough of Hackney at present, a meeting is held every two months between senior officers from the area and representatives of the local community (called the Hackney Police/Community Consultative Group). This is an open forum where views on policing issues affecting local people can be expressed. It allows the community an opportunity to question the policies of local senior officers and is open to the general public.

There have also been a number of other more localized initiatives to improve the degree of police liaison with the communities they serve, especially in difficult urban areas. Also in Hackney, on a smaller scale, every four months, residents are invited to attend a sector forum to discuss with their local Inspector and permanent beat officers issues of concern to them. Community liaison is in some ways a police-sponsored alternative to the more radical initiatives by local people to become involved in policing issues which developed from the late 1970s onward. These often had what has been referred to as a "critical and reformist" agenda, seeking to monitor police activity, provide information for the media, investigate deaths and injuries in police custody and during the policing of industrial disputes and disturbances (Uglow, 1988, p. 131). Many of these bodies were funded by local authorities, especially in Labour-controlled boroughs in the inner cities; as such they were not, themselves, necessarily totally representative of the views of all, or even a majority, of local inhabitants.

Some critics have argued that police-sponsored meetings are largely cosmetic and merely "talking shops" not having any measure of real control. However, they do provide a valuable avenue of communication between police and public. It appears that this is increasingly appreciated by the police as well as residents. For example, at a local level, in Hammersmith in West London, a superintendent observed that a recent marked fall in recorded crime in the area was linked to "targeting the right crimes, identifying problem areas and policing them effectively." He felt that "Hammersmith police officers listen to what people tell us and concentrate on what local residents want us to do." Pursuant to this policy, the division sent 12,000 questionnaires to homes in the area during September 1996, asking residents to suggest future directions for police strategy (*Hammersmith and Fulham Gazette*, August 30, 1996, p. 2).

PROFESSIONAL CHALLENGES

Individual Professionalism

Occasional brutality and dishonesty by police officers certainly occurred in the 1940s and 1950s, but much of the public, especially its respectable elements, appear to have been oblivious to this. In a letter to *The Times* (March 24, 1991), a retired solicitor, who earlier in his career had practiced extensively in the Metropolitan Magistrates Courts, suggested that a degree of brutality and the fabrication of evidence had been widespread in the London area for years and that stipendiary magistrates had been largely indifferent to it. The eminent retired Court of Appeal Judge, Sir Frederick Lawton (1991, p. 2), has also observed that when he practiced before those same courts as a barrister, from 1935 to the early 1960s, he was often instructed by his client that the police had "put on the verbals" (fabricated admissions) and even that the accused had been assaulted after arrest.

Whatever the reality, there was a widespread mystique about and belief in the sterling qualities of the British Police in the 1950s among many (probably most) sections of British society. The "typical" British "bobby" was seen as having honesty, integrity, fairness, impartiality, avuncularity, straightforwardness and courage. This remembered image continues to provide an easily accessible (and nostalgic) model for a disillusioned modern public to look back to as a point of comparison. Although the image did not always correspond with reality, the community trusted its police (Graef, 1989). There was a widespread and high level of approval of the police, something which appears to have lasted until the mid-sixties. Opinion polls of that era revealed that public confidence in the police was high compared to other national institutions and that the police enjoyed cross-party political support.

Since the 1950s, all this has changed. At the start of the 1990s, one respected British expert on policing felt able to state quite explicitly that in Britain the "police are in trouble." The identified symptoms of this trouble, he felt, were a tide of scandals involving allegations of fabricated confessions and perjury by officers and of records being destroyed or altered, as well as claims of "fitting up" and brutality by the police. Other symptoms were trial juries allegedly manifesting a marked new reluctance to convict on unsupported police evidence, the large sums paid out in damages as a result of civil actions for wrongful arrest and assault by police officers, and the apparent inability of the Independent Police Complaints Authority to penetrate the wall of silence it regularly met when police officers under investigation closed ranks. The same writer also felt that there were other, more concealed, indicators of a deep-seated malaise. Among these, for example, were the thousands of working days lost by the service as a result of men going sick through stress or psychosomatic illness, an apparently high rate of resignations among experienced policemen and university graduates in the force and a shortage of potential candidates willing to take the sergeants' examination for promotion (Graef, 1991).

Many of these symptoms were a reflection of the fact that at the beginning of the 1990s the police in Britain were facing an unprecedented loss in public confidence and in their perceived "legitimacy." To many forces around the developed world, this would not have been a novel situation. In Britain, it was the culmination of a fall from an almost unique status that had been gathering pace since the 1960s. The fall resulted in the police being seen sometimes as a potential source of abuse of power and as a threat to freedoms and liberties (and thus democratic society) rather than their trusted upholder. This view was linked to a perceived decline in individual police professionalism. Concerns about this decline had been manifest in a number of areas in recent decades, particularly about the levels of police corruption, brutality, and the giving or fabricating of false evidence by officers to secure convictions. All three issues were present in the decline in public esteem of the police from the 1960s onward.

Corruption

Perhaps as a harbinger of things to come, in 1959 the Chief Constable of Worcestershire was convicted of misappropriating police funds and received a sentence of eighteen months' imprisonment (Emsley, 1991, p. 162). In the following decade, in 1969, reporters from *The Times* newspaper tape-recorded London Criminal Investigation Divisions (CID) detectives offering colleagues a bribe on behalf of an accused criminal. The allegations made by the newspaper were investigated by Frank Williamson, one of the official Inspectors of Constabulary, who later opined that there were three types of police officer at Scotland Yard: those who were corrupt, honest men who knew about the corruption and did nothing, and those too stupid to appreciate that there was any corruption (Campbell, 1994, p. 200).

Corruption appeared to grow inexorably. In particular, it spread to the Flying Squad (which deals specifically with serious armed robbery) and the Vice and Drugs Squads. It eventually became apparent that in the 1960s and 1970s, in some specialist sections of the police and especially in the Metropolitan Police in London, there was an almost endemic level of corruption. Ultimately, even the very senior officer at the head of the Drug Squad, Detective Chief Inspector Kelaher, was convicted of conspiring to pervert the course of justice and imprisoned. As John Alderson was to remark, although the British Police had possessed "one of the most outstanding and well-deserved reputations for a robust and sturdy integrity" this was all the more reason "to regard the existence of a considerable amount of corruption in the CID of the Metropolitan Police as a major setback—an illustration of how a police subculture can go wrong" (Alderson, 1979, p. 67).

After a number of belated and ineffective false starts, effective steps were taken to combat this grave problem, especially during the Commissionership (of the Metropolitan Police) of Sir Robert Mark. Reforms resulted in large-scale resignations by, and disciplinary actions against, literally hundreds of suspect officers and produced, at least in respect to corruption, a very much improved and cleaner service by the 1980s. But much damage to the public image had been done. Today, although a much smaller problem than twenty years ago, similar cases periodically crop up. In part, corruption will often occur because of the necessarily intimate relationship between detective officers and the underworld, in Britain as anywhere else in the world, such officers nursing informants for information by turning blind eyes, etc.

Some have even argued that when interviewing occurs at a police station recent additional controls may have encouraged corruption (Hobbs, 1988, p. 226). One serious alleged case in the 1990s became the focal concern of a special police investigation ("Operation Jackpot"). Between 1991 and 1993, there were allegations that the police in Stoke Newington (a rundown part of north London with a high Afro-Caribbean, Turkish and other ethnic minority populations) were involved in recycling crack cocaine and other drugs seized from dealers in the notorious Sandringham road area. A police officer from this station, a Detective Constable, was also, separately, convicted of stealing 2,000 pounds worth of rare books, in his capacity as scene of the crime exhibits officer, from the home of the victim of a brutal murder. He was later sentenced to eighteen months' imprisonment (Campbell, 1994, p. 211).

More recently, in Scotland, a convicted drug dealer giving evidence in court claimed that a Grampian Police officer, who was accused of three corruption charges involving the dealer (allegedly accepting 6,000 pounds among other payments for information about police intelligence and anti-drug surveillance operations), was a scapegoat for other corrupt officers on that force (Smith, 1995). Also in 1996, in one of the worst individual cases of police corruption in England since the 1970s, a Detective Constable from the Metropolitan Police was sentenced to eleven years' imprisonment after he demanded 60,000 pounds to destroy evidence about a drug dealer, and subsequently spent the bribe on a champagne lifestyle. An Old Bailey jury convicted him of four counts of corruption and one of perverting the course of justice. He was described at his trial by the prosecutor as steeped in corruption and dishonest to the core.

At a much less serious level, there has been some public concern in recent years about the apparently high number of police officers, especially senior ones, who are Freemasons, and consequently there have been calls for membership of this theoretically secret society to be forbidden to serving officers. Several Home Secretaries have also expressed concern about police membership of Masonic lodges but have resisted banning it. In 1985, Scotland Yard formally advised new officers that it might be "unwise" to be a Mason. A police handbook further advised that a shrewd policeman would "probably consider it wise to forego the prospect of pleasure and social advantage in Freemasonry so as to enjoy the unreserved regard of all those around him." In 1993, the Police Federation, representing Britain's 125,000 police officers, narrowly defeated a conference motion (made by a Police Inspector) declaring that Masonic membership was not conducive to the independence of the police (*Glasgow Herald*, December 7, 1993).

Police Brutality

Despite the successes of the 1970s against organized police corruption, other types of policing problems either developed (or became more overtly apparent) and took their place in generating public concern. These included allegations of police brutality. Again, a harbinger of this may have occurred in 1962, when a rhino-whip was apparently used on a suspect held in a police station in Sheffield. During violent political disturbances and disorder in 1974, a student died as a result of a blow from a police baton while protesting against a National Front rally in Red Lion Square in London. Four years later, in a similarly fierce confrontation, Blair Peach, a New Zealand teacher and political activist, died at a rally in Southall, apparently as a result of a blow from a police Special Patrol Group radio being used as an improvised baton.

Outright, deliberate police brutality (as opposed to periodic excesses during disturbances) and planned attacks on detainees at police stations or people in the streets appear to be very rare in Britain (certainly compared to some other developed countries). However, it has occurred from time to time. One of the worst such incident of police brutality in recent years occurred in August 1983, when three police vans from the Metropolitan Police Force's District Support Unit were patrolling in North London. One van, containing eight officers, encountered a group of five black youths, ages 13 to 16, walking down a side street near the Holloway Road. A number of the officers left the van and assaulted the youths in what the trial judge described as a "brutal, bullying, unprovoked attack upon innocent schoolboys" (East, 1987, p. 1010). One 16-year-old boy had his nose broken as a result of being punched in the face while others received lesser injuries. Other than some shouted abuse by another group of teenagers earlier in the evening, there appears to have been no reason at all for the attack. After the incident, the officers involved returned to their colleagues in their patrol van, which then drove away. None of the other policemen in the van reported the incident to their superiors.

Following the victims' complaints, the Complaints Investigation Branch of the Metropolitan Police carried out an initial investigation and interviewed the members of all three vans operating in the area. This investigation was unsuccessful, partly through misfortune (the investigators felt the descriptions of the police involved provided by the victims better fitted members of another of the three vans), but largely because the investigators met a wall of silence from the police officers of all three vans. It later transpired that in the week after the attack, the crew of the van involved met in a London park and agreed among themselves not to cooperate with the police investigation. As a result, the investigation lasted for over two and a half years with no success. The Director of Public Prosecutions decided (quite reasonably on the evidence available to him) that there was insufficient evidence to prosecute any officer at that time. By April 1985, all twenty-two officers in the three vans had been interviewed twice and all had maintained their silence on (or at least ignorance of) the issue.

The Police Complaints Board, to whom the investigators also reported, was replaced in April 1985 by the new Police Complaints Authority (which had greater lay participation). This body, in February 1986, published a damning report on the lack of progress and claimed that a prosecution was unlikely because of a cover-up by the officers involved. This led to a powerful a media campaign, led by the *London Evening Standard* (the main paper for the capital) and a former policeman and editor of an independent police magazine, the *Police Review*. The magazine's editorial on February 7, 1986, contained a strongly worded attack on the "five bastards serving in the Metropolitan police force." In part prompted by this, Sir Kenneth Newman, then Metropolitan Police Commissioner, set up another inquiry, this time with the promise of immunity from prosecution for those officers not involved in the assault. The necessary information was promptly provided by one of the officers in the van and the guilty men were arrested and charged. In July 1987, the five officers involved in the "Holloway Road Transit case" were convicted, three of them for assault and conspiring to pervert the course of justice, and were sentenced to four years' imprisonment each. Another officer was convicted on the conspiracy charge and sentenced to eighteen months, while the sergeant in charge of the transit van in which they were operating was sentenced to three years for his part in the conspiracy and for failing in his duty as an officer to stop the assaults. The trial judge told the officers when sentencing, "you behaved like vicious hooligans and you have lied like common criminals (East, 1987, p. 1011)."

The incident prompted some structural changes. District Support Units were disbanded and replaced by Territorial Support Groups (larger units with a greater degree of senior supervision and training). Explanations for the officers' conduct that night have ranged from the bad apple theory to the strains of policing (the officers had apparently been on duty in the van, in high summer, for thirty-two of the previous thirty-six hours and one was receiving psychiatric treatment after being stabbed) to the allegation that the Sergeant had lost any effective control of his men (East, 1987). The slow pace of the investigation and the inability to reach a speedy conclusion also raised concern about available methods of investigating such misconduct; the wall of silence encountered by the investigators has been a regular experience of those investigating other allegations of police misconduct.

Although this incident occurred thirteen years ago, fresh (less serious) incidents still occur from time to time. In Scotland in 1996, a Police Constable (also a former amateur heavyweight boxer and karate black-belt), was jailed for three months after being convicted of assaulting a youth of 19 during a strip search in a cell at the headquarters of the Grampian Police force in Aberdeen. The victim was left bleeding from a series of blows to his head. Though the trial Sheriff (judge) accepted that the victim was a drug user who had earlier stolen something from the police officer concerned in the assault, he also observed in sentencing that "I would be failing in my duty if I were not to take a serious view of police officers who assaulted people in their custody" (Urquhart, 1996).

Fabrication of Evidence

Since the end of widespread and organized corruption, police concoction of evidence to secure convictions has probably been the subject which has caused most alarm in the general public. Again, the first major cases came to light in the 1960s. In 1965, a mentally disturbed Detective Inspector, "Tanky Challenor," in the Metropolitan Police Force, was found to have regularly planted evidence on alleged criminals and protesters, being finally exposed after planting a brick on a demonstrator at the visit to London of the Queen of Greece. The Inspector was subsequently found to be insane (though this had apparently gone undetected by his superiors for a considerable period), but three of his colleagues were convicted of conspiring to pervert the course of justice and jailed for a total of eleven years.

Even worse, however, were a number of well-publicized *causes célebres* involving the fabrication of evidence which came to light in the 1980s and 1990s, involving apparent police misconduct, some of it committed much earlier. These included scandals associated with the West Midlands Serious Crime Squad, which was disbanded in 1989 after widespread allegations (well founded) of producing fabricated evidence against suspects, which prompted several external investigations and disciplinary actions against officers. These were especially disturbing because they occurred in what was (after the Metropolitan Police) the second-largest force in England and Wales, and because they indicated a major institutional problem caused by poor management and control by senior officers. Although an extreme case, the problems in the West Midlands force can be seen to reflect wider policing difficulties throughout the country at that time.

At roughly the same period, there occurred the dismissal of a Detective Sergeant and the "in-house" disciplining of half a dozen of his more junior colleagues, for the manipulation of crime statistics in the Medway towns in Kent; the release of the "Guildford Four" (a group of people who were convicted of Irish Republican Army (IRA) bombings in the 1970s) from prison on appeal because their convictions may have been partly based on fabricated evidence (at their own trial the officers were acquitted by the jury); and the release on appeal, on similar grounds, of others convicted of IRA bombings in the 1970s, such as those in the "Birmingham Six" case. In his judgment in the Birmingham case, Lord Justice Farquharson of the Court of Appeal stated that at best some of the police officers involved in the case were lying in court about their notes being a continuous record of their interviews of the suspects, while at worst they had conspired to fabricate part of the interview. However, some have argued that these cases were merely the publicized (and most serious) tip of an iceberg of police dishonesty.

During the 1980s, there was a growing level of doubt about the reliability of much police testimony, especially that relating to oral admissions or "verbals" at police stations. In some respects this was merely an awareness on the part of the public of something that had been privately suspected and talked about by some lawyers for many years. Thus, in 1986, an experienced criminal practice barrister observed of his career that "with the passage of time it became apparent to me that police perjury occurs with great frequency in London (where I mainly practice)." From conversations with other lawyers he felt that this "was regarded as a commonplace." The lawyer conducted a crude straw poll among his colleagues at the Bar, consulting fifty-five experienced barristers practicing mainly in the London area, most of whom often appeared for both prosecution and defense (as is common in England). From his own experience, he felt in-court perjury by police officers had been proved, on the balance of probabilities, to have occurred on average in three out of every ten trials that he conducted, in both the summary and in the higher (Crown) courts. His sample of barristers produced a similar result: 74.55 percent replied that in their experience this was a realistic estimate with which they would agree (Wolchover, 1986, p. 181).

Sir Frederick Lawton attempted to assess why some police officers committed perjury or fabricated evidence and identified several potential reasons for police malpractice. He felt that some of the most serious cases tended to occur when Criminal Investigation Divisions detectives were investigating particularly horrific crimes, such as the Guildford and Birmingham bombings (which resulted in several dozen deaths) or even serious offenses like armed robbery committed by experienced criminals, which prompted a feeling on the part of the investigating officers that such people were enemies of society who must be put away. This feeling was especially strong when coupled with a belief that they knew who the culprits were but were frustrated in bringing the suspects to trial by the rules of evidence and procedure. He strongly believed that police officers do not set out to bring about the conviction of anyone whom they knew to be innocent. Additionally, Sir Frederick felt general "over-zealousness" was another important cause. This zeal was reinforced by the widespread belief among officers that their promotions depended upon having a record of successful prosecutions (something that has ramifications for the setting of external performance targets by central government). He also felt that police standards of behavior would never improve as long as officers thought that those responsible for discipline looked the other way when there were accusations of malpractice. Yet he had always been surprised at the seeming lack of interest shown by chief constables in allegations made in the course of a trial that police officers under their control had been guilty of malpractice. To his knowledge, there had been many cases in which juries had acquitted after such allegations had been made yet there had been no disciplinary proceedings afterwards (Lawton, 1991).

As a result of these revelations, by the end of the 1980s one respected (and by no means radical) commentator expressed the view (of the Metropolitan Police) that it is difficult sometimes to resist the conclusion that there are more rotten apples than fresh ones (Levin, 1990). This loss of trust in the police was potentially disastrous. As the British Crime Surveys have repeatedly indicated, 91 percent of crimes known to the police are reported to them by the public. The police were present at the scene of a crime, when it occurred, in only 3 percent of cases. Thus, the loss of perceived legitimacy threatens their most vital source of information and cooperation.

Non-Partisanship and the Policing of Civil Disturbances

Other developments in the 1970s and early 1980s also eroded confidence in police even-handedness on the part of at least some sections of the community. Non-partisanship, the avoidance of becoming overtly involved in the wider political debate, had been a well established facet of British policing up to the early 1970s, with senior officers careful not to say anything that might be interpreted as supporting a party political position or as favoring a particular political party. However, this reluctance became frayed in the 1970s as senior officers, and rank-and-file organizations such as the Police Federation (the professional representative body for junior police ranks), began to be more overt in giving opinions about the state and problems of law and order in contemporary Britain. Often, it has been alleged, these comments reflected a Conservative perspective. Indeed, after the new government of Mrs. Thatcher came to power, the Police Federation broke with its previous tradition of appointing its parliamentary adviser from the opposition benches and reappointed an outspoken Conservative MP, Eldon Griffiths, to the position. At a less overt level, a few Chief Officers, such as James Anderton in Manchester, regularly gave their opinions about the general state of the nation's morals. There were also public advertisements, paid for by the Police Federation, calling for the restoration of the death penalty (which was abolished in Britain in 1965).

At the same time, cross-party consensus on policing was further eroded by the emergence of highly radical elements in the Labour party who increasingly demonstrated profound hostility towards many routine and traditional forms of policing. In the May 1981 Council Elections in England's urban areas, many radical Labour councils were elected which provided a power base for the overt and critical questioning of police policies. As a result of this and other processes, between 1981 and 1985 there were repeated and highly public clashes between some of these authorities and their local Chief Constables, particularly in Manchester and Merseyside (Reiner & Spencer, 1993, p. 4).

Such conflicts reached a climax in public order policing, with the policing of industrial disputes in particular prompting fears of an American-type politicization of the police. The police have faced two major public order challenges in the past twenty years. The first came from inner city riots (with substantial racial overtones), with particular widespread disorder occurring in 1980, 1981 (on a much wider scale), 1985, as well as (to a lesser and more localized extent) in 1995. The second challenge came from policing disturbances associated with a series of serious industrial disputes in the late seventies to mid-eighties, disputes which necessarily had heavy political aspects to them.

This was particularly so with regard to the prolonged miners' strike between 1984 and 1985. This strike led to the largest and most sustained mobilization of police resources since the General Strike of 1926. The use of the police's National Reporting Centre (NRC) to deploy resources on a nationwide basis, as well as the very nature or style of the policing of the disputes itself were particularly controversial in this process. The NRC function was (and is) to keep information on available police resources among the forty-five forces throughout England and Wales for times of emergency. It was (and is) administered by the President of the Association of Chief Police Officers, an officer appointed on a yearly basis. At points during the strike, the NRC deployed up to 8,000 officers from the police support units (effectively mobile reserves) of forty-three English forces. For example, the Nottinghamshire Constabulary received assistance in this way from thirty-seven other forces. Nottinghamshire is a coal-mining area where there were some fierce disturbances.

There were numerous allegations of violence by police involved in the operations; many of these were well documented. Serious violence was also used against officers by strikers and demonstrators. More than 400 formal complaints against the police were made. There was also widespread concern, from a civil liberties angle, about the extensive use of police road blocks to prevent travelling (flying) pickets reaching pits long before they neared them. By one estimate, 164,508 presumed pickets were turned away from Nottinghamshire alone in the first twenty-seven weeks of the strike. There were also worries about the deployment of large numbers of police to shepherd strike breakers to and from their work. At its most extreme, on one occasion, an estimated 1,500 officers escorted one man into the Corton Wood pit in South Yorkshire. The Police Federation argued that the success of the police operation proved that the UK could manage without a paramilitary riot squad akin to the French CRS. Ordinary police officers, albeit with special training and equipment, were sufficient. Critics contended that the specially-trained and equipped support units were tantamount to having such a French-style force.

Unlike some Continental or American police forces, the traditional approach of British Police to policing civil disorder has been to avoid an excessively formidable appearance and to cultivate the image of the unarmed and non-aggressive British Bobby. As they replaced the army and militia's involvement in such disputes in the nineteenth century, the use of bayonets (and occasionally even guns) was replaced by baton charges which in turn were later replaced by cordons of officers, with linked arms, holding back the disorderly elements. When violence erupted, the police would "win by appearing to lose," in the words of a former Metropolitan Police Commissioner Sir Robert Mark, by being popularly seen and portrayed as the victims of violent crowds or mobs rather than their oppressors (cited in Reiner, 1992, p. 64). However, as has also been observed, this situation was premised on the fact that the police did eventually win.

In the 1970s this approach started to fail and changes to traditional methods began to be introduced (Waddington, 1987, p. 39). A turning point in police attitudes to such disturbances appears to have begun during the policing of an earlier industrial dispute at the Saltley coke depot in 1972, when 800 police officers had been confronted by 15,000 strikers (from places other than the depot concerned, this being a case of secondary picketing). The police agreed to close the depot in the face of this mass pressure and fear of widespread and violent disorder (i.e., the police "lost"). The disorderly and violent end to the West Indian Notting Hill carnival in 1976, when policemen were seen on television shielding themselves from missiles with dustbin lids taken from the streets, led to the first moves towards what some have termed paramilitarism. The large-scale inner-city riots of 1981 accelerated this process, with the police acquiring flame-proof overalls, perspex shields, stocks of CS gas and baton rounds (plastic bullets). As yet, these bullets have not been used outside Northern Ireland, though they were deployed at the serious riots in Tottenham in 1985.

When striking miners tried to close the Coke depot at Orgreave in 1984, in an attempted "rerun" of Saltley, they were confronted by officers from a dozen forces, uniformly equipped in helmets with visors, large shields and long batons. These officers periodically opened their ranks to let out lighter groups of mounted police officers and officers on foot carrying small round shields to take offensive action against the crowds. These officers were backed up by a new and highly controversial Manual on Crowd and Riot Control Techniques. At Orgreave, unlike Saltley, the police won and the depot was not closed. However, police conduct and apparent violence during the incident contributed to a situation in which court cases subsequently collapsed against every miner arrested and charged with riot (Emsley, 1991, p. 173).

It must be acknowledged that, qualitatively, the level of violence shown toward the police in such disturbances has increased enormously in recent years. In Tottenham in 1985 this included attempted shootings with illegally held pistols and other firearms and the brutal hacking to death of a community police officer by a mob. The tactics that had worked and gained widespread admiration during the anti-Vietnam War demonstrations in Red Lion Square in the 1960s (cordons and linked arms) would have been potentially suicidal in Tottenham in 1985. Additionally, although at first sight unattractive, it must also be acknowledged that traditional approaches to policing severe modern civil disorder using untrained, poorly supervised and ill-equipped officers, can themselves sometimes be highly counterproductive. For example, studies of the St. Paul's riot in Bristol in 1980 (the first major disturbance of the decade) produce an image of initial chaos as the police rushed to the scene of the riot in a largely unorganized manner—which may itself have encouraged further violence. Unorganized responses can also be counterproductive in other ways. Unable to deal effectively with the main (and most dangerous) source of the trouble, ill-prepared officers will be tempted to arrest those most readily available even if only peripherally involved. Ill-protected police officers are also under potentially greater pressure to take violent pre-emptive action for fear of personal injury. In some circumstances, an ineffective response may make situations worse. In the Tottenham riots of 1985 officers were withdrawn from the area during the build-up to the riot in an attempt to avoid provocation, but in fact appear to have merely provided an opportunity for the riot to develop (Waddington, 1987, p. 43). Well-protected, disciplined and, most important, trained police can sometimes exercise a greater degree of both control and restraint.

Damage to the Police Image

As a result of the apparent scandals and despite significant and relatively successful efforts to improve individual professionalism and training in the mid- to late 1980s and throughout the 1990s, public esteem of the police in Britain has seldom been lower than at present (though it may have improved in the most recent years). A leading scholar of British policing has referred to this as a "cruel paradox." Although in many ways the professional standards, integrity and conduct of the police service have never been higher, or police training and accountability more thorough, public support and confidence in them is at a level that is lower than at any time since the decade after their formation in the first half of the nineteenth century. Virtually all surveys of public opinion have registered a sharp decline in trust for officers in recent years. As Professor Reiner has noted, "Chief Constables might be forgiven some puzzlement when they find that the medicine has not improved the patient's condition. Public support continues to plummet, though professionalism increases." Perhaps even more worrying for the service, this decline in confidence appears not to be limited to the young males of what has been termed the "urban underclass", who have traditionally been the focus of a high degree of police attention (the term used by the police, "police property", accurately reflects the mutual sentiments), but has extended to much of the educated middle classes, formerly a bulwark of support for the police, but a social group now much more inclined to question its conduct and efficiency. Whatever the reality might be, the police are now perceived by many sections of society as more abusive and corrupt and less efficient than before (Reiner, 1991b).

Of course, there are several explanations for this process other than the tarnished image occasioned by well-publicized scandals. Some of these are purely practical. Sir Frederick Lawton has traced part of the decline in public faith in the police to the wide availability of motor vehicles in modern times (now spread well beyond the richer classes) and the ensuing and regular contact the general (i.e., non-criminal) public have had with the police over motoring matters (in *The Times*, January 16, 1990).

In part the decline of legitimacy was probably inevitable. For better or worse, Britain is a much less unquestioning and deferential society than was the case forty years ago. It is a country whose inhabitants are more aware of their rights than previously and who also have greatly improved access to legal redress for the infringement of those rights. On this issue, the analysis of Sir David McNee, the former Metropolitan Police Commissioner, is probably valid. Noting that actions against himself and his officers had risen from 16 in 1967 to 182 in 1982, he attributed the increase to a number of causes, but in particular to "a greater knowledge by individuals of their rights and of the high monetary rewards sometimes made by civil courts, together with the monitoring activities of professional bodies concerned with individual rights and of the media" (McNee, 1983, p. 181). Since 1982 there has been a further dramatic increase in civil complaints filed.

OPERATIONAL CHALLENGES

Crime Levels

According to one expert on policing, Britain has become "a more crime-ridden, disorderly and lawless society than at any time since the middle of the last century" (Reiner, 1991b). This development is significant to policing a democracy because rising crime has produced a general atmosphere of crisis and social breakdown as well as recent debate about police effectiveness and the need for giving them greater powers. An atmosphere of crisis is not conducive to a healthy democracy; crime, or fear of crime, can be as much a constraint on freedom as can government. Rising crime also means that despite large increases in officers, the numbers available in some areas to deal with emergencies or simply to patrol are often inadequate. Inevitably, questions about police effectiveness tend to overlook the fact that the police play a wider role than simply dealing with crime, and that police tactics, in whatever form, probably have a relatively small impact on levels of crime.

A major increase in crime in Britain since 1945, at least until very recently, is borne out by official statistics. Initially, some comfort was taken from the belief that this apparent rise was substantially a reporting and recording phenomenon, and that the more extreme assertions by some commentators were based on ignorance about the real situation in the past. There is undoubtedly still some truth to this analysis. All such statistics are flawed to a considerable degree. However, since the early 1980s, the results of the British Crime Survey (essentially victim report surveys) have eroded much of the confidence behind this assertion, as they indicate a progressively rising level of unreported crime (if occasionally less dramatic in extent than the official police figures).

The 1950s saw the major initial increase in recorded crime, a trend that was sustained, at varying rates, until 1992. At first this may have been partly linked to the new post-war advent of mass consumerism which created numerous easy and potentially lucrative targets for theft. However, the rise in crime has continued steadily. The total number of notifiable offenses (those collated and published in the annual Criminal Statistics) for 1992 stood at 5,594,000, and amounted to a more than tenfold per capita increase since 1950, having grown from just over 1 offense per 100 people in 1950 to 10.6 per 100 in 1992. Not only had the absolute numbers of recorded offenses increased enormously but their nature had changed. In part, this reflected obvious social changes. In 1950 car ownership in England was quite rare and inevitably this resulted in comparatively few crimes involving such vehicles. By the 1990s crimes involving such vehicles, whether theft of or from them, made up about 25 percent of the total (high by the standards of most European countries). Other changes reflected more worrying social changes. Crimes of criminal damage (vandalism), also very rare in 1950, now make up about 15 percent of the total crime picture. Crimes of violence against the person now greatly exceed cases of fraud or sexual cases, something that was not the case in the early 1950s.

In 1950, a conviction for a criminal offense was very rare among men in nearly all social groups. Today, cohort studies of men born in the early 1950s reveal that by the age of 31 a third of them will have acquired at least one criminal conviction for an indictable offense (the figure for women, at 8 percent, is much lower). Despite this, many of the total of recorded offenses are attributed to a relatively small proportion of British society. According to the Home Office cohort studies, up to 65 percent of court convictions accumulated by members of the cohort were distributed among 7 percent of its members, each acquiring six or more such convictions (Maguire, 1994, p. 233). This is itself, perhaps, telling evidence of the reemergence of a criminogenic underclass.

Some recent signs have been slightly more hopeful. In 1995, recorded crime fell by 2.4 percent. This followed falls of 1 percent in 1993 and 5 percent in 1994. Of the 5.1 million crimes recorded by the police in England and Wales in 1995, 93 percent were property offenses, with half of all offenses being either burglary or car crime. These categories, burglary and vehicle crime, fell by 1 percent and 4 percent respectively from 1994 to 1995. The actual theft of motor vehicles fell by 5 percent or 24,600 offenses between 1994 and 1995. The fall in vehicle crime is significant as it was the third consecutive annual fall, with 1995 levels of recorded vehicle crime 15 percent lower than in 1992 (a fall of 226,000 offenses). This is especially important as the great majority of this type of crime is reported to, and recorded by, the police (something that is necessary for insurance purposes), and hence are highly reliable numbers. Car crime is also an area where Britain has had one of the highest rates in Europe. Theft as a whole fell by 4 percent (98,000 offenses), having also fallen for three consecutive years. Smaller falls occurred for fraud and criminal damage.

Violent crime (including violence against the person, robbery and sexual offenses), although it makes up only 6 percent of recorded crime, is most feared. It continued to rise by 2 percent (an increase of about 5,400 offenses) from 1994 and 1995, though this was a slightly slower rise than the average for the previous decade. Within this category some types of crime actually fell: offenses of violence against the person fell by 1 percent, and recorded sexual offenses by 5 percent. However, this decline was offset by a marked increase in the number of recorded robberies (up by 14 percent), with particularly heavy increases in London.

In response, the Metropolitan Police Commissioner Sir Paul Condon launched "Operation Eagle Eye," which was specifically aimed at catching street muggers in order to address a growing level of fear among Londoners of being attacked and robbed in public. Sir Paul Condon observed that in the first six months of 1995 alone there were around 20,000 such muggings in London. Pursuant to "Eagle Eye" the police planned to employ the augmented use of intelligence and the surveillance and targeting of suspects as well as the extensive use of undercover police officers.

Although the 1995 recorded crime figures in England and Wales saw the third consecutive annual fall in recorded crime after decades of swift increases, some police officers and academics have claimed that the improving figures are heavily massaged, and even a "sham," arguing that fear of losing insurance no-claims bonuses deterred some people from reporting crimes from vehicles and that a series of offenses (such as burglaries from a group of homes in a block) might be recorded as a single crime (Henry, 1996). There is considerable scope for such manipulation. As a police officer observed in the late 1980s, "if you've got a Chief Superintendent who is looking for more resources, one man stealing ten things goes down as ten crimes. If you've got a Chief Superintendent who is looking to become a Commander, then it's only one crime." It is likely, though, that computerization has reduced the scope for fiddling with the numbers (Graef, 1989, p. 275).

Yet, available victim surveys (especially the British Crime Survey and the General Household Survey) indicate that during recent years a decreasing proportion of offenses have been reported to the police, largely out of concern that it would make it more expensive to arrange insurance coverage for cars and homes. The decline in crime could also be a consequence of government attempts to monitor police productivity more tightly, tempting some officers to do disappearing tricks with unpromising cases. Certainly, the whole of the T.I.C. scheme (offenses admitted to by the defendant and "taken into consideration" when sentencing but not charged separately, thus qualifying as having been "cleared up" for police records) has been open to abuse for a long period. One officer noted that "that's the way it's been done since time immemorial" (Graef, 1989, p. 274). However, these ploys appear to have been in use long before 1992, hence should not affect the currently reported crime rate. The reduction in car theft must be seen as particularly significant.

If an optimistic approach is taken, 1995 crime levels were 8 percent lower than those in 1992 (a fall of 468,000 recorded offenses). This is a reduction matched in this century only in the years 1913–1915 and 1952–1954. Crime fell in thirty-five out of the forty-three police forces in England and Wales, with parts of Wales and Durham showing particularly large reductions. At the least (reports suggest that the fall may not have been sustained recently) the decline is an indication that the huge increases in recorded crime in recent decades are levelling off even if not falling appreciably. The overall clear-up rate for such reported crimes in England was unchanged from 1994 at 26 percent (though it was lower in London). Clearance rates for violent crimes were at about 65 percent. Offenses of violence against the person have a clear-up rate of 77 percent, and 92 percent of homicides and 89 percent of attempted murders were cleared up in 1995.

A relatively low homicide rate has been a marked feature of English society in the past two centuries. In the 1890s, it stood at 1 per 100,000 a year and declined even further in the Edwardian period (Gatrell, 1980, p. 286). This low rate survived a number of major social traumas, such as World Wars I and II, with often only very small variations (Gartner and Archer, 1981, p. 79). Even today, the English homicide rate is still fairly low by some European (for example Italian), let alone North American, standards. In England and Wales in 1993 there were 675 offenses initially classified as homicide compared to 682 in 1992 (HMSO, 1995, p. 71). However, the homicide rate in England and Wales has been rising slowly yet steadily for much of the period since 1961 (when 265 cases were initially recorded as homicide), something that was especially worrying given the improvements in medicine in this period (HMSO, 1995, p. 76). This increase has levelled off since 1986 and the average has been fairly constant (in the high 600s). Few homicides in England are the result of shootings, partly as a reflection of the country's tight gun control. In 1994 in Great Britain (including the separate legal jurisdiction of Scotland), 75 people were killed and 2,512 injured in attacks involving the use of firearms. The relatively low number of deaths to injuries suggests that most of the injuries were pellet wounds from shotguns and air guns (readily available) rather than rifle and pistol bullets. However, in the wake of the Dunblane tragedy in Scotland in 1996, when a madman murdered sixteen small children in a school, another firearms amnesty was held in England, Wales and Scotland between June the 3rd and 30th of that year, to encourage people who held unlicensed weapons to hand them in to the police without fear of prosecution. (Many of these guns are often elderly relatives' war souvenirs).

Although aware of the risks of romanticizing the past, there was a general feeling on the part of many British Chief Police Officers that at even the lowest level, violence and disorder in Britain in the 1980s and 1990s was worse than in earlier decades. A typical Chief Constable recalled "when even a diminutive chap like me could be called [to a pub disturbance] and people would freeze and everything would stop. Now the likelihood is that they would all turn on the police and assault him" (Reiner, 1991a, p. 167). However, although fatalities among serving police officers have increased slightly, they remain relatively low in Britain. (Northern Ireland is a very different situation). There were five police deaths in the five years preceding May 1995. In the London area (the largest and one of the most dangerous) eleven serving officers from the Metropolitan Police died violently in the course of their duties in the ten years from 1984 to 1994. Earlier, in 1983, three died from terrorist incidents connected with the IRA. A significant proportion of police deaths (unlike most murders) were by shootings.

Migration

As an island on the Western extremity of Europe, Britain has not had a history of migrants or temporary guestworkers arriving from the east of the continent. However, as a result of large-scale and permanent immigration, Britain, and especially the larger English Urban areas, has become a multi-racial society since the end of the 1940s, with large numbers of people arriving from the Caribbean from the 1950s on and from the Indian subcontinent since the 1960s. Initially, many police forces lagged behind the rest of society in reflecting these changes, resulting in areas that were largely made up of members of ethnic minorities being policed by officers who were overwhelmingly white. The move to a multi-racial cultural society has also produced additional policing problems by the inevitable reduction of commonly held social norms. Examples include the widespread and socially accepted use of cannabis among some sections of West Indian youth or different attitudes about the role of women among some Muslims.

There have also been allegations, sometimes well substantiated, that the police force itself contained considerable numbers of racially prejudiced officers and even a culture of such prejudice. In 1979 Sir David McNee, the then-Metropolitan Police Commissioner, proposed that a detailed study examining relations between the Metropolitan Police and the community be carried out. The subsequent report, from the Policy Studies Institute, raised a number of awkward issues with regard to police/ethnic minority relations. The researchers found that "racialist language and racial prejudice were prominent and pervasive" in the force and that many officers were preoccupied with racial differences and often believed that West Indians were inherently prone to deviance (Emsley, 1991, p. 67). It is, however, important to note that although the "canteen culture" of the police may include elements of racial prejudice, attitudes do not necessarily translate into discriminatory conduct on the street. This disjunction between culture and behavior in the field was caught by the Policy Studies Institute Survey which stated, after observing groups of police officers, that racialist language and racial prejudice were prominent and pervasive. Yet they also found that these officers, as they went about their work, often had "relaxed and friendly relations with members of racial minorities" (PSI, 1983).

Although there were periodic problems between ethnic minority groups (especially Afro-Caribbeans) and the police from the late 1950s on, these became particularly pressing as a result of large inner-city disturbances and riots since the start of the 1980s. In 1980 there was serious trouble in the St. Paul's area of Bristol. The Metropolitan Police also faced regular allegations of harassment and abuse of the notorious (and now replaced) "sus" law, a nineteenth-century provision whereby police were allowed to arrest people on the suspicion of their intent to commit a crime. In "Operation Swamp," carried out in Brixton in 1981, police, including elements of the Metropolitan Force's Special Patrol Group, were brought in to stop a serious (and rapidly growing) level of street crime. In this process they stopped over 1,000 mainly black youths for questioning and made over 100 arrests (many for resisting the police). The resulting tensions in the area appear to have fuelled the subsequent Brixton riots.

Lord Scarman, in his inquiry, found that the riots that occurred from April 10th to 12th in 1981 had been sparked by an incident involving an injured young black youth and two police officers. Some members of the community felt the incident was harassment (the officers were actually trying to assist the youth) and a large crowd of black youths swiftly gathered. Subsequently there was widespread rioting in which 279 policemen (and 45 members of the public) were injured in the violence; there was also widespread looting and nearly thirty buildings were burnt down. Lord Scarman felt that the police were partly to blame for a breakdown in relationships with the local community, and that there had been cases of racial harassment among junior officers on Brixton's streets. Rioting in other parts of Britain's inner cities (usually those with high ethnic minority populations), such as Toxteth, followed later that year, sometimes of an even more intense kind. Moss Side in Manchester, Chapeltown in Leeds, and Handsworth in Birmingham saw major outbreaks, while there were smaller-scale disturbances in Wood Green and Hackney in London and in Nottingham and Coventry. The police were the targets for much of the violence. In Toxteth, iron railings were pulled out of fences and used as improvised spears against the police, along with petrol bombs, rocks and bricks. A large section of Liverpool's inner city area was also burnt. Police equipment and tactics at the time were clearly inadequate. The CS gas which was used by the Merseyside force (all that they had on hand at that time) were barricade penetration cartridges fired from shotguns, not special riot gas canisters lobbed from launchers. The cartridges could produce serious wounds if they hit individuals.

In the aftermath of these disturbances, Lord Scarman was commissioned to conduct an inquiry and produce a report. The main recommendations of his report included the introduction of statutory consultative committees to make the police more accountable, a better arranged and organized plan and accompanying action to deal with inner-city social and economic problems, a greater amount of independent oversight of the police complaints procedure, more thorough police recruit training on dealing with and preventing the development of public disorder, greater effectiveness in dealing with racially prejudiced conduct by police and more recruitment to the police from ethnic minorities. Lord Scarman supported a proposal by the Commission for Racial Equality that racially prejudiced or discriminatory behavior should be included as a specific offense in the police disciplinary code and also stated that it should be made clear to all police officers that the normal penalty for racially prejudiced behavior is dismissal (Scarman Report, 1986).

Social conditions can obviously encourage crime and disorder. The Scarman Report of 1986 also concluded that ethnic minorities had suffered disproportionately from the problems faced by all inner-city residents. "Unemployment and poor housing bear on them very heavily." This was a problem which Sir Paul Condon's controversial (and widely publicized) letter on mugging in London in 1995 (*Daily Telegraph*, July 7, 1995) addressed, explicitly stating that "very many of the perpetrators of mugging are very young black people, who have been excluded from school and/or are unemployed." Housing is poor in the large council estates where much of the black section of the community lives. The local schools' performance is much below the national average. One school, the Dick Shepherd in Brixton's Tulse Hill estate, was closed in recent years after its students achieved only a quarter of the national average in GCSE (school examination) passes. But low achievements have not always been a simple matter of a lack of resources. The local (and radical) council to Brixton (Lambeth) has had a mixed record in addressing the area's problems, although considerable sums of government money have been spent. A major reduction in many services, including housing and education, has recently occurred there, with serious allegations of managerial incompetence, mismanagement, wasted resources and corruption (Valleley, 1995).

Some of the almost insoluble problems in this area of policing can be seen in the reasons for, and the reaction to, the Commissioner's much-publicized letter on

London mugging in 1995. He made public what many police officers had been saying in private for a long time—that disproportionate numbers of black youth in the capital were involved in this type of crime. In the year preceding June 1995, there were 22,000 recorded street robberies in the capital with (according to victim statements) 80 percent being carried out by blacks. Scotland Yard figures revealed that 60 percent of those subsequently arrested for street robbery were black. Fear of antagonizing the black community had previously prevented public comment. However, as a major police initiative was planned to combat such (swiftly rising) street crime ("Operation Eagle Eye"), senior officers feared that without announcing publicly the reasons behind such a move, an anti-racism backlash prompted by the number of arrests involving young blacks suddenly increasing might occur. However, by going public on this (in a carefully worded letter), in an attempt to prevent the type of misunderstanding that had accompanied "Operation Swamp" in 1981, the Commissioner brought about a barrage of criticism and allegations of racial prejudice from minority groups.

There is a considerable degree of distrust between some minority groups and the police, with potentially serious consequences. A survey in the mid-1980s revealed that only 62 percent of West Indians in London would be willing to give evidence in a street robbery case compared to 84 percent of Londoners generally (Robillard & McEwan, 1986, p. 11). Because few young working class blacks will cooperate with the police, it has been hard in London and other big cities to arrange identity parades (where the victim picks out a suspected criminal) which will meet the standard of legal admissibility. As a result, the Metropolitan Force often resort to putting their suspect on a busy escalator in the underground system while the victim travels in the opposite direction to that of the suspect, to see whether he or she can identify him. Members of the ethnic minorities (as well as younger people) who were the victims of crime were also significantly more likely to feel that the police had not dealt adequately with, or demonstrated sufficient concern about, their cases. Asians were particularly unlikely to report that the police had been very polite (HMSO, 1994, p. 63). Yet local victim surveys, such as the Islington Crime Survey, have indicated that ethnic minority members are disproportionately the victims of crime. Some research indicates that Afro-Caribbeans are 50 percent more likely to be stopped by the police in vehicles and four times more likely on foot (Leishman et al, 1996, p. 200). In the poorer and rougher areas where they are disproportionately concentrated, the rates for "local" whites are also probably much higher than for the white population in general.

Steps have been taken in recent years to counteract minority resentment by attempting to increase the numbers of serving officers from the ethnic minorities as well as incorporating race relations in the training of police recruits. Crimes based on or motivated by racial harassment (such as assaults or the defacing of houses) have also been given greater priority by the police. By the 1980s there were also recruitment drives specifically aimed at recruiting potential officers from the black community, something that was highlighted by the BBC television series *Black in Blue*, broadcast in 1990 (Emsley, 1991, p. 176). Although their numbers are still relatively small compared to the size of the minority community in the urban areas (policing is still an "overwhelmingly white occupation"), recruiting posters in cities "invariably feature the face of a black or Asian officer" (Leishman et al, 1996, p. 210). Undoubtedly, much has been achieved in this area yet there is still a considerable amount to do. For example, also in 1990, an Industrial Tribunal decided that a police officer of Asian ethnic origin had been discriminated against when he tried to become a detective in the Nottinghamshire Police. The Tribunal's written decision noted that there was no overt racism in the force, but was critical of some officers and of the racial banter directed at black officers. Additionally, some black officers will face hostility from elements in their own communities. But in some policing environments even the best efforts by the police, however well motivated, will be wasted. Thus an officer engaged in community policing (and accompanied by a researcher) in an inner-city area, noted that there were "hopeless situation(s) in which there were no grounds to attempt to cultivate relationships." This officer, when newly appointed, had attempted to visit a group of workshops, mainly run by Rastafarians (and suspected of heavy involvement in altering stolen cars) to say "hello" and attempt to build a relationship. However, in response, "no one said hello and some looked right through him." There was a tangible atmosphere of hostility to the policeman (Fielding, 1995, p. 144).

It is also important not to generalize about ethnic minorities as a whole, but to distinguish among individual minorities or even sections of minorities. For example, an analysis of local London Crime Surveys such as the Islington Survey suggests that although the police are perceived as being unfair and treating groups differently by age, race or gender by a much higher proportion (77.5 percent) of young (16–24) male blacks (i.e., Afro-Caribbeans) than any other racial group, the figure for similar-aged white youths who believe this to be the case (46.5 percent) is significantly higher than that for Asians (32 percent).

RESPONSES TO CHALLENGES

Administrative Reforms

The Reversal of Politicization

There have been a number of concerted attempts by senior officers in recent years to improve standards of police professionalism and to regain public support. Beginning in the late 1980s, the process of politicization has been reversed to a considerable extent. Labour Party political leaders are now assiduous in attending police conferences, and in being interviewed by police journals. Senior police officers are increasingly distancing themselves from an excessively close relationship to the Conservative (or any other potential) government (Reiner, 1992, p. 96). Additionally, since the end of the Wapping dispute (involving printworkers) in 1987, Britain has not seen industrial disputes with anything like the bitterness of earlier years. Civil disturbances have been confined to the inner cities and to some large Northern housing estates. Nevertheless, the legacy of Conservative repressions of strikers and a lingering distrust remains, especially in the areas where the problem was most acute (such as former mining villages).

Lay Involvement in Policing

Another way in which the police have sought to promote a sense of public support in recent years has been to encourage a degree of voluntary involvement by the public in their own policing, and thus to reduce the police/public division. The most obvious example of ordinary people becoming involved in policing is the Special Constabulary program, which enrolls people who give up some of their spare time on weekends and in the evening to work, without pay, as part-time policemen. The government has encouraged this development, which has also been supported by senior officers, especially when other civilian self-policing agencies, such as the Guardian Angels (imported from the United States with fairly limited success in the late 1980s), have appeared to threaten their own position. The use of these part-time officers has varied. Traditionally (they are not a new innovation) they have played a very minor, even cosmetic, role, helping out at public events, etc. But some forces have increased their numbers and given them an expanded role, though this has raised serious questions as to their level of competence and training (the latter inevitably rather limited) to deal with emergencies. In the Metropolitan Police Special Constabulary there were 1,550 Special Constables in March of 1995; encouragingly, a relatively high proportion of these, 170, were from ethnic minorities. They performed a total of 407,102 hours of operational policing that year, ranging from going on divisional patrol and assisting regular officers at large sporting and ceremonial events (such as F.A. cup matches), to giving instruction to the public in crime prevention techniques. The Home Secretary, Michael Howard, announced in 1996 that a significant expansion of the Special Constabulary, from its present strength of 20,000 nationwide, would play a vital role in the necessary "partnership" between the police and the community to fight crime. A further ten million pounds was promised in special grants for police forces to increase their numbers of special constables, funds that were to be supplemented by local commercial sponsors.

At a more widespread level, Neighborhood Watch (NW) schemes have proved to be a success in terms of membership numbers. There are now over 143,000 schemes in England and Wales covering some 6 million households. Under these schemes organized groups of people report suspicious or unusual local activities to the police. Recently, the Home Office has sanctioned a limited role for NW street patrols under supervised conditions. NW supporters believe that these schemes have made a significant contribution to reducing crime and have also helped to reduce the fear of crime by reassuring residents of the neighborhoods in which they are situated that local people are watching out for them. However, the scheme has also received criticism from other commentators who feel that it is largely ineffective and cosmetic. An analysis of the early days of the scheme suggested that it did reduce the fear of crime, as well as improve local social cohesion, but that there was little evidence of a significant improvement with regard to public reporting of suspicious incidents to the police or subsequent police clearance rates (Bennett, 1988, p. 241). It has also been found that in high-risk areas, where relations with the police are poor, schemes are more likely to fail. These areas are also the most dissatisfied with the resources available from the police to back up the NW scheme. Although introduced as a community-based crime prevention scheme that would require little support from the police, it has become apparent that it requires an appreciable and active police involvement if interest in such schemes is not to flag (Husain, 1988, p. 44).

Improvements in Police Training and Recruitment

Police training since the 1980s has stressed an increased need for sensitivity on the part of officers. New recruits to the police have traditionally been drawn largely from white (though this is slowly changing) working-class and lower-middle-class males, who quickly fit into a distinct rank-and-file police or "canteen culture" which, it has been alleged, is steeped in prejudice. Within the ranks, cultural norms have been manifest in a resistance to women police officers with perceived "modern" attitudes and to homosexuals (though there is now a gay and lesbian police officers' association). These prejudices have also influenced contact with the wider public. In Manchester in the 1980s, the Chief Constable, James Anderton (admittedly in no way a typical Chief Constable), publicly described AIDS as God's curse on homosexuals who were "drowning in a cesspit of their own making." In the middle years of the same decade, another policeman observed that "you hear remarks about poofs, Pakis, lezzies, women, students, the rich, the media, politicians, all foreigners, the Scots, the Irish—you name it. We hate everybody." However, in the past decade the police have made significant efforts to improve their levels of "human awareness" (to employ a police college term) and to address some of these prejudices during the fifteen weeks' initial training at the college. In 1993, a Chief Superintendent was ordered to resign his position after allegedly referring (rather foolishly to a local newspaper reporter) to women delegates attending a conference on domestic violence as "a bunch of lezzies" (Graef, 1993).

Many more women officers have been recruited and promoted to command positions (now making up, for example, 20 percent of the total Manchester force and undertaking all police jobs from motorbike patrols to firearms and public order). In 1993, 13.2 percent of English officers were women, compared to 8.6 percent in 1981. In 1994, three women were made deputy chief constables and in 1995 there followed the appointment of the first female Chief Constable, Pauline Clare (of the Lancashire Constabulary), who has herself commented on the extensive changes since she joined, a period when "the Constabulary was segregated; policewomen belonged to their own department and had their own conditions of service in what was perhaps regarded very much as a man's world" (*LancCon*, 1996).

Some urban forces are also now attracting significant numbers of ethnic minority recruits, while in 1990 there was the formation, by homosexual officers, of the Lesbian and Gay Police Association, one of the aims of which was specifically to "work toward better relations between the police and gay community" as well as to promote the position of homosexual officers within the service (Leishman et al, 1996, p. 202). Police officers have started liaising more heavily with the large gay communities of London and some other cities. Indeed, a police liaison officer has been appointed at Bootle Street station specifically to look after the interests of Manchester's local "gay village" (Graef, 1993).

Some commentators have alleged that such efforts are often little more than window dressing or PR campaigns that make little or no difference to the average patrol officer on the beat unless local commanders support such policies as fully as the Chief Constables who introduce them and who head their forces. However, their very existence is in itself significant of attempts to bring the service-up-to-date. Other informed observers have remarked that in many areas there is a distinct change in the style of policing in Britain in recent years.

Community Policing: Sensitivity and Styles

Police sensitivity was something that was expressly encouraged by Lord Scarman (following the widespread inner city riots of 1981, particularly in high ethnic-minority-populated areas such as Brixton) in his 1981 report, and has clearly had some, albeit limited, impact on the ground. Scarman urged the application of traditional principles of policing based on extensive consultation and cooperation with the communities being policed. This is sometimes described as community policing and has had the support of senior officers such as John Alderson. In Lord Scarman's report it was stressed that the primary police aim of preserving public calm and avoiding disorder should have a higher importance than the strict enforcement of the law at any cost. Also indicative of a desire to regain some of the esteem that the police of the Dixon of Dock Green era had enjoyed was the introduction of the Plus Programme by the Metropolitan Police, under the then-Commissioner Sir Peter Imbert (following the Wolff Ollins report of 1988) in September 1990. The program was promoted by a "Statement of Common Purpose" prominently displayed in all police stations and emphasized the service nature of policing (Emsley, 1991, p. 177). However, some commentators, including former policemen, felt that it was likely to "run aground ... because of the strength of beliefs at canteen culture level, and even in the management ranks" in the same way that initiatives by Imbert's predecessor Sir Kenneth Newman had failed. Indeed, it was even pessimistically reported that "such attempts at radical change will always fail" because of internal attitudes (Goldsmith, 1991, p. 395). This form of community policing was also expounded by John Alderson, the influential former Chief Constable of Devon and Cornwall.

A survey of encounters between the police and the public in English cities concluded that hostility on either or both parts was present in one in ten encounters (though in only one in fifty encounters was hostility present in both parties). To an extent, this was inevitable. "It is unrealistic to suggest that all such hostility could be eliminated, or even should be: There are times when the officer needs to be assertive or aggressive in order to establish his authority or bring a wrongdoer under control." The important thing is to eliminate unnecessary or gratuitous hostility in police-public encounters (Southgate, 1987, p. 1). Police attitudes in this respect have matured in the past decade. Thus, senior officers in the Hackney Division of the Metropolitan Police have noted that although Hackney is a poor and ethnically mixed part of London there were no "No-Go" areas for the police in the Division. The police were able to patrol everywhere, including the rougher public housing estates. However, the senior officers also accepted that there was always a potential for disorder in such an environment. As a consequence, "we ask officers to act circumspectly and sensitively while carrying out their duties." This included an emphasis on displaying a "correct attitude at all times" as an "abrasive" attitude or careless remark could transform an apparently simple situation into one involving confrontation and possible injuries (MetPol, 1996).

This approach to policing has not always gone unchallenged either within the police services in Britain or by some outside commentators. Such attitudes can be found in the die-hard, or traditional, school of policing. But other critics are thoughtful individuals who genuinely worry that the reality (if not the theory) of community policing is sometimes policing at a level that is tolerable to the most aggressive and least law-abiding sections of the community in urban areas. Furthermore, the growth in certain types of crime has meant that some forms of sensitive or even traditional policing may sometimes be inappropriate. Indeed, Lord Scarman (1986) himself expressly recognized this in his report, as far back as 1981, stating that there will continue to be circumstances in which the use of hard policing methods including the deployment of the Special Patrol Group is appropriate, even essential. Perhaps reflecting some of these concerns, in 1994, the Home Secretary Michael Howard ordered police forces in England and Wales to reduce their use of cautions given by a senior and uniformed police officer at a police station, after a person has admitted an offense, in lieu of prosecution. His concern was that the increasing use of cautions for more serious offenders and the use of repeated cautions for minor offenders might lead juveniles to think they could get away with lawbreaking.

Police sensitivity in other areas has also been improved. The Home Office issued a "Victim's Charter" in 1990 which described how victims of crime could expect to be treated when having their cases investigated by the police, along with what other services the victims were entitled to receive from the Criminal Justice System. Police handling of rape cases has been made much more sensitive, from the victim's perspective, in part following the much publicized and televised and highly insensitive handling of a reported rape by the Thames Valley Constabulary in 1982. There are also changing perspectives as to what sort of behavior should be the subject of police intervention. Sometimes this means increased involvement in delicate areas that the police have often shunned in the past. For example, traditionally and until the early 1980s, involvement by the police in domestic disputes (between spouses and co-habitants) was often limited to trying to pacify the disturbance and effect a reconciliation between the parties. However, increasing agitation from women's groups and others about domestic violence in London resulted in the formation of a Working Party in 1984, at the behest of the then-Metropolitan Police Commissioner, to examine police practice in such cases. Its report was published two years later and made several recommendations in favor of a change in police approach from one characterized by mediation to one of active intervention. This could be achieved by, for example, the arrest of the apparent assailant (often the husband or boyfriend) and a greater degree of support and advice for the victim. Guidelines for the Metropolitan Police were prepared using the Working Party's recommendations, which lowered the threshold at which police intervened actively in domestic disputes and emphasized the need for victim support.

It has been argued that the police in recent decades have developed an excessive emphasis on the detective work of the Criminal Investigation Department (CID) at the expense of foot patrol and its attendant scarecrow function (though some doubt about the effectiveness of deterrence in a sophisticated modern society has also been expressed); and there has been a proliferation of specialist departments. Uniformed patrol is being increasingly carried out by young and inexperienced police officers or by police officers who have failed to make promotion to other areas of work or to the CID. This distinction has been found to be reflected in the police culture, where such routine work is sometimes held in low esteem. During the period from the 1960s onward, especially following the introduction of "unit beat policing," officers increasingly started to use patrol cars (rather than foot patrols or bicycles) to cover their beats, a change that was highly conducive to breaking links with local people.

Within the police force itself, by the late seventies, salary levels had also fallen significantly in real terms. Although pay was subsequently improved again in 1978, after the Edmund-Davies Inquiry recommended substantial salary increases and a guarantee that future rises in pay should be index-linked to the cost of inflation, one particularly unfortunate result of low pay in the mid to late 1970s was that it was often difficult to attract and retain high quality recruits. Many experienced officers left the service, sometimes resulting in inexperienced and excessively aggressive young officers, not properly commanded by senior men, operating on the streets, especially in the combustible ethnic minority inner-city areas of the large English conurbations, areas such as Toxteth, Brixton and Moss Side in Liverpool, London and Manchester. This was something that was to cause problems in the early 1980s. Additionally, some officers received promotions who were less than ideal for command positions.

The British Crime Survey of 1992 has made it clear that for the majority of the public seeing police patrolling on foot is "related to satisfaction across a broad set of performance measures," despite the fact that empirical work has also demonstrated that it was not very effective in making arrests and clearing up crimes. To this extent, the former Metropolitan Police Commissioner, David McNee, may be incorrect in believing that "the confidence of the public in the police depends as much on effectiveness as on a comforting attitude and a smiling, kindly face" (McNee, 1983, p. 180). In 1992, 61 percent of those surveyed felt that foot patrolling by the police was inadequate in their areas (compared to only 25 percent for vehicle patrols). Such patrols played an important role in reducing fear of crime, although they were expensive in terms of resources and tended to increase response times to emergency calls (HMSO, 1994, pp. 64–65). Interestingly, this preference for high visibility policing was found within all ethnic groups other than Afro-Caribbeans, though it was also less common among young males generally than for the greater population. Although some specialization is inevitable (obviously sophisticated terrorism and organized crime cannot be effectively dealt with by foot patrol), this argument asserts that the process has gone too far and that there is a need to return to the basics of policing. Some forces, such as the Metropolitan, have tried in recent years to put a greater emphasis on uniformed patrol.

International Influences

As the world becomes smaller due to transport and communications revolutions, crime has become ever more international. There have been allegations that Russian and Italian Mafia money launderers have been exploiting loopholes in the British banking system to move millions of pounds through London's financial centers. Luigi Palmieri, head of Interpol's Organized Crime Task Force, stated that "one of the problems for the UK is the banking sector," and suggested that the laws should be more rigid. Palmieri said it was relatively easy to transfer money in and out of London; although transactions by individuals might be scrutinized, companies acting as fronts for organized crime could evade checks (cited in Tendler, 1995). Transnational crime has necessitated international cooperation by police forces; as crime has become internationalized so has policing. Britain has had a longstanding and active role in Interpol. A British policeman and former Scotland Yard detective, Raymond Kendall, is currently secretary general of Interpol at Lyons in France.

Britain has been (sometimes cautiously) involved in many of the important policing developments occurring within the European Community. These countries are committed to many common values and standards, allowing a relatively speedy development of close contacts. Yet at present, the only criminal laws for which the EU can require standardization among its members are in the areas of insider dealing, computer crime, money laundering and environmental crime (Schwarzenegger, 1996, p. 142).

In the treaty on the European Union signed at Maastricht in 1992, the member states of the European Community took the decision (Union Treaty, article K1(9)) to establish a European Police Office (Europol). European police forces were seeking to draw up plans for a Community-wide system of intelligence to counter the threats posed by terrorism, drug trafficking and other serious forms of international crime. This proposal was supported by the Declaration on Police Cooperation made in conjunction with the Union Treaty (Cullen, 1992, p. 383). By 1995 a Europol computer had been set up at The Hague in Holland to allow information sharing and liaison officers from the member states, including Britain, had started to work there.

Additionally, a special dedicated intelligence unit, the European Drug Intelligence Unit (EDIU) has also been set up, with British participation, to combat the growing threat from international drug dealing. Operational since 1994, it is (like Europol) based at The Hague. It also serves primarily as a clearing center for national police forces, providing information on drug trafficking and attendant crimes such as money laundering. At present it has no investigative powers.

Britain has refused to join the seven-nation Schengen agreement on open borders. Schengen allows citizens of many European Economic Community (E.E.C.) countries to travel between their states without border checks. The British government believes that such checks are necessary in the fight against international crime, illegal and economic immigration, rabies, etc. As an island, unlike the land-bordered countries in the Schengen agreement, border controls are probably also more effective in Britain, perhaps explaining the reluctance to relinquish them. Other countries have experienced problems with the agreement. France initially withdrew from the agreement following concerns that there were insufficient checks against drug smuggling from Holland. Holland responded by reinstating checks on its border with France.

There is considerable debate at present whether Europol will be given real powers to deal with (international) crimes in Europe, by investigating, arresting, etc., in the same way that national forces do. The German government supports this idea. Also in 1994, the European Union was a party to the Berlin declaration of 1994 (with the US, Canada, Eastern and Central European countries, Switzerland and Morocco) proposing greater mutual support in crime control (Schwarzenegger, 1996, p. 38).

Other influences (especially on British police equipment) have come from North America. These have included concepts such as Neighborhood Watch and some types of weapons. For most of its history, down to the present day, British policemen, other than those engaged in embassy or diplomatic protection, VIP bodyguards, etc., have been limited to their traditional truncheons. This issue has been reopened, especially in London and other big cities, and is partly linked to a perceived increase in the use (or at least carrying) of guns by criminals. As a partial measure to increase their police power without issuing them pistols, police officers have recently (and controversially) been issued with American-type CS gas sprays, and the traditional short truncheon (in use for over a century) has been replaced with the larger (and heavier) plastic side-handled baton (also an import from the US). This baton was chosen because trials suggested that it was popular with officers and also led to fewer police casualties in confrontations. However, a recent death as a result of a blow from such a baton has prompted concern, and led the coroner at the deceased man's inquest to recommend that more training be given to officers in the employment of the weapon.

Laws

Police Powers

Law empowers the police yet also controls them and provides the rules that the police must attempt to enforce. With regard to rules of conduct there has been some concern at the perceived overeach of the criminal law. Some have argued that in areas such as the use of soft drugs, status offenses, etc., the law should be changed. Recently, former Scotland Yard detective Raymond Kendall, now Secretary General of Interpol, urged decriminalization for people who use drugs, including hard drugs such as cocaine and heroin, but not for traffickers who deal in the drugs. He felt that drug abuse was more a social, health and welfare problem rather than a police problem. Already, in England, the reality is that a first offense of possession of small amounts of cannabis for personal use is met by a police caution (if admitted) rather than formal court action. However, most politicians including the Prime Minister have made it clear that they are strongly opposed to decriminalization, even of soft drugs. In 1996, some British Chief Constables also recommended the legalization of brothels.

The law regulates the British police in a number of different ways. Foremost among them are that they can be prosecuted if they commit criminal offenses, and sued if they commit civil torts, such as false imprisonment or wrongful arrest. Another important way in which the law, via the courts, can regulate police conduct in the investigation of crime is by clearly defining their powers and by punishing breaches and abuse of such power by excluding any resulting evidence from the court. Aspects of the *Police and Criminal Evidence Act of 1984* (PACE) have provided the judiciary with markedly increased control of policing via the ability to exclude evidence. There has been a marked trend since 1986 for a more active judicial intervention.

It has been asserted that according to the tenets of a liberal society policing remains acceptable to the wider public only when the police are in possession of carefully and precisely defined legal powers to interfere in the lives of the public. The process of generalizing powers and making them rest on the individual officer's discretion "removes a central constraint" on the police (Uglow, 1988, p. 52). Prior to the advent of the *Police and Criminal Evidence Act of 1984*, police powers and responsibilities, whether patrolling the streets or on arresting and detaining suspects, were, as the Royal Commission on Criminal Procedure explicitly observed in its 1981 Report, rather confused and unclear. They were governed by a mixture of common-law rules and the recommendations contained in the non-binding Judge's Rules, judicial guidelines on good practice, periodically issued from 1911 onward. One of the aims of the Royal Commission was to codify and clarify the law in these areas.

The PACE 1984 Act and its associated Codes of Practice was partly designed to regulate precisely what police officers could and could not do in most given situations, as well as to introduce safeguards for those detained and/or questioned by the police. The impact of the new Act has been significant in the ten years of its operation (it came into force on January 1, 1986). Codes of Practice which provide firm and relatively clear guidance for police conduct in a variety of areas are issued pursuant to section 66 of the Act. Breaches of the codes are themselves specifically admissible in evidence (under section 67 [11] of the Act) and can be considered by the courts in determining questions of evidential admissibility. The five Codes cover the treatment of detainees (accommodation, standards, provision of rest and refreshment, etc.), the right of access to free legal advice (guaranteed by section 58 and code C, section 6, of the Act, and the duty solicitor scheme), the videotaping of interviews, the identification of suspects (e.g., the correct procedure for "I.D. parades"), etc. So extensive are these provisions that some judicial (and much police) comment has suggested that the balance of the investigative process has, as a result of PACE, tipped too far towards the suspect,[2] while inevitably the documenting of the necessary safeguards provided by PACE has added to the burdens imposed by an increasingly bureaucratic service. In May 1993, the Home Secretary, having become concerned, from apparently anecdotal evidence, about the level of paperwork following an arrest, even commissioned a study into the matter which resulted in Police Research Series Paper. This has contributed to a situation in which it has been alleged that a typical medium-sized force of 2,500 officers can often only keep 125 men 'on the streets' when those doing administrative work, detective work, undergoing training, on leave, or at headquarters are accounted for (cited in Burke, 1996).

PACE has provided many safeguards which were notably absent in the 1970s and early 1980s (a period which, in retrospect, produced a number of embarrassing miscarriages of justice based on police malpractice), as well as firm guidance to correct police conduct. By itself this might not have been enormously significant. However, the Act also provided the judiciary with new discretion to exclude evidence that was unfair, something which provided teeth in controlling police abuse of these safeguards. Unlike the American experience, where decisions such as *Miranda v Arizona* appear to have encouraged the exclusion of tainted evidence, in England the Courts took a progressively more restrictive view of the existence of such a discretion, culminating in the landmark decision in the 1980 House of Lords' case of *R v Sang* [1980] AC 402.

Although the case concerned an agent provocateur, the court took the opportunity to issue an opinion on the wider topic of the court's discretion to exclude evidence that had been obtained in dubious circumstances. The tenor of the Court's decision was summed up by Lord Diplock when he stated that although a judge in a criminal trial had a discretion to refuse to admit evidence if in his or her opinion its prejudicial effect outweighed its probative value and, with regard specifically to confessions and evidence obtained from the accused after commission of the offense, if these were obtained improperly, he or she had no general discretion "to refuse to admit relevant admissible evidence on the ground that it was obtained by improper or unfair means." Thus, the Lords upheld the view of the lower courts that the judge had no power to exclude evidence simply because it emanated from the activities of an agent provocateur.[3]

However, section 78 of PACE 1984 gave the courts a new statutory (and general discretion) to exclude from criminal trials evidence which has been obtained "unfairly" (Stone, 1995). This provision was in addition to the court's statutory and common-law powers to exclude specific types of evidence (such as that expressly governing confessions contained in section 76 of the Act). Section 78 was a late amendment to the 1984 Act, having been put forward by the government in the face of an attempt by the Lords (acting in their parliamentary capacity) and led by the (former) judge Lord Scarman, to insert a stronger provision. The section stated:

78 (1) In any proceedings the court may refuse to allow evidence on which the prosecution proposes to rely to be given if it appears to the court that, having regard to all the circumstances, including the circumstances in which the evidence was obtained, the admission of the evidence would have such an adverse effect on the fairness of the proceedings that the court ought not to admit it.

This provision has resulted in much case law since it came into force in 1986. Although sometimes it has been difficult to identify the principles being applied in its use, it is clear a lack of probity by and bad faith on the part of the police with regard to serious breaches of the Codes (but not simple inadvertence on the part of the police) have been of crucial importance in deciding whether to bring the discretion into operation. Bad faith in this context includes deliberately (or knowingly) exceeding police powers, failing to meet the requirements of the 1984 Act and its Codes, or indulging in generally dishonest or deceitful behavior.

An illustration of this is the case of *R v Mason* [1987] 3 All ER 481. Mason was arrested on suspicion of committing arson. The police had no direct evidence linking him with the crime. Nevertheless, they told the defendant and his solicitor (the court was particularly concerned at this aspect of the deceit!) that they had found fragments of glass from a bottle of petrol near the fire, and that his fingerprints were on the fragments. This was totally false. The accused, in the face of this allegation, confessed, and the trial judge allowed the evidence of the confession to be admitted against him. On appeal it was held that the deceit practiced on M. and his solicitor was very reprehensible, and since this confession was the only significant prosecution evidence, the conviction would be quashed. Although this obvious trick would probably have led to exclusion even under the common law rules (as set out in *Sang* [1980] AC 402), other cases have gone further and have suggested that the simple fact that the police are aware that they are acting beyond their lawful powers may in itself constitute sufficient bad faith to justify exclusion (see on this *Matto v DPP* [1987] Crim LR 641).

Significantly, when confirming subsequently (in *R v Alladice* [1988] Crim LR 608) the admissibility of a confession (despite wrongful police denial of access to a solicitor), the Court of Appeal expressly noted that there was no evidence of deliberate bad faith on the part of the investigating officers and that it was clear that the defendant had known his rights and that it had made no difference to the outcome. Despite this latter fact, the Court also stated that "if the police had acted in bad faith, the Court would have had little difficulty in ruling any confession inadmissible under section 78." Thus bad faith on the part of the police, whether deliberate trickery or an awareness of exceeding their powers, is a significant factor in favor of the exclusion of evidence. Of course, even with the 1984 Act, the codes or the audiotaping and now videotaping of interviews in operation, there is no ground for complacency, as was illustrated by the interviews in the "Cardiff Three" case (*R v Miller, Paris and Abdullah* (1993) 97 Cr App Rep 99). These were described by the Royal Commission of 1993 as having been conducted in a "loud and aggressive way." There are still concerns about the competence of some legal representation at police stations, though the Law Society (representing England's solicitors) has attempted to improve this.

Not everything that might be considered by laypeople to have involved a trick on the part of the police will necessarily be regarded by the courts as justifying exclusion. In *R v Christou* [1992] 3 WLR 228, police undercover officers set up a mock shop, called "Stardust Jewellers," where they pretended to deal in secondhand jewelry in an area with fairly high crime levels. The transactions with customers were secretly recorded on hidden video cameras. The defendant was filmed dealing with stolen goods and also making incriminating statements as to the provenance of the goods. It was held by the Court of Appeal that the evidence was admissible because the police had not actively encouraged or incited the commission of offenses, the defendant had "applied himself" to the trick, and there was thus no unfairness. Similarly, in *Williams v DPP* [1993] 3 All ER 365 the police left an apparently unattended lorry parked in a road, apparently containing cartons of cigarettes, which the accused subsequently stole; the court considered that there was no unfairness, as again the defendant had not been actively encouraged.

Failing to allow such evidence would prevent many undercover operations. In *R v Bryce* [1992] Crim LR 728, however, an unrecorded conversation between a plainclothes police officer and the defendant about a car which the defendant was selling was excluded partly because it had deliberately been intended to produce incriminating statements and should thus have been preceded by a formal caution (which would have alerted the defendant). More controversially, in *R v Bailey; R v Smith* (*The Times*, 22 March, 1993), evidence was not excluded. The officers investigating a crime and the custody officer at a police station acted out a conversation in front of the defendants in which the custody officer, pretending to act against the wishes of the investigating officers, insisted on placing the two defendants in the same cell (the cell was bugged). The defendants, lulled into a false sense of security by this, subsequently had a conversation which contained damaging admissions (and which was recorded). The Court of Appeal found nothing so wrong in what the police had done as to justify exclusion, even though it was a way of circumventing the rule that they could not themselves question the defendants any further (because they had both already been charged with offenses) (Stone, 1995).

Other areas in which the police have come in for robust judicial censure have been in the use of suspect profiling and entrapment. A clear recent illustration concerned the investigation into the brutal murder of Rachel Nickell while she was out walking with her baby on a London common in 1992. This crime prompted a large police investigation and raised serious questions about the use of undercover investigative techniques. The initial investigation failed to obtain any significant leads on the identity of the killer. A forensic psychologist was then employed to provide a psychological profile of the murderer. Information given to the police about a local man's sexual fantasies by a former pen friend led the police to suspect him of possible involvement (he had also been fairly close to the murder scene at the time). An undercover operation was then set in operation by the police with a view to obtaining admissions linking him to the murder. A police woman corresponded with the man (Colin Stagg) and encouraged him to send her examples of his sexual fantasies. When Stagg, after some prompting, produced fantasies which were similar to those employed in the method and place of the murder, he was arrested and charged. The defense argued at trial that the fantasies were largely induced by the undercover officer and heavily encouraged by her promise of a sexual relationship to a lonely man, and that thus the evidence should be excluded under section 78 PACE 1984. These arguments were accepted by the trial judge, Mr. Justice Ognall, who also made a fierce, and heavily publicized, attack on the police methods employed. He termed the investigation "a misconceived and deceptive operation" and urged particular care in the use of psychological profiles. The judge also stated that "if that route involves trespass into the territory of impropriety, the court must stand firm and bar the way." He excluded the evidence both on the basis of a circumvention of Code C (governing the questioning of suspects), and on the grounds of general unfairness (Doherty, 1994).

The 1984 Act also established the Crown Prosecution Service (CPS) as an independent body to review cases and oversee the conduct of prosecutions, taking responsibility (and pressure) from the police in this area.

Civil actions against the Police as a Means of Accountability and Control

Dissatisfaction with the existing complaints system and the desire for redress have encouraged the growth in actions against the police in the civil courts. Obviously, and without dispute, the police are accountable to the law and the courts in the exercise of their powers and duties. They can be prosecuted for crimes resulting from their misconduct (such as assault—always quite difficult because of the standard of proof and the circumstances of most incidents make it hard to verify such allegations "beyond reasonable doubt"), and they can be sued for damages in the civil courts if it is alleged that they have committed a tort against members of the public (such as false imprisonment). The courts have taken a very much more robust position in recent decades, while the public appears to be much more rights conscious and willing to litigate, leading some commentators to observe that "whilst local democratic accountability has declined both in substance and as a political issue, police accountability to the law for the use of their legal powers ... has been enhanced" (Reiner & Spencer, 1993, p. 13).

Since 1979 there has been a significant increase in civil litigation against the police for actions such as unlawful arrest, assault, wrongful imprisonment and malicious prosecution. These are much easier to establish than in a criminal court, because of the lower standard of proof (the balance of probabilities). A fair number of these suits have been successful, sometimes producing substantial damage rewards, and constitute an increasingly important aspect of accountability. An early example was the 8,000 pounds paid out in the case of *George v The Commissioner of Police for the Metropolis* (*The Times*, March 31, 1984). The awards became much higher as the decade proceeded. In 1990, a black former world champion boxer, Maurice Hope, having been arrested by two Metropolitan policemen, was subsequently charged with smoking and eating cannabis, assault and obstruction of the police. At court the Crown Prosecution Service offered no evidence (i.e., it abandoned the case) and Mr. Hope was told by the trial judge that he left the court "without any stain on his character." He subsequently sued the police, claiming exemplary damages for false imprisonment and malicious prosecution. While denying liability (normal in such settlements) the Metropolitan Force offered him 50,000 pounds in damages by way of an out-of-court settlement. More than 230,000 pounds in damages were awarded in the High Court against the Metropolitan Police Force alone in 1989. Many other awards have been made against provincial forces. In one case, West Midlands police agreed to pay a record 70,000 pounds damages to a man who said that detectives of the disbanded Serious Crime Squad had fabricated his confession to armed robbery. The police force also agreed to pay his 250,000 pounds costs to settle the civil action (*Daily Telegraph*, October 14, 1993).

The police have not only been sued for abuse of their positions and powers. They have also been on the end of civil actions from victims or relatives of victims of crimes alleging tortious negligence in their investigation or suppression of those crimes, in cases such as *Hill v Chief Constable of West Yorkshire* [1989] AC 53 (an allegation of negligence in the hunt for the Yorkshire Ripper serial killer) and more recently in *Alexandrou v Oxford* [1993] 4 All ER 328. The courts have usually shown little enthusiasm for upholding such complaints.

Media

Crime and policing are preoccupations of the modern media, both in newscasting and entertainment. In recent years, the police in Britain have become much more sophisticated in their own handling of journalists and television. Occasionally, this has even extended to inviting reporters and film crews to cover major operations in advance. Thus, in 1989, a large operation to arrest twenty suspected drugs dealers on Tottenham's notorious Broadwater Farm estate was accompanied by previously briefed journalists and cameramen. The reasons for this were partly to show that the police still had control of an estate where one of their officers had been brutally murdered in 1985, to provide independent evidence that the operation was conducted properly and to avoid the rumor mill that accompanied some police operations in Black areas earlier in the 1980s, sparking major disturbances. One senior police officer commented "I believe the raid showed the Metropolitan Police as an open, honest, caring organization." Another, Commander Stevens, the officer in charge of strategy for the operation, was explicit in explaining the reasons for allowing journalists to accompany the police: "The stakes were too high to allow this operation to be distorted or misrepresented. There was a very serious risk of riot and disorder in Liverpool, Manchester, Brixton, Bristol—you name it" (Silverman, 1991). Some have criticized such manipulation, others welcomed the element of independence provided by such public exposure. Recently, and perhaps indicative of greater police sophistication in this area, the Metropolitan Police have published their own partnership magazine, *The Link*, aimed at the public and produced on a quarterly basis (24,000 copies are distributed in London) to strengthen ties between the police and the public, to provide a forum for the exchange of ideas and information and to keep the public up-to-date with developments in policing.

To an extent, life imitates art. It has been alleged, probably with some validity, that (usually) fictional television accounts of policing from abroad (almost entirely the United States), such as the 1970s series *Starsky and Hutch* and some Clint Eastwood films, which normally focus on the policeman as a macho action man—one who bangs car doors shut rather than simply closing them, always runs even when he could walk, and is not adverse to a bit of rough handling of dubious characters—contributed to the growth of a police "canteen culture" which stressed a reactive and aggressive form of policing. This culture was exemplified by groups of officers cruising around in transit vans as a "fire-fighting" force and implicitly degraded many of the more mundane but nevertheless essential policing tasks (such as school visiting) as not being proper policing.

This claim is impossible to assess. It is certainly the case that American TV policemen rapidly spawned their counterparts in British television, for example Inspector Jack Regan of the very popular 1970s TV program *The Sweeny*. Regan was a man who had little time for legal or procedural limitations that stood in his way in the fight against hard-core criminals; and he was always subsequently shown to be correct in his analysis of the situations he faced, at the expense of the "by the book" careerists who were his superiors. Certainly, he was in a radically different mold from P. C. Dixon of *Dixon of Dock Green* who arrested his criminals by saying "you'd better come with me, my lad" and finished each episode by giving a short homily on the moral of the episode and wishing viewers a "good night all."

CONCLUSION

The state of democratic policing in Britain provides grounds for both concern but also optimism. It is probably still fair to say that Britain enjoys one of the best and most accountable systems of policing in the world. Police abuse of power still has the power to shock, rather than simply anger, something that is not universal even in the developed democracies. Many of the better traditions built up in former times have survived, if sometimes in an attenuated form. For example, there still exists a marked reluctance among the great majority of officers in all British forces, including the Metropolitan Force, to the routine carrying of firearms, making Britain the only large developed country where this occurs. The results of a survey of its members published by the Police Federation in May 1995 indicated that 79 percent of their members (a somewhat smaller proportion in London) opposed the routine arming of the police, though 83 percent did support an increase in the number of occasions when officers were armed for special purposes (as has already happened in practice with armed response vehicles in urban areas).

Serious problems developed in the 1970s and 1980s (though many of these were probably always present and were merely identified and publicized for the first time). Many of these problems are still present to a degree (and probably always will be), not least because no organization that numbers over 125,000 can fail, regularly, to produce a quota of corrupt, brutal or dishonest policemen. However, the 1990s have seen marked steps towards addressing many of these issues, with changes to training, tactics, recruitment and policing priorities and the safeguards provided for those involved with the police contained in the *Police and Criminal Evidence Act 1984* and its attendant codes. Together, these probably mean that the young, ill-educated (and now quite often black) males who traditionally have constituted police property are faced by a police force that is closer to the public image of the service than for many years.

The police will never, and can never, regain the apparent widespread acceptance that they had in the 1950s. For better or worse, Britain, from the South of England to the North of Scotland, is a less deferential, more rights conscious and more legalistic society, one in which all arms of government are under intense and permanent scrutiny. While it is easy to assert that Britain, and England in particular, has become more fragmented and divided as a society in recent years, many facets of this phenomenon lie in aspects of modern life (and even permissiveness) and in changes in the wider society (rather than, for example, specific economic policies). Such matters are under the control of the government (let alone the police) in only a very limited way. England is now a multi-racial and multi-cultural society with other large and identifiable minorities as well, whether identified by sexuality (in particular homosexuals) or by the plethora of new and alternative religions and lifestyles. Such a society is inevitably more fragmented than the homogenous, cohesive and comparatively close-knit, but arguably more socially restrictive, society of the immediate post-war years. In such a society a sense of wider community and social solidarity is almost bound to be reduced. It has been observed, with considerable truth, that although the "Report of the Royal Commission on Police Power and Procedure" in 1929 noted that consent, "a proper and mutual understanding between the police and the public," was necessary "for the maintenance of law and order," it is at least arguable that Britain is now so culturally diverse that "there is no unified consent that could be given to the police" (Robillard & McEwan, 1986, p. 12).

In his important (and influential) book *Policing freedom*, John Alderson (1979, p. xi) felt that one of the objectives of policing a "free, permissive and participatory society" was to "provide leadership and participation in dispelling criminogenic social conditions through cooperative social action" and that it was necessary for the police to "contribute toward liberty, equality and fraternity in human affairs." His views were supported by Metropolitan Police Commissioner Kenneth Newman in the mid-1980s when he stated that police should collaborate with other agencies "to develop solutions which address the root cause rather than the symptoms of crime" (Leishman et al, 1996, p. 105). But it is sometimes hard to see how these fine ideals can readily be achieved. Many of the social problems that appear to encourage crime are completely beyond police control (though they take much of the blame for them) and may well be beyond government's ready control as well. While it has been observed that the breakdown of the post-war political consensus with its accompanying atmosphere of "political Anglicanism" based on full employment and ever-rising prosperity has "thrust the police into the centre of a series of conflicts, particularly, though not exclusively, in the arenas of industrial relations and inner city tensions" (Lustgarten, 1987, p. 23), it is not easy to see how this consensus could be restored. Much has been said about government's responsibility for the economic causes that have (according to some accounts) fuelled the increase in crime. In particular these critiques have focused on the apparent emergence of a demoralized inner city proletariat. One noted academic even observed that the Conservative government's "economic and social policies have expanded and demoralized the urban underclass, reversing a century-long process of social integration" (Reiner, 1991b). There might be some truth to this, yet there are also many other factors at work that are totally independent of government. The same writer has observed that change was also fuelled by the development of a less deferential culture which both weakened inhibitions against offending and the authority of the police themselves (Reiner, 1995).

Another problem with such an analysis is that high levels of unemployment and racial tensions have been a Europe-wide phenomenon. Indeed, Britain currently ranks comparatively low in some of these respects when compared to other countries (for example severe and often violent racial tensions in Germany and exceptionally high unemployment in Spain). Britain currently has the lowest level of unemployment of any major EEC country. Despite much rhetoric from government during the 1980s about the need to cut spending, this did not occur in practice (though many people believe that it did). State expenditure still accounts for about 43 percent of the GNP and, again despite much rhetoric, spending on welfare has increased inexorably in real terms since 1979. With an aging and entitlement-conscious society and a revolution in health care, demands on government expenditure have been relentless. As a consequence, policy options are much more limited for any government (of whatever political hue) than is sometimes suggested. The simple lack of a consensus among academics as to the cause and cure of Britain's social problems is itself indicative of these difficulties.

For example, as an extreme contrast to those who emphasize the redistributive potential of the welfare state, the controversial American social philosopher, Charles Murray (1990, p. 1), has also focused on Britain as a classic example of the emergence (or, remembering the early Victorian period, reemergence) of a growing underclass in a modern state. This is a phenomenon that he has attributed, in large part, to high levels of illegitimacy rather than inadequate welfare provision, this in turn being both a symptom and a cause of the destruction of family life. Murray felt that rates of illegitimacy and crime were closely connected; until the 1950s, illegitimacy in Britain had roughly remained stable for centuries, at about 5 percent (sometimes less) of the population. By 1976, it was about 9 percent; by 1992, 31 percent and still growing. This rise coincided with the great increase in crime. According to Murray's analysis much welfare spending is even counterproductive, encouraging the emergence of what he has termed (very harshly) the "New Rabble" in British society.

To an extent, policing theory has recently turned full circle in England. After years in which some critics saw a considerable role for the discretionary application of the substantive law against petty crime and social nuisance, especially as part of a community approach to policing, the "broken windows" theory of urban crime, pioneered in the 1980s (Wilson & Kelling, 1982) has arrived from America in the form of zero-tolerance initiatives. According to this analysis, although policing does not have a major *direct* influence on rates of crime, in promoting order by dealing firmly with minor infractions and public nuisances (such as drunks), it can, indirectly, have a major influence on levels of more serious crime in the policed area, by promoting community solidarity and, in turn, preventing respectable elements from moving to pleasanter areas. Zero tolerance has received limited application in a few small Northern cities and in the Kings Cross area of London (a popular location for prostitution). Critics, mindful of the early 1980s, feel that it will damage police/community relationships, especially in multi-racial areas, if attempted on a wider scale. Proponents cite the apparent success of such a strategy in New York.

Whatever the reasons for the growth in crime since 1945, it has placed the police (and the wider criminal justice system) under considerable stress, even with the greatly increased resources available to them. Added to this have been other major challenges in the types of crime being faced (for example, Britain faces the highest level of terrorist threat in Western Europe) or institutional challenges (such as the growing perceived threat to their position from alternative forms of private policing, whether private or municipal, provided by Britain's large security industry and local councils). There is no room for complacency. Despite these, and many other continuing problems, Britain is ending the century with a police force that is in better condition than it has been for a considerable number of years and is substantially able to face the challenges of policing a modern democracy.

NOTES

[1] There are three separate legal jurisdictions in the United Kingdom: England and Wales, Scotland, and Northern Ireland. Each has its own police system. Because of the special nature of the problems in Northern Ireland and the necessarily specialist response of much of the Royal Ulster Constabulary (R.U.C.), Northern Ireland will only be touched on when its police practices influence policing in the rest of the United Kingdom. This chapter will focus on England and Wales, by far the most populous section of Britain, but will also periodically refer to the Scottish experience.

2   See, for example the speech on "balance of fairness" by then-Lord Chief Justice, Lord Lane, in *R v Alladice* [1988] Crim LR 608, p. 380.
3   That there is still no automatic exclusion in such situations post-PACE was confirmed by the Court of Appeal in *R v Smurthwaite; R v Gill* [1994] 1 All ER 898. It was accepted, however, that the fact that evidence had been obtained in this way might be a factor in deciding whether to exercise the discretion to exclude under section 78.

REFERENCES

Ascoli, D. (1979). *The queen's peace: The origins and development of the Metropolitan Police 1829–1929*. London: Hamish Hamilton.
Alderson, J. (1979). *Policing freedom*. Estover: MacDonald and Evans.
Baker, K. (1993). *The turbulent years: My life in politics*. London: Faber and Faber.
Bennett, T. (1988). An assessment of the design, implementation and effectiveness of neighbourhood watch in London. *Howard Journal of Criminal Justice, 27*, 4. 241–255.
Burke, J. (1996, September 15). The thin blue line. *The Sunday Times*.
Campbell, D. (1994). *The underworld*. London: BBC Books.
Constabulary Journal. Various issues.
Cornish, W. R. & Clark, G. D. N. (1989). *Law and society in England: 1750–1950*. London: Sweet and Maxwell.
Cullen, P. (1992). The treaty on European Union—Maastricht and beyond? *Journal of The Law Society of Scotland, 37*, 383–389.
Doherty, M. (1994, November 4). Watching the detectives. *New Law Journal, 144*, 1525–1526.
East, R. (1987, October 30). Police brutality—lessons of the Holloway Road assault. *New Law Journal, 137*, 1010–1012.
Emsley, C. (1991). *The English police: A political and social history*. London: Harvester/Wheatsheaf.
Fielding, N. (1995). *Community policing*. Oxford: Clarendon Press.
Gartner, G. & Archer, R. (1981). Homicide in 110 nations: The development of the comparative crime data file. In L. Shelley (Ed.). *Readings in Comparative Criminology* (pp. 78–99). Carbondale, IL: Southern Illinois University Press.
Gatrell, V. A. C. (1980). Theft and violence in England 1834–1914. In B. Lenman & G. Parmer (Eds.), *Crime and the law* (pp. 238–337). London: Europa.
Goldsmith, A. J. (Ed.). (1991). *Complaints against the police: The trend to external review*. Oxford: Clarendon.
Goldstein, H. (1977). *Policing a free society*. Cambridge, MA: Ballinger.
Graef, R. (1989). *Talking blues: The police in their own words*. London: Collins Harvill.
——. (1991, March 18). What's gone wrong with the police? *The Independent*, vol. 88, no. 17.
——. (1993, December 20). Can the police ever root out their bigots? *Evening Standard*.
Hain, P., (ed.) (1979). *Policing the Police*. London: John Calder.
Henry, I. (1996, March 31). Crime figures a sham say police. *Sunday Telegraph*.
Hobbs, D. (1988). *Doing the business: Entrepreneurship, detectives and the working class in the East End of London*. Oxford: Clarendon Press.
Holdaway, S. (Ed). (1989). *The British police*. London: Edward Arnold.
HMSO (1991). *A study of the police complaints system*. London: Home Office.
——. (1994). *Contacts between police and public: Findings from the 1992 British crime survey*. London: Home Office, Research Study Number 134.
——. (1995, 1996). *Criminal statistics for England and Wales*. London: Home Office.
Home Office (Various). *Circular*. London: Home Office
——. (1993). *Police reform: A police service for the twenty-first century*. London: Home Office, CM 2281.
Hughes, S. (1989, August 4). MPs back "best in the west" police complaints authority despite growth in civil actions. *New Law Journal*, p. 1062.
Husain, S. (1988). *Neighbourhood watch in England and Wales: A locational analysis*. London: Home Office, Crime Prevention Unit Paper 12.
LancCon (Lancashire Constabulary). WWW homepages at http://www.ehche.ac.uk/community/bluelight/forcei.htm

Law and Society Gazette. Various issues.
Lawton, Sir F. (1991, May 8). Tarnished police evidence. *Law Society's Gazette, 88,* No. 17, 2.
Leishman, F., Loveday, B., & Savage S., (Eds.). (1996). *Core issues in policing,* London: Longmans.
Levin, B. (1990, December 10). Come and get rotten apples, only 50,000 pounds a go. *The Times.*
Lustgarten, L. (1987). The police and the substantive criminal law. *British Journal of Criminology, 27,* 23–30.
Maclean, B. D. (1993). Left realism, local crime surveys and the policing of racial minorities. *Crime, Law and Social Change: An International Journal, 19,* 1, 51–86.
Maguire, M. (1994). Crime statistics, patterns and trends: Changing perceptions and their implications. In R. Reiner, M. Maguire, & R. Morgan (Eds.), *The Oxford Handbook of Criminology* (pp. 233–291). Oxford: Oxford University Press.
McLaughlin, E. (1992). The democratic deficit: European Union and the accountability of the British police. *British Journal of Criminology, 32,* 473–487.
McNee, Sir D. (1983). *McNee's law.* London: Collins.
MetPol (Metropolitan Police, Scotland Yard), on the Worldwide Web at http://www.open.gov.uk/police/mps/home.htm.
Murray, C. (1990). *The emerging underclass.* London: Institute of Economic Affairs.
New Law Gazette. Various issues.
PSI (Policy Studies Institute). (1993). *Police and people in London.* London: Author.
Reiner, R. (1991a). *Chief constables.* Oxford: Oxford University Press.
——. (1991b, September 5). Unhappiest lot of all. *The Times.*
——. (1992). *The politics of police.* (2nd ed.). Toronto: University of Toronto Press.
——. (1994). Policing and the police. In R. Reiner, M. Maguire, & R. Morgan (Eds.), *The Oxford Handbook of Criminology* (pp. 705–772). Oxford: Oxford University Press.
——. (1995, October 12). Selling the family copper: the British police. *The Independent.*
——. & Spencer S. (Eds.). (1993). *Accountable policing: Effectiveness, empowerment and equity.* London: Institute for Public Policy Research.
Roberts, R. (1973). *The classic slum.* Harmondsworth: Pelican.
Robillard, St. J. & McEwan, J. (1986). *Police powers and the individual.* Oxford: Basil Blackwell.
Scarman report (1986). Harmondsworth: Pelican Edition.
Schwarzenegger, C. (1996). Borderless Europe, borderless crime. *Hosei Riron: the Journal of Law and Politics (Japan), 29,* 20–40.
Sheehy Report (1993). *Report of the inquiry into police responsibilities and rewards.* London: Home Office, CM 2280.
Silverman, J. (1991, October 4). Raids, riots and rumour control; the police are getting better at handling the media. *The Independent.*
Smith, G. (1995, December 7). Grampian officer is scapegoat, dealer tells High Court. *The Glasgow Herald.*
Southgate, P. (1987). Behaviour in police-public encounters. *The Howard Journal, 26,* p. 153–163.
Spencer, S. (1985). *Called to account: The case for police accountability in England & Wales.* London: National Council Civil Liberties.
Stone, R. (1995). Exclusion of evidence under section 78 of the Police and Criminal Evidence Act: Practice and principles. *Web Journal of Current Legal Issues, 3,* 10–30.
Stephens, M. (1988). *Policing: The critical issues.* Hemel Hempstead: Harvester/Wheatsheaf.
Tendler, S. (1995, May 5). Mafia launders cash in London. *The Times.*
Uglow, S. (1988). *Policing liberal society.* Oxford: Oxford University Press.
Urquart, F. (1996, April 16). Judge jails policeman for "serious and unprovoked attack" on teenager in cells. *The Scotsman.*
Vallely, P. (1995, The Independent). How Brixton became a byword for disorder. *The Independent.*
Waddington, P. A. J. (1987). Towards paramilitarism? Dilemmas of policing civil disorder. *British Journal of Criminology, 27,* 1, 37–46.
——. (1991). *The strong arm of the law: Armed and public order policing.* Oxford: Oxford University Press.
——. (1996). Public order policing: Citizenship and moral ambiguity. In F. Leishman, B. Loveday, S. Savage. et al. (Eds.), *Core issues in policing* (pp. 114–130). London: Longmans.

Wilson, J. Q. & Kelling, G. (1982, March). Broken windows. *Atlantic Monthly*, 29–38.
Wolchover, D. (1996, February 28). Police perjury in London. *New Law Journal*, February 28.

Cases cited
Alexandrou v Oxford [1993] 4 All ER 328
George v The Commissioner of Police for the Metropolis (*The Times*, March 31, 1984).
Hill v Chief Constable of West Yorkshire [1989] AC 53
Matto v DPP [1987] Crim LR 641
R v Alladice [1988] Crim LR 608
R v Bailey (*The Times*, March 22, 1993).
R v Bryce [1992] Crim LR 728
R v Chief Constable of Sussex ex parte International Trader's Ferry Ltd ([1995] 4 All ER 364
R v Christou [1992] 3 WLR 228
R v Gill [1994] 1 All ER 898
R v Mason [1987] 3 All ER 481
R v Miller, Paris and Abdullah (1993) 97 Cr App Rep 99
R v Sang [1980] AC 402
R v Smith (*The Times*, March 22, 1993).
R v Smurthwaite [1994] 1 All ER 898
Williams v DPP [1993] 3 All ER 365

11

Challenges of Policing Democracies: The Dutch Experience

DANN VAN DE MEEBERG and ALEXIS A. ARONOWITZ

INTRODUCTION

The Netherlands (Continental Europe), the Netherlands Antilles and Aruba (both in the Caribbean) together form the Kingdom of the Netherlands. The Netherlands is situated in Europe along the North Sea and shares boundaries with the Federal Republic of Germany to the east and Belgium to the south. The country is basically flat, with a few hills in the southern areas around Maastricht. About half of the country lies below sea level. The ground surface of the Netherlands covers 41,160 square kilometers. Administratively, the country is divided into twelve provinces.

As of January 1,1994, the population in the Netherlands reached slightly more than 15.3 million inhabitants, with a population density of 452 persons per square kilometer—making the Netherlands the most densely populated country in Europe. Only two cities, Amsterdam and Rotterdam, have over one million inhabitants (Central Bureau of Statistics, 1995). The economic structure of the Netherlands is based on agriculture (dairy products, flowers); industry (high tech, chemical, heavy); trade; and transport (sea ports, railway). Two thousand five hundred kilometers of motorways and 4,000 kilometers of major waterways make transport possible to every part of the country. The average income per capita is approximately fl. 30,000 Dutch guilders (approximately US $19,000).[1]

The Netherlands is a democratic, constitutional monarchy. One of its most important features is that the rights and obligations of both citizens and the government are spelled out in laws and in the Constitution. Decisions concerning laws, rules and policy are made on the basis of the democratic principle that a majority of the body representing the people decides. At the national level, this is the Parliament (comprised of two Houses), at the provincial level the Provincial legislatures, and at the municipal level the population is represented by local councils.

In order to implement the democratic state principle optimally and to guarantee that certain bodies do not usurp power, the polity has been divided into a legislative power (Parliament), an executive power (the sovereign and the ministers) and an independent judiciary (Ministry of Foreign Affairs, 1990.) This division is also known as *trias politica*.

THE CONCEPT OF DEMOCRATIC POLICING IN THE NETHERLANDS

Prior to the introduction of the revised Police Act on April 1, 1994, the police in the Netherlands were divided into 148 municipal police forces, which were mainly responsible for the large urbanized municipalities, and a nationally-organized force, the National Police Force, which operated in more rural areas. A debate over changing the organizational structure of the police advanced rapidly when a new government came into power in 1989. The reasons were the following:

First, the organizational structure of the police was no longer considered adequate; a decentralized police force was judged to be inefficient and therefore not effective, which in turn affected the quality of police performance.

Second, the organizational structure of the police could no longer respond quickly and effectively to increases in and to the new forms of crime, such as organized and international crime. As crime increased and changed in form, the clearance rate continued to decrease.

The government report, *Een nieuwe politiebestel in de jaren' 90* [A New Police Structure in the Nineties] (1990), provided a blueprint for the new police organization. On the basis of this report and further elaboration and completion, the reorganization was begun and a new Police Act was developed in Parliament. One aim of the reorganization was an increase in the scale and an improvement in the management of the police, which in turn would make them more capable of performing their tasks adequately.

With the imposition of the 1993 Police Act (on April 1 of 1994), the Netherlands was divided into twenty-five territorial forces, so-called regions, which are closely linked to judicial districts and the administrative regions. Each region has a regional police force. A number of police services, which traditionally covered the entire nation, such as Traffic Police, Waterways Police, the Protection Service of the Royal Family and the National Criminal Intelligence Service were organized at the national level and became part of the National Police Agency. Together, the twenty-five regional forces and the National Police Agency make up the Dutch police. In all, the Dutch police comprise about 40,000 police officers and civil staff (Fijnaut, 1994; Strooper, 1994).

Being an administrative body with far-reaching powers which can encroach upon the citizen's liberties and integrity, the police must observe the principles of the democratic state. Most important, the laws to which the citizen is to adhere apply to the police as well. The police are especially bound to abide by legislation, regulations, constitutional law and case law. As the administrative body which sees to it that the citizens abide by the laws, thus countering crime and maintaining public order, the police should not disregard laws and regulations themselves.

Apart from having to abide by the laws, the police are accountable to the competent authorities as regards the execution of their duties. The police must answer to both the Ministry of Justice and the Ministry of the Interior (see Figure 11.1). The question of which government body or figure maintains authority over the police depends on the duty performed by the police.

If police action is aimed at maintaining public order, the mayor in the municipality concerned is the competent local authority. The mayor is accountable to the local elected body, the council, for the execution of her/his powers and for oversight of the police. At the provincial level the police must answer to the Queen's Commissioner; at the national level the Minister of the Interior is the competent authority. The Ministers of Justice and of the Interior are responsible for appointing

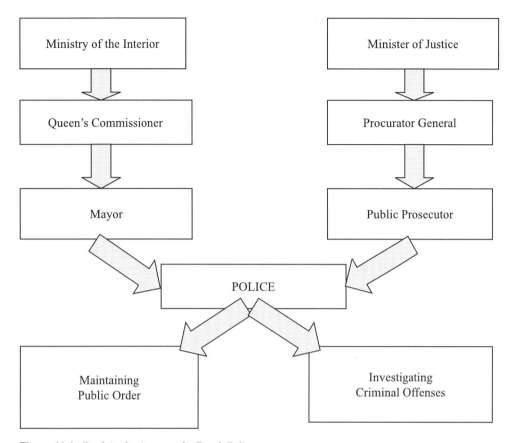

Figure 11.1. Dual Authority over the Dutch Police

the regional police chiefs. The appointments are conferred by Royal Decree (Ministry of the Interior, 1994).

In the case of criminal proceedings or where the police execute their duty of maintaining legal order within the framework of the criminal law, the public prosecutor is the competent authority over the police. The public prosecutor is not directly accountable to a representative body. The public prosecutor is responsible to the Procurator General (one of five procurators general at the Appellate Court level), who is accountable for the execution of his or her authority to the Minister of Justice, who, in turn, is accountable to the national Parliament (Ministry of the Interior, 1994). This guarantees that police action can be checked by Parliament. If necessary, the actions of the police can be rectified by those in authority by a democratic decision made by the members of the House of Representatives.

Priorities in police duties are set in legally-based tripartite consultations between one of the mayors, who is the administrator of the regional police force, the public prosecutor and the chief of police. This system of providing various checks and balances is typical of policing in a democratic society. Aside from whether the police are capable of "policing their own behavior" (addressed in a latter section in this chapter), the control over police authority exercised by various other branches of government ensures that police uphold democratic ideals in Dutch society.

OPERATIONAL CHALLENGES

This section addresses a number of operational issues linked to crime and migration. First, we discuss the problems of crime in general and organized crime in particular, characteristics of the offender population, society's assessment (and fear) of crime, the rise in and alternative forms of crime prevention, detention in the Netherlands and the policy implications of crime for democratic policing. Second, we describe recent patterns of migration and their suspected and real impacts on crime.

The Crime Problem

Normal Crime

Crime in the Netherlands has become an increasing problem. Social norms, whether confirmed in criminal law or not, are frequently disregarded. Regardless of the great effort invested, neither society nor the authorities have been able to respond to crime adequately. The picture of crime in the Netherlands is becoming increasingly clear. Apart from crime data supplied by the police, there are national victims surveys among the population and the business community, municipal and regional population studies (the so-called Police Monitor), victim research, self-report delinquency studies and offender analyses and national inventories of organized and group crime.

From the 1950s to the present day, recorded crime has increased fivefold and the clearance rate has decreased. While the clearance rate was 51 percent in 1965, it fell to 30 percent in 1980 and to 19 percent in 1993 (Eijken, 1994).

Official statistics indicate a sharp increase in violent crimes[2] between 1985 and 1990, from 37,100 to 50,300. The following increases were registered: 51,400 (1991); 58,500 (1992); 61,400 (1993); and 65,900 (1994). Property offenses[3] increased from 840,700 in 1985 to 851,100 (1990); 872,200 (1991); 950,100 (1992); 951,700 (1993); and 977,300 (1994) (Centraal Bureau voor de Statistiek, 1996, Table 9, p. 415). Official statistics also reflect an increase in victims of crime. Annually, 35 percent of the population[4] and more than 40 percent of businesses become victims of crime (Eijken, 1993).

Crime, in particular street robberies, robberies, car and bicycle thefts, occurs most frequently in the four largest cities in the Netherlands: Amsterdam, Rotterdam, The Hague and Utrecht. Crime types such as vandalism, bicycle theft and making threats comprise a major share of the crime problem, and recently burglaries and car thefts have been on the rise. Damages and losses as a result of victimization have increased greatly (particularly as a result of car theft and burglaries).

Internationally, the Netherlands ranks high in victimization statistics, which may be due, in part, to the large number of bicycle thefts, which create the appearance of unusually high levels of overall victimization (Council of Europe, 1995). Victimization studies indicate that the rate of burglaries, in 1991, is comparable to that of neighboring countries, and car theft ranks relatively low (Eijken, 1994). Shoplifting and burglaries are the types of crime occurring most frequently in the business community. In the catering industry, crime committed by employees ranks high (Centraal Bureau voor de Statistiek, 1996, p. 5a). (See Table 11.1)

The Offender Population

Turning to the offender population, one sees strong gender and age differences. Based upon official police statistics, adult males are responsible for 73 percent of

Table 11.1. Crime in the Netherlands

Year	Number of crimes	Clearance rate
1980	705,697	30%
1985	1,093,726	24%
1990	1,150,185	22%
1994	1,308,724	19%

(Source: Central Bureau of Statistics/Scientific Research Department, Ministry of Justice, 1994)

registered offenses, juvenile males for 15 percent, adult females for approximately 10 percent and juvenile females for approximately 2 percent. Based upon population statistics, annually, approximately 3.3 percent of adult males, 5.5 percent of juvenile males, 0.45 percent of adult females and 0.5 percent of juvenile females have had contact with the police. This means that a higher percentage of juvenile males come into contact with the police than other population groups. There has been a gradual rise in the criminal activities of youths between 12 and 18 years of age; after that, delinquency decreases with age. Young people rank highest for a variety of crimes, with the exception of drunken driving and tax fraud, where middle-aged groups are predominant (Eijken, 1993).

Almost all research shows that the core offenders are characterized by minimal education, little work experience and high rates of recidivism, often in combination with an addiction to drugs. They have a more systematic way of operating and their criminal career ranges from petty offenses to serious forms of crime.

Organized Crime

A problem of growing concern in the Netherlands is the existence of organized criminal groups. Periodically, an inventory is made of the number of criminal groups. While experts disagree as to the number of organized criminal groups[5] operating in the Netherlands, a 1995 police inventory identified 450 active criminal groups, the majority of which were involved in the hard drugs (66%) and soft drugs (39%) trade. In total, there are approximately 100 highly organized criminal groups operating in the Netherlands. Of the highly organized groups, almost two-thirds are homogenous: forty-four are groups of Dutch origin, ten groups are of Turkish origin, and the other eleven groups are groups stemming from Morocco, Surinam, Iran, China and Eastern Europe (Van Traa, 1996). Violence, both internally and against others in the criminal world, is commonplace in these highly organized ethnically-homogeneous criminal groups. Turkish groups rank particularly high. A quarter of the criminal groups have a nucleus composed of both Dutch and ethnic minorities.

These groups consist of twenty members on the average, but some have over 100 members. Most of the highly organized criminal groups operate on an international level. Three-quarters of them operate throughout Europe and one-third focuses on Central and Eastern European countries (Van der Heijden, Tham, & Nierop, 1995).

These new developments present particular problems to law enforcement officers. The international nature of such associations, language barriers and the clannishness of homogenous and foreign organized groups have made penetration by law enforcement agencies virtually impossible. This has required law enforcement to devise new ways of approaching the growing problem of organized crime.

In recent years, crime investigation in the area of organized criminal groups has been intensified. Seven special police teams have been formed and charged with investigating (inter)national organized crime. At present, they are administered by joint regional police forces; however, in the future they will probably be administered at the national level.

Another attempt to control or limit organized criminal groups in the Netherlands, in addition to but apart from a repressive approach, is the strategy based on the adage "follow the money." Crime suppression requires intensive cooperation between the police, the prosecution service, the various ministries (Justice, Finance, and Interior) and other authorities such as the tax service. Therefore, asset-stripping teams have been formed and, based on law, have been able to seize monies which were obtained as a result of criminal activities. The teams' asset-removing powers have provided funding for such investigations; consequently organized crime (or rather the funds seized as a result thereof) funds its own suppression.

It is expected that in the coming years the situation involving crime and public security will witness the following changes (Stichting Maatschappij en Politie, 1991):

(1) A strong increase in the less visible forms of insecurity whereby the victims will predominantly be the authorities and the business community;
(2) A stabilization or decrease of less serious forms of frequently committed crime, such as vandalism and shoplifting;
(3) An increase in violent crime, especially by socio-economically underprivileged youths;
(4) Greater risks of public order disturbances, such as increases in public protests;
(5) A decrease in the rate of road safety improvement, due to sharply increasing mobility; and
(6) Increased feelings of insecurity.

Fear of Crime

It is conspicuous that feelings of insecurity remained reasonably stable during the past decade. Between 1981 and 1989 approximately 40 percent of the public reported being afraid of crime. This figure increased between 1990 and 1994 to over 60 percent (Eijken, 1994, Figure 7). The fear of crime is influenced more by the seriousness than by the frequency or the increase of such offenses (Van de Heijden, 1986). Respondents living in Amsterdam report the highest levels of insecurity and fear (Eijken, 1993).

While interviews and surveys showed that the public's apprehension of crime remained relatively stable and low during the past decade, the public's insecurity began and continues to increase in the 1990s. Crime is seen as the principal social issue. Surveys show that the Dutch have become increasingly pessimistic about controlling the growth of crime (Ministerie van Binnenlandse Zaken, 1994). Yet in spite of such fears, crime is not so high that it would deter foreign tourists from visiting the Netherlands.[6]

Crime Control Policies

Crime Prevention

Crime prevention has not yet reached a saturation point. Currently, two-thirds of all households are secured with locks and almost the same proportion have a light on when the occupants are absent. Half the houses are equipped with outdoor lights and 6 percent with burglar alarm systems (Central Bureau Statistics, 1995, Table 3, p. 405). The number of safety locks and alarm systems is on the increase. The Dutch also assign high priority to the preventative task of the police (Eijken, 1993).

The private security sector boomed in the 1980s and 1990s. Other forms of functional supervision have also increased, such as that of the city watchmen (*Stadswacht*). They are uniformed, unarmed civil staff, employed by a municipality and charged with maintaining order on the streets and assisting the public. Their number has increased greatly since 1989, the year in which they were introduced. The number of private security guards (e.g., employed by housing corporations) has risen as well.

Detention

Since 1980, the number of detainees has risen by 45 percent. Prison sentences have grown longer and there has been a clear increase in prison sentences imposed for violent crime, property crime and drug offenses. In 1990, 80 percent of the convictions for drug offenses and about 50 percent of the convictions for violent crime and property crime resulted in the imposition of a sentence of imprisonment (Eijken, 1993). The average duration of a prison sentence has doubled in that period. Still, the number of prisoners per 100,000 inhabitants is low in comparison with neighboring countries (Council of Europe, 1995, Table 3b). On December 31, 1994, the prison population in the Netherlands stood at 8,467 inmates. Of these, 3,999 (47%) were persons convicted of an offense, 3,361 (40%) were awaiting trial and 1,107 (13%) were staying outside of the institution (Centraal Bureau voor de Statistiek, 1996, Table 48, p. 435).[7]

Imprisonment is not usually imposed for less serious offenses; however, punishment for conviction occurs in a high percentage of cases. Almost 90 percent of the convictions for traffic violations and 70 percent of the convictions for vandalism are punished with a fine or a period of incarceration in combination with a fine.

A comparison between recorded crime and crimes committed by those detained shows remarkable differences. Thirty-six percent of the detainees had perpetrated violent crimes and 28 percent of those detained were sentenced for property crimes, whereas violent crimes make up just about 2 percent and property crime no less than 75 percent of crime recorded by the police (Central Bureau of Statistics, 1996, Table 38, p. 425).

In the Netherlands there are still insufficient cells in detention centers and prisons. This is due not only to an increase in the length of prison sentences but is also a result of the detention capacity allocated to illegal and undesirable aliens awaiting deportation hearings.

The Netherlands at one time boasted one of the most progressive correctional systems in the world. It was characterized by short prison sentences, small, modern facilities and sufficient cell capacity. Unfortunately, the Netherlands is following in the footsteps of other democracies: higher crime rates, more violent offenses accompanied by an increasing fear of crime, longer prison sentences and problems of overcrowding in prisons. The challenge to the Netherlands, as in most democracies, is to find alternatives to prison sentences. The Netherlands has risen to the challenge by allowing police officers to impose administrative fines (*transacties*), thus reducing the need to process cases through the courts and impose penal sanctions. Other alternatives to incarceration have been recently introduced as penal sanctions.

While crime has increased and the clearance rates have decreased to approximately 20 percent, the actual percentage is even lower if the estimate of real crime figures (the dark figure of crime not recorded in official statistics) is taken as a basis. Despite lower clearance rates, the offender, once apprehended, is more likely to be dealt with than in the past. This is due, to a great extent, to the police now being able to impose administrative fines in misdemeanor offenses, instead of having to process the case through the court and hope for a penal fine in non-serious offenses. Offenders have less chance than before of being sentenced to imprisonment due to the police's ability to impose administrative fines and to alternative punishments. The terms of imprisonment imposed, however, have become relatively longer.

In recent years, the general public's attitude towards sentences has changed. Citizens prefer legal action to result in a change of behavior through "assignment penalties" (e.g., restitution or repairing damages caused by the offender, while under the supervision of the probation department) which have been imposed increasingly over the past few years. The labor penalty (community service) is the most frequently used. Still, they serve little to relieve overcrowding in the prison system and their results in terms of recidivism are not considered spectacular: they do not always prevent convicted people from committing similar offenses again (Van der Heijden, et al., 1995).

A novelty in the Netherlands is an experiment combining house arrest with electronic monitoring. Attached to the leg of a convict is a band which is linked up to a central computer. The convict's free movement is restrained and in this way he or she serves a term of detention at home and can maintain contact with her/his family, job and society in general.

Although new prison construction is under way, the Netherlands has succeeded in implementing alternative sanctions. It remains to be seen whether crime prevention programs will succeed in reducing the crime problem and thus eliminating the need for new prisons in the future.

Implications of Crime and Crime Control for Democratic Policing

How do crime and its detection present challenges to the police? One of the most important functions of the police is to maintain order and provide a safe environment for the public. This implies that the police task is to prevent and solve crimes. Crime rates are generally seen as an indication of how well the police are doing their job, and when crime rates rise it is often the police who carry the burden of the blame. An increase in crime rates, however, may not indicate poor police performance but may be the result of a number of other factors.

According to a researcher at the Central Bureau for Statistics (CBS) in the Netherlands, an increase in crime in the form of reports made to the police is due partially to an improvement in the computerized registration of offenses which has resulted in better (and possibly more) crime statistics. The hiring of new police officers will also ultimately result in more registered crime. A third and more controversial explanation is that increased reports to the police may be the result of the public's expanded trust in and willingness to talk to and work with the police. An increase in crime as the result of any of these explanations is actually due to improvements in the police department.

The CBS reports that the Netherlands is no more or less safe or dangerous than it was ten years ago. An increase in crime has been matched by an increase in the population. When one looks at the population most likely to be victimized—those between the ages of 15 and 39—there were 98 victimizations per 1,000 in the population in 1984 compared to 102 per 1,000 in 1994, an increase of only 4 percent. Once again, the CBS warns that this increase may not be due to an actual increase in crime (Van Zijl, 1996).

The most recent police statistics released by the Ministry of Justice, however, are promising. According to the Minister, crime in the Netherlands (in 1995) has declined for the first time since 1945. The number of *processen-verbaal*, or official police reports, reflect a 5 percent decline over the 1994 statistics (Van Zijl, 1996). As crime statistics may be an inaccurate indicator of police performance, however, measures other than crime rates must be used to appraise police success.

Another challenge presented to police departments operating in democratic societies is the need to create a balance between protecting society by solving crimes and arresting offenders and the need to protect the rights of the accused. This issue came to the forefront in the Netherlands in a recent controversy surrounding an interrogation method, the Zaanse interrogation method, used by the police in Zaanstad. An uncooperative murder suspect in Groningen was put under excessive psychological pressure[8] until a confession was obtained.

In this method, intensive interrogations take place in a small room, often in the presence of five investigators. Photos of the victim cover the room. Interrogations are often videotaped, sometimes without the knowledge of the suspect; hidden cameras may be used. Despite the fact that this method has proven successful, resulting in confessions, forensic and legal experts are appalled at the manipulative, "psychological choke-hold" of this interrogation technique (Rombouts, 1996). The Minister of Justice has asked the Investigative Advice Commission (*recherche-adviescommissie* or RAC) for advice concerning this interrogation method which has been in use in the Netherlands over the past ten years.

Good policing demands that crimes be solved. It also, however, requires that the police protect the rights of the accused and remain within the boundaries of the law. This conflict illustrates the difficulty of policing in democracies, yet its importance must not be underrated. In theory, by abridging suspects' rights the police could solve most crimes. Yet the price they would pay in terms of public fear of the police and lack of community support and cooperation is immeasurable. Only by using fair and legal practices will the police maintain good public standing and achieve success in their endeavors to prevent and solve crime. The police in the Netherlands, as well as in other democratic societies, will continue to be confronted with the problem of creating a balance among such issues as maintaining good public relations, creating a safe environment, preventing and solving crime, and protecting the rights of the accused.

Migration

Migration to the Netherlands has resulted in the country having to deal with two very distinct foreign populations encompassing different nationalities with very different problems. On the one hand are the political asylum-seekers whose numbers increased drastically from 20,346 (1992) to 35,399 (1993) and peaked at 52,576 (in 1994) before decreasing to 29,258 in 1995 (Ministerie van Justitie, 1996). From diverse backgrounds, they often arrived in the Netherlands with little more than the clothes on their backs. They speak no Dutch, are prohibited from working and are often isolated in detention centers and camps until the Immigration and Naturalization Service and/or the courts have determined their future status.

The second population of immigrants to the Netherlands consists of those from previous colonies (such as Indonesia and Surinam) or others seeking work or a better life (such as Moroccans and Turks) (Ministerie van Binnenlandse Zaken, 1994). Immigration almost doubled from 46,200 in 1985 to 81,300 in 1990. In 1993 the Netherlands registered 87,600 immigrants (Central Bureau Statistics, 1995, Table 32, p. 63).

The Netherlands has traditionally been a society characterized by ethnic diversity. While the Netherlands has been known for its tolerance and widespread acceptance of foreign populations, it has been faced recently with increasing problems, particularly among young ethnic minority males. The number of criminal youths from ethnic groups is relatively high.[9] In the big cities, ethnic minority youths are responsible for a disproportionate part of youth crime (de Haan & Bovenkerk, 1993) and account for as much as 70 percent of all juvenile crime in the four largest cities in the Netherlands (Eijken, 1993).

On the whole, Moroccan youths account for a large part of the offenses—over 40 percent of thefts and burglaries, as well as violent crimes (theft with violence, threats and assault).[10] They rank lower only for vandalism. Ethnic minorities are almost exclusively responsible for violent crime and theft with violence (mugging). Young females show a similar pattern on a much smaller scale. Moroccan youths are represented much more strongly than all other ethnic minority groups: almost half the male youth who have come into contact with the police for suspicion of involvement in juvenile crime are of Moroccan origin. A quarter of all Moroccan male youths in the big cities between the ages of 12 and 17 have dealings with the police every year. Juveniles are mainly involved in theft and burglary. For girls this applies in 90 percent of the cases (Ministerie van Binnenlandse Zaken, 1995).

The very high number of repeat offenders among Moroccan youths is a matter of serious concern. There are strong indications that police and judiciary have adopted a new policy. Increasingly, first offenders are sent home or referred to social workers or probation officers after a warning or reprimand. If the formal justice system is only able to respond after the second or third offense, it is critical that social workers intervene prior to or after the first offense to prevent further involvement of these youths in the criminal justice system.

While their numbers are relatively small compared to the native population in the Netherlands[11], foreigners comprise a large percentage of the jail and prison population. Thirty percent of detainees are Moroccan, Turkish, Surinamese or Netherlands Antillan.[12] A strikingly high proportion of the non-Dutch detainees, especially those older than 40, were convicted on drug charges (Eijken, 1993; Van der Heijden, et al, 1995). (See Table 11.2)

The number of victims of crime among the ethnic population in the big cities is, not surprisingly, high. Research involving victimization in the three largest cities indicates that foreigners[13] are more often than the Dutch the victims of violent crime (Loef, 1992). Furthermore, ethnic minorities more often fall prey to multiple victimizations. Many Moroccans are also the victims of crime in medium-sized cities. Surinamese and Antillan immigrants stand out as victims of car crime (Ministerie van Binnenlandse Zaken, 1994).

The challenge to the Netherlands exists in finding ways of reducing both the crimes committed and the victimizations experienced by ethnic minorities. It must be remembered that this country is dealing with two distinct groups of foreigners: legal immigrants and asylum-seekers with uncertain status. The two groups, and their various problems, must be viewed separately.

Asylum-seekers, during the time that they are awaiting a decision on their status from the Ministry of Justice, receive social welfare support. If their applications are rejected, all welfare payments cease. If the land from which they came does not have a repatriation agreement with the Netherlands, they cannot be forced to return. In essence, then, they are released onto the street with little money, no health or social benefits, and no living or work permits. There are few alternatives available to them: they may either live with relatives, work illegally or turn to prostitution or other forms of crime to support themselves. Lack of a legal status may necessitate turning to crime as a means to basic life support.

Table 11.2. Overview of ethnic origin of detainees in 1993

Continent	Percentage	
Europe,	64	
incl. Netherlands,		52
Germany,		1
Turkey		6
Africa,	11	
incl. Morocco		8
Asia,	4	
incl. Middle East		2
South America,	19	
incl. Surinam		11
Dutch Antilles		6
Rest of America	1	

(Source: Central Bureau of Statistics, Prison Statistics 1994)

The motivation for the participation of other ethnic minorities (Moroccan, Surinam, Netherlands Antillan or Turkish origin) in crime is different. The problems facing these groups have less to do with legal status than with integration into the society. Social problems facing these particular groups have more to do with poor education, limited work experience and job opportunities, discrimination and prejudice.

Regardless, however, of whether the problems have to do with legal status or lack of equal opportunities, the challenge lies in socially integrating foreign populations and providing educational, work and social opportunities. A number of special projects focusing on Moroccan youths have been initiated. Their aim is to reduce the high rates of criminality among this population. The success of these and other programs would ensure not only a possible reduction in crime, but may also temper feelings of discrimination against ethnic minorities.

RESPONSES TO CHALLENGES

International Influences

Over the past years, formal police cooperation in Europe has intensified. A reason for this was the fact that national law enforcement was increasingly hampered by cross-border aspects of public order and safety: increased freedom of movement of persons within the European Union and immigration from countries bordering the European Union. Also, international organized crime has experienced a strong growth in power and scope. Since the disintegration of the East bloc and its legal orders, the threat of crime from these countries has caused (and continues to cause) great concern to the countries in Western Europe.

One of the first steps toward international police cooperation was taken as a result of the threat of terrorism in Western Europe in the 1970s. At their meeting in Rome in 1975, the Council of Ministers decided to establish the Trevi consultations. Later, issues such as organized crime, drug trafficking, illegal immigration and asylum became items for these meetings. Trevi was not an organization, but a platform for regular consultations comprising *ad hoc* working groups. In spite of the loose structure, Trevi consultations have paved the way toward a more substantial and coherent structure of police cooperation at the European Union level.

During the Trevi consultations, countries were found to have differing views on the significance and implications of the free movement of persons in the European Community. A number of member countries, beginning with the Benelux countries (Belgium, the Netherlands and Luxembourg) in the 1980s, decided to abolish checks at their common borders. In 1985, the Schengen Agreement was signed by participating countries[14] (*Overeenkomst* ..., 1985). As a result of the agreement, the Schengen Information System, a European data collection and information search system, was introduced. This database mainly contains information aimed at protecting the outer borders by preventing illegal immigration.

The Union Convention (*Dossier Maastricht*, 1993), *inter alia*, provided for police cooperation (Europol) aimed at preventing and countering terrorism, drug trafficking and other serious forms of international crime and, where necessary, for the inclusion of certain aspects of cooperation among Customs services. Europol is mainly concerned with transnational police cooperation involving the exchange of confidential criminal intelligence information.

After becoming completely operational, Europol will offer important advantages in comparison with other large systems for exchanging information. Different from the Schengen Information System, the Europol database will contain search data on criminal offenses. Furthermore, its important advantage lies in the fact that it will also exchange so-called "soft" information—personal data—under the observance of rules and procedures regarding privacy which still have to be established.

Another medium for international police cooperation is Interpol. It is the most important medium for the exchange of police information. Interpol operates worldwide and is a network of national police organizations, not of countries.

The task of coordinating preliminary consultations and the Dutch position with regard to police cooperation lies with the Ministries of the Interior and Justice. Also very closely involved, by their police professional knowledge and general coordinating policy responsibility, are the National Criminal Intelligence Division—CRI—of the National Police Agency and, at higher levels of decision-making, the Ministry of Foreign Affairs. Specific areas, such as terrorism and traffic, require the input of other agencies such as the Inland Secret Service and the Ministry of Traffic and Waterways. Parliament plays a democratically controlling role.

The CRI also has an international network of liaison officers. In many places in the world, a police liaison officer is stationed at Dutch embassies in foreign countries.

International cooperation, with the best of intentions, remains problematic. In addition to language barriers, cooperation is formalized and governed by international treaties and European Court of Justice decisions. While international cooperation is guaranteed in cases involving serious offenses (terrorism, murder, etc.), obligations to cooperate become more obscure when the offenses fall into the fiscal or "political" arenas. A recent study conducted by the Research and Documentation Centre of the Dutch Ministry of Justice documented international tax fraud in the European Union (Aronowitz, et al., 1996). Legal requirements often restrict a simple exchange of information. An anticipated increase in international tax fraud has forced investigative agencies in the countries to begin establishing informal ties to other law enforcement and tax agencies to expedite and aid in the investigation. Fiscal liaison officers in embassies is one way of facilitating the exchange of information in international fiscal frauds.

The challenge to all law enforcement agencies is to exert influence on law-makers to modify existing regulations. An alternative is to create more informal avenues for the exchange of information so that valuable time is not lost, which further hinders investigations.

Professionalism

The public's acceptance of police action is closely linked to the way in which police operate. Acceptance and support are of particular importance in democratic societies because the police have to cope with independent citizens whose acceptance of the actions of the police depends heavily on their legitimacy.

In the growing complexity of modern society, rules and regulations are decreasingly adequate means for social control for the individual police officer, who must be constantly flexible in and adaptable to new situations. In a concrete situation, an officer can only meet the needs of the modern, democratic citizen if the officer provides quality service and that requires, first of all, professionalism on his or her part. In this respect, professionalism is the expert execution of his or her duty while being aware of personal accountability for decisions made and actions taken. Professionalism, however, is not only defined by the actions of a single officer. The degree to which a police officer is able to act as a professional is, in part, dependent upon the degree to which he or she is supported by a professional organization.

A professional organization aims at providing quality services. Quality features can be divided into mission features and cultural features. Mission features include a focus on raising the level of security and the quality of life in a society, being externally directed, assuring itself of material and formal legitimacy, and effectively and efficiently using the available means. The cultural features of quality include reliability, credibility, customer orientation, helpfulness, problem solving ability, flexibility and enthusiasm.

A Dutch proverb says "Craftsmanship is mastery." Essentially that is what professionalism is all about. It applies both to the individual police officer and to the police organization. Mastery reflects itself in quality and it is also linked to leadership.

Both quality and leadership are enhanced in the Netherlands by a three-tiered system of education and training, monitoring and quality control. Owing, *inter alia*, to pressure from the police unions, a centrally standardized system has been developed for education and training. Basic police training in the Netherlands is eighteen months.[15] Supervised practical periods are part of the course and the nationwide police examination is government controlled. The recruits' training encompasses knowledge of laws and social as well as police skills. Traditionally, professionalization and the democratic legal system in which the police perform their duty are given emphasis in police education.

A professional police force is one which demands the dedication of its members. Dedication to and pride in the organization are, in part, a result of recruiting quality candidates, providing excellent training and a salary commensurate with a professional career.[16] An indication of the dedication to a police force may be measured in the attrition rate. Annually, the attrition rate among police officers nationwide amounts to approximately 1–2 percent (Jansen & van Sluis, 1990).

External monitoring is accomplished by systematic, large-scale surveys (conducted by an external consulting agency) during which people are interviewed concerning their experiences with the police. Ample attention is given to security, service, availability and visible presence.

Currently, a centrally standardized and directed system of quality control is being developed to complete the recent restructuring of the Dutch police. This system, introduced in the form of a Bureau of Quality, includes visitations by fellow police colleagues. Auditing for efficiency and effectiveness takes place by analyzing performance indicators (Van de Meeberg, 1996).

Integrity

A democracy presupposes a community, and integrity presupposes loyalty to that democratically-ruled community. The personal moral identity of the individual police officer is a question of adjustment to the norms which apply in a democratic community and of loyalty to the other members of that democratic society. It is one of the characteristic aspects of professionalism that loyalty and integrity are inextricably bound up with one another. Therefore, a professional police officer must be incorruptible.

In order for the police officer to be professional and incorruptible, he or she requires the permanent backing of his or her force, which in its turn should also be professional and incorruptible. But merely good intentions are not sufficient. They should be structured in policies and their administration. It is also of the essence that the police organization and its work be transparent.

The measures that are taken in this framework aim at reducing risks from, reinforcing resistance against, and reducing the vulnerability to corruption and illegal conduct of an organization and its members. Limiting the issue of integrity to fraud, corruption and the dangers of organized crime, however, does not do justice to the problem. The diversity and quantity of ethical issues force a broadening of intellectual and ethical horizons so as to be able to judge the overall coherence and determination of priorities for policing. At the same time, the practical application of ethics is predominantly sought in the operationalization of values and norms and must not degenerate into moralizing. Operationalizing norms and values is directly linked to leadership, to the culture of the organization, and to training and educating the staff. As it applies to leadership, the emphasis is on consistency and on effective (internal) security management, because these are conditions for maintaining a credible system of norms and values in the police force. Culture applies to clearly formulating and pursuing an organization's values and norms and not so much to a summation of what is and what is not permitted. As far as education and training are concerned, the essence is ethics and defining, clarifying and acknowledging moral dilemmas.

The question "who controls the controller" is ageless. Faith is good, but control is better and this holds true in a democracy as well. In the Netherlands, that control is executed by three types of bodies: separate investigative bodies, independent judiciary and supervision by democratically elected/nominated authorities.

Complaints about the Police

As mentioned before, the use of police powers may infringe on the rights of the citizens. Although legally based, the question is whether the use of powers is lawful under all circumstances. There are times in which the police are required to take rapid action. In complex situations, when action is required, it is not always possible to form a reasoned opinion about the use of certain powers which can have a massive impact on the lives of citizens involved. Constitutional rights may be violated or (excessive) violence may have been used.

To allow the citizen the right to express a violation of his or her rights, the 1993 Police Act provides that each force shall establish a procedure to deal with citizens' complaints. Such a procedure enables the citizen to file a complaint with the police chief and to request a judgment against the police officer's conduct.

With regard to the complaints procedure, the 1993 Police Act provides that an independent commission be installed which advises the chief of police about the settlement of complaints. The reason to opt for an independent commission to handle the complaints procedure lies in the experience of dealing with complaints against the police prior to the implementation of a complaints procedure provided in the Police Act. Prior experience with an internal police review board taught us that citizens felt that an internal board could not be trusted and that complaints were only dealt with superficially. It was thus felt that an independent commission would inspire more trust and complaints would be dealt with more seriously.

Not every complaint will lend itself to treatment by and advice from the complaints commission, hence only complex and important complaints are eligible for submission to and advice by the independent commission. The Police Act further provides that the complaints procedure must contain rules concerning the time period within which a complaint should have been dealt with, the registration of complaints filed, subsequent decisions taken by the chief of police and their periodical publication in annual reports.

The aim is to solve the problem which has arisen between the citizen and the police and consequently to restore the citizen's faith in the police. The police organization, too, benefits from the complaints procedure. Complaints received give an indication of—possibly structural—shortcomings in the functioning of the police. Complaints, if evaluated and sustained, provide a picture of the quality of the functioning of the police. The police administration can benefit from this and the appropriate department can take measures to improve the functioning.

Furthermore, the Police Act provides that a periodical overview be published of complaints received for each force and the decisions subsequently taken. Thus, there is some indication as to whether, and to what extent, certain complaints are a measure of structural shortcomings of the police functioning and where, if necessary, attention should be paid to the means to remove these shortcomings.

The citizen's ability to file a complaint concerning the action of a police officer during the execution of his or her duties should be distinguished from a disciplinary procedure or criminal proceedings with regard to the officer concerned. Although a complaint and the subsequent decision by the chief of police may result in a disciplinary measure or the institution of criminal proceedings, this will not automatically be the case.

Not every well-founded complaint against a particular officer concerning his or her action is regarded as one in which disciplinary measures or criminal proceedings are required. The sanction taken against the officer should correspond with the egregiousness of the conduct displayed. Thus, a disciplinary or criminal procedure is generally instituted if a police officer has violated administrative procedures or committed a punishable offense.

If citizens do not agree with the decision of the chief of police, they have the right to file a complaint with the National Ombudsman, a legally established independent office. Citizens can also address the ombudsman in relation to the conduct of an administrative body (which encompasses all authorities). On the basis of her/his findings, the National Ombudsman will prepare a report in which he or she indicates whether or not an individual's or administrative organ's conduct was deemed acceptable.

The National Ombudsman will not make an inquiry into a complaint until the complaint has been conveyed to the force which employs the police officer involved (the so-called information requirement). In practice, this means that the complaint will first be investigated and dealt with by the force concerned before the National Ombudsman becomes involved in the matter.

The administrative bodies responsible for the police fall within the competence of the National Ombudsman. In 1994, 20 percent of all inquiries made by the National Ombudsman pertained to the police. In 1995, 358 complaints to the National Ombudsman involved police action, frequently involving the use of excessive force. In approximately half of the cases in 1995 the National Ombudsman ruled in favor of the complainants and against the police (*Volkskrant*, 1996). The guarantee that the National Ombudsman independently assesses police action has, since this office was established in 1983, become an indispensable element in the democratic state of which the police form a part (Doomen, 1996).

In closing, because the success of police work relies so heavily upon community support, and because community support is lacking when police professionalism is absent, it is essential that the police stress professional behavior through high ethical standards and fair treatment of the public. When this fails, police departments must make measures available to the public to air their complaints. This the Netherlands has successfully done.

Legislation

Legislation has been instrumental in creating new bodies or offices to deal with the changing nature of crime in the Netherlands. In the late 1980s, investigations into criminal organizations were reinforced by establishing Financial Support Bureaus and Criminal Intelligence Services within the police and by creating more intensive cooperation with the Inland Revenue. Additionally, systematic economic crime analysis was developed and a funding arrangement was created for supra-regional enquiries (Regulation on Police Special Investigation Costs). In the western part of the country, an inter-regional investigation team has been operative since 1989. Six other such teams were established, and 1996 saw the formation of such a team on the national level, as part of the National Police Agency.

New bills have been introduced. These have given the police greater powers to investigate criminal groups, including organized crime. Also, a bill is in the making whereby preparatory activities, under certain conditions, become punishable offenses. A bill is also underway extending the power of the police to tap not only telephones but also other modern forms of communication.

A study is being made of the policy regarding the issuing of licenses for waste processing. In the framework of countering money laundering, a law has come into force under which banks and other financial institutions are obliged to report unusual and suspect financial transactions over the sum of fl. 25,000 (approximately US $13,000) to the Unusual Transactions Disclosure Centre, which is administered by the National Police Agency. Currently, accountants are being consulted with regards to setting up a similar center to which unusual and suspect cases involving accountants can be reported.

On the basis of the Civil Code, proposals have been made to check prior history and records when new limited or incorporated companies are established, with a view to preventing criminal organizations from infiltrating legitimate businesses and society.

In 1996, changes in the legal bases of police and prosecutorial investigations were debated and decided. The issue concerned the legal use of police powers. Various investigation methods were being used without an explicit legal basis. Until recently, these actions were condoned by the courts. Within the recent past, however, court cases have failed to result in a conviction due to the use of evidence unlawfully obtained through investigation methods not explicitly stated by law. A parliamentary commission, *Enquetecommissie opsporingsmethoden*, more commonly known as the Commission Van Traa (Van Traa, 1996),[17] conducted hearings into, among others,[18] the legality of police investigative methods. The weeks-long hearings resulted in conclusions and recommendations for formal regulations with regard to investigation methods and tighter guidelines with regard to application and control. It remains to be seen whether, to what degree and how quickly the recommendations of the Commission Van Traa will be adopted. This remains a challenge as the Commission, while seeking to hold police and prosecutors accountable to basic legal standards in conducting investigations, must balance the needs of the police to fight crime (and particularly organized crime) against the needs of the community to be protected from abusive police investigative methods.

Another area of concern, and one in which Dutch culture and legislation come into stark conflict with its neighbors, is the Dutch policy of permissiveness with regard to the sale of soft drugs in so-called coffeeshops. The Dutch drugs policy focuses on suppressing all drug traffic and production, provides for the medical and social treatment of (hard) drug addicts and condones cannabis production at home and sale by coffeeshops on a small scale. This policy has come under foreign pressure,[19] particularly from France which had, at one point, cut off bilateral talks with the Netherlands. The Dutch policy of tolerance was being blamed for the drug problem in French border towns. The Netherlands has come under criticism from internal sources as well, who argue that we cannot condone the sale of "small amounts" of hashish and marijuana in coffee shops, which must maintain a large stock to satisfy customer demand, yet outlaw the import and distribution of the drug. Will the Netherlands be able to control the trafficking, production and distribution of drugs and still permit the sale in coffeeshops? And, perhaps more important, can the Netherlands continue to uphold a unique policy of tolerance yet also maintain positive working relations with its European neighbors and international allies? In an ever-increasing march toward a European Union, the Netherlands may find itself isolated by its policy of tolerance toward soft drugs.

Relations with the Media

In the Netherlands, the media monitor the authorities with keen interest. Each day will see an article or broadcast about the police and its performance and action. The results are given much attention. The media also have an obligation to inform the public independently and in doing so expose wrongdoings by a democratically controlled government and, in particular, its police force. The police are used to this practice. For years, the police have been a journalist's source for matters which take place in our densely populated society. The police have important tasks as concerns criminal investigations, public order disturbances and rendering assistance. Society is informed of what goes on in our country through the media. Consequently, the adage for the Dutch police is: a democratic police is as open to the media and the public as much as possible, unless current operations or the privacy of victims or defendants is jeopardized.

The Netherlands adheres to the principle that the authorities operate transparently and openly, pursuant to the Openness of Public Administration Act. This Act does not provide sanctions, but it clearly shows how openness should be realized in this country. Freedom of speech and freedom of the press are constitutional rights.

It can be said that the police in the Netherlands publicly account for their activities and are held publicly accountable through the media. But this principle, too, has its limits. Research journalism, i.e., journalists who do their own investigations and publish their findings, poses risks to the police. Journalism may expose questionable police practices but it can also jeopardize an investigation.

The media must walk a fine line. On the one hand, they have a responsibility to the public to keep them informed. This includes reporting on positive as well as negative police behavior. But the media must balance the right of the public to access to information against the right of the police to protect an ongoing investigation.

The police are sometimes the center of media attention. The parliamentary hearings conducted by the Commission Van Traa concerning the investigation methods used by the police to combat serious and organized crime were televized daily. Special talk shows were broadcast during the hearings and newspapers and magazines provided daily and weekly coverage.

A number of recent court hearings and trials have also focused public attention on police activities. In the past, the police, at times with and at other times without the knowledge of the department of public prosecution, infiltrated criminal circles through the use of informants in order to apprehend top level drug dealers. Criminal police informants could then make a career within criminal organizations. With the knowledge and protection of the police, these informants were able to actively bring large quantities of drugs onto the Dutch market. Slowly these facts began to surface as lawyers of accused drug defendants began challenging police operations and the use of informants. The public nature of court trials made it possible for the media to give extensive accounts of these activities.

There is more. The house of the public prosecutor who was in charge of the investigation of one of the largest and most important recent drug cases was burglarized and a number of diskettes containing highly confidential information were stolen. The diskettes also contained information on the controversial investigation methods used by the police. The criminals' intention had apparently been to discredit the public prosecutor and to undermine the authority of the government agencies involved, the public prosecution service and the police. The diskettes were offered to the lawyers of the persons charged and also to a number of publishers. The defense attorneys used the information to seek acquittals for their clients and the press published some sensational articles claiming that the police break the law in order to enforce it.

Disregarding, for a moment, the actions on the part of the police, two questions remain unanswered. Can a lawyer use stolen information to defend a client? Should a journalist be allowed to publish truthful information that was stolen? Debates are ongoing in the Netherlands concerning these issues. Will allowing such behavior result in an agreement by the media not to publish this kind of information which would otherwise prove to be counterproductive to justice and also be an instrument used by criminals, who in their pursuit to undermine and corrupt the authorities will use any means possible? Despite the fact that this type of information is counterproductive to the police and prosecution service, it could be argued that this is only so because the police used "questionable" methods to infiltrate such organizations and collect enough evidence to result in arrests. Had the police adhered to legally approved methods, had the proper authorities, up to and including the Minister of Justice, been made aware of the situations, the information on the diskettes would have been of no value to the lawyers of the defendants nor of any great interest to the media.

Despite the fact that the media have been highly critical of the police and prosecution department within the recent past, they do serve the positive role of watchdog, of assuring that the authorities abide by the rules and procedures spelled out in a democratic society. After all, success in fighting crime cannot be measured by arrest alone, nor by arrest and conviction when that same conviction is later overturned by an appellate court due to illegal methods of operation employed by the police. If police adhere to the rules, the game can, if not easily, at least be fairly won.

The media should continue to scrutinize police behavior. They help to assure the public that the police are acting legally. No one wants to see the police involved in illegal action. But even more disheartening (to the authors) is watching guilty defendants exonerated on the grounds that the police acted in violation of the law. The media serve as a constant reminder that if the police are unable to adhere to the rules of the game, this knowledge will be made public.

CONCLUSION

One of the most important ideals in any democracy is the concept of all people being treated as equals. This right is spelled out in Article 1 of the Dutch Constitution. It guarantees that "Everyone in the Netherlands will be dealt with in the same manner under the same circumstances. Discrimination based upon religious belief, convictions, political orientation, race, sex or upon any other ground is prohibited."[20] As a constant reminder to the police, this article is printed in huge letters covering an entire wall of the entrance and reception area of the detective school in Zutphen.

The Netherlands is a democracy. The police in the Netherlands are bound to abide by the law and the regulations as they provide a service to members of society. The Netherlands, however, is also a tolerant country. This means that its inhabitants have a great degree of individual freedom and that, in principle, there is space to exercise that freedom for everyone.

Within recent years, friction has arisen between these principles and the reality of the world in which we live. The police and justice authorities have gradually been faced with a changing crime picture. Much of what the police do is aimed at tracking down drugs and countering drug-related crime. Some people think the war on drugs is lost; others would like to crack down harder on drugs. Policing in a democratic society requires that police abide by certain laws, regulations and restrictions which often reduce the immediate, short-term effectiveness of the police. This is particularly so in the battle against drugs and organized crime operations which are privy to large amounts of money and sophisticated measures for protecting their operations.

Policing in a democracy requires abiding by the laws and allowing some criminals to escape because under the current system it is impossible to legally obtain evidence against them; or, it requires changing the laws through democratic processes to expand police powers to allow police to more easily infiltrate such organizations using more creative, avantgarde methods. In our view, the area of tension between the tasks of the police and their powers, including the investigation methods, and the rights of citizens and their demands for protection and security will become greater.

The Dutch police and justice system (as well as the social welfare system) has within the last seven years begun facing another problem. The Netherlands has become (even more) a multi-ethnic society. Although within the last year the numbers have been receding, many people have sought (and continue to seek) asylum in the Netherlands while many others reside in the Netherlands illegally. Policing a multi-ethnic society requires understanding and being able to relate to that society. In order to help improve relations between minorities and the police, the police have actively attempted to recruit more minorities and women as officers. A police force which more nearly mirrors its population in terms of ethnic background, race and gender often has a better working relationship with the public.[21]

Another recent area of concern, facing not only the Netherlands but all European Union nations, is the movement toward a single Europe. With the fall of internal EU borders and border controls, the smuggling of contraband and persons over the borders has become increasingly easier. Keeping a check on the movement of people requires increased communication and cooperation between various agencies within these European countries. This cooperation is governed by various treaties and the European Court of Justice. In addition to formalities, communication is complicated by differences in language, skills and computerized systems. The Dutch police must establish networks and work within the parameters and guidelines established by these governing treaties and court cases. This problem is slowly being overcome in the area of international value-added tax fraud (Aronowitz, et al., 1996). It remains to be seen how effective Europol will be in aiding in the fight against international crime.

A further concern facing the Dutch government in the move toward harmonization of laws and policies in the European Union is its independent, and sometimes unpopular, drugs policy. Will the Schengen member countries pressure the Netherlands to abandon its drug policy in light of the need to harmonize regulations, as remains the hope of conservatives, or will the Netherlands be able to continue to maintain open borders and open communication with its neighbors and still maintain its liberal domestic policy (Van Zwol, 1996)? This, too, is a question of time.

The media serve as a watchdog and publicly hold the police accountable for their behavior. The public is also informed of police behavior through public reports issued through the office of the National Ombudsman. It is not enough, however, to monitor police behavior when things go wrong. There must be a way of measuring police performance without having to rely upon complaints by citizens or the media uncovering a scandalous story. Terlouw, et al (1994) propose a model of measuring police performance based upon hard and soft indicators.[22] The researchers go so far as to suggest that funding would be based upon the performance of the force. Such a concrete measurement tool would aid in improving the working methods of police departments and to hold them accountable not only for solving crimes but for interacting with the community and making citizens feel safe.

The new Bureau of Quality was implemented in May 1996. This Bureau measures the quality of police services based upon the evaluation by the customer (the prosecution service, city council and the citizens), the personnel (the officers on a particular force) and society (organizations within the society). The bureau also examines the management of processes, administration and strategy, management of personnel and hardware, and leadership. In this way the police organization in the Netherlands is attempting to improve the quality of police services to ensure an effective, efficient and professional police force.

The recent media circus surrounding the Van Traa parliamentary hearings focused attention upon the inadequacies of the police. The recent shift in the type of crime police are facing clearly indicates that the practices which the police are required to follow are sorely inadequate for combatting more organized forms of (and large-scale) drug crimes. The parliamentary hearings are just the beginning. They will have a great impact upon the future of policing in the Netherlands. But if the laws are inadequate, then they must be changed. This, however, is not the responsibility of the police. In a democracy, it is the task of the politicians to supply the police with good guidelines for the execution of their duties.

NOTES

[1] Since the exchange rate fluctuates, this is an approximate figure. In August 1997, since the value of the dollar had risen compared to many currencies, the proper amount would be about $15,000.

[2] Violent crimes include rape, assault, other sexual offenses, threat against life, abuse, imputable death and bodily injury, theft with violence and extortion.

[3] Property offenses include forgery, larceny, burglary, other aggravated theft, embezzlement, fraud and receiving stolen goods.

[4] With slight fluctuations, and differences in age and gender, this figure has remained fairly consistent between the years 1980 and 1992 (Eijken, 1994, Table 7 in the Appendix, p. 104).

[5] In testimony to the parliamentary Commission Van Traa, investigating, among others, organized crime in the Netherlands, the figures varied from 30–35 groups to over 100.

[6] After having visited the Netherlands, tourists are usually positive about feeling safe.

[7] Due to shortages of cell space, those awaiting trial and sometimes convicted persons are recorded in the prison system but are not incarcerated and stay outside of the institution. When a prison cell becomes available, they must report to the prison to begin serving their sentence.

[8] Within a period of 72 hours, the suspect underwent 32 hours of intensive interrogations. He was promised a lighter sentence if he cooperated and later threatened and told that his wife and child could only be given police protection if he confessed.

[9] This is based upon official police statistics. In contrast, results of an international self-report study (Junger-Tas, Terlouw & Klein, 1994) show that the rate of criminality among Moroccan and Turkish youths in the Netherlands is lower than among Dutch, Surinam and youths of other ethnic minority groups.

[10] Dutch male youths are especially involved in thefts, burglaries and vandalism.

[11] In 1995, the foreign population comprised approximately 5 percent of the total population (Ministerie van Justitie, 1996).

[12] Over half the detainees (52%) are of Dutch origin, the remainder are from other European countries or those excluding the four previously-named groups.

[13] Foreign residents of ethnic minority status are referred to in Dutch as *"allochtonen."* We are speaking about this group here as opposed to foreign tourists, who are also often the victims of crime, particularly of car burglary or robbery (Ommens & Van Wijngaarden, 1992; Willemse, 1993).

[14] Schengen member states are Germany, France, the Netherlands, Belgium, Luxembourg, Italy and Austria.

[15] In addition to an initial extensive educational and training period of eighteen months, further training is provided throughout a police officer's career. A four-year management school allows officers with university education to enter into management positions. Police administrators often do not rise from within the ranks as in the US, but are university graduates with backgrounds in management or law. As part of the four-year education, they are required to complete internships within various police departments.

[16] Beginning annual salary for a police officer, as of August 1997, is approximately 35,000 Dutch guilders (or $18,000).

[17] The Commission Van Traa was named after its chairman, M. Van Traa.

[18] The Commission also addressed such issues as organized crime in the Netherlands, the Delta-method (allowing drugs to enter into the country with the knowledge of the police and prosecutors office as part of a controlled investigation), the use of informants and undercover officers, the use of information and international investigations, and the actions of various investigative organizations.

[19] Certain German federal states have supported the Dutch policy and have asked the Netherlands not to give in to international pressure.

[20] Article 1 of the Dutch constitution: *"Allen die zich in Nederland bevinden, worden in gelijke gevallen gelijk behandeld. Discriminatie wegens godsdienst, levensovertuiging, politieke gezindheid, ras, geslacht of op welke grond dan ook, is niet toegestaan"*.

[21] One Dutch police officer in Amsterdam has spent the last ten years studying Mandarin Chinese. He serves as the police/liaison officer for the entire Chinese community in the city. The citizens go to him with problems concerning social welfare as well as crime.

[22] Hard indicators are the number of recorded crimes and the clearance rate, fines, etc. Soft indicators are measures of police service, defined as: judgment of the public on overall police performance, police availability, willingness to report crimes, victimization rates, and feelings of safety.

REFERENCES

Aronowitz, A. A., Laagland, D. C. G., & Paulides, G. (1996). *Value-Added tax fraud in the European Union*. Amsterdam: Kugler Publications.
Centraal Bureau voor de Statistiek (1996). *Statistisch jaarboek 1996*. Voorburg/Heerlen.
Centraal Bureau voor de Statistiek (1994). *Criminaliteit en strafrechtpleging 1993/1994*. Den Haag: SDU/CBS Publikaties.
Central Bureau Statistics (1994). *Prison statistics*. Voorburg/Heerlen.
Central Bureau Statistics (1995). *Statistical yearbook 1995*. Voorburg/Heerlen.
Council of Europe (1995). *European sourcebook of crime and criminal justice statistics*. Strasbourg:
de Haan, W., & Bovenkerk, F. (1993). *Moedwil en misverstand: Overschatting en onderschatting van allochtone criminaliteit in Nederland. Tijdschrift voor criminologie, 35*, 277–300.
Doomen, J. (1996). *Behoorlijk blauw: Politieoptreden in Nederland*. Utrecht: Contact.
Dossier Maastricht: verdrag betreffende de Europese Unie (1993). Den Haag: SDU Servicecentrum Uitgeverijen.
Een nieuwe politiebestel in de jaren '90 (1990). Tweede Kamer der Staten-General, vergaderjaar 1989–1990, 21, 461, nr. 1. Gravenhage: SDU uitgeverij 's.
Eijken, A. W. M. (1993). *Criminaliteitsbeeld van Nederland: Aard, omvang, achtergronden, spreiding en preventie van criminaliteit, 1980–1992*. Den Haag: Ministerie van Justitie, Divisie Criminaliteitspreventie.
Eijken, A. W. M. (1994). *Criminaliteitsbeeld van Nederland: Aard, omvang, preventie, bestraffing en zorg voor slachtoffers van criminaliteit in de periode 1980–1993*. Den Haag: Ministerie van Justitie, Divisie Criminaliteitspreventie.
Fijnaut, C. J. C. F. (1994). *Politiewet 1993*. Nijkerk: Callenback.
Jansen, T. G., & van Sluis, V. (1990). Verloop bij de politie: het beland van gestructureerd exit-onderzoek. *Tijdschrift voor de politie, 52*, 491–495.
Junger-Tas, J., Terlouw, G.-J. & Klein, M. (Eds.). (1994). *Delinquent behaviour among young people in the Western World: First results of the international self-report delinquency study*. Amsterdam/New York: Kugler.
Loef, C. J. (1992). *Alloctonen als slachtoffer van onveiligheid, achtergrondstudie voor de integrale veiligheidsrapportage*. Den Haag: Ministerie van Binnenlandse Zaken.
Ministry of Foreign Affairs (1990). *Netherlands in brief*. The Hague.
Ministry of the Interior (1994). *Police in the Netherlands*. The Hague.
Ministerie van Binnenlandse Zaken (1994). *Integrale veiligheidsrapportage 1994*. Den Haag.
Ministerie van Binnenlandse Zaken (1995). *Veiligheidsbeleid 1995–1998*. Den Haag: SDU Servicecentrum.
Ministerie van Justitie (1996). *Het vreemdelingenbeleid kent z'n grenzen*. Den Haag.
Oomens, H. C. D. M. & van Wijngaarden, J. J. (1992). *Toerisme en onveiligheid*. SMP–Cahier (2)–Arnhem: Gouda Quint.
Overeenkomst ter uitvoering van het tussen de regeringen van de staten van de Benelux Economische Unie, de Bondesrepubliek Duitsland en de Franse Republiek op 14 juni 1985 te Schengen gesloten akkord betreffend de geleidelijke afschaffing van de controles aan de gemeenschappelijke grenzen [Schengen Agreement], (1995). Den Haag: Ministerie van Justitie.
Rombouts, R. (1996). *Vernietigende kritiek op Zaanse verhoormethode, Het Parool*, 15 August.
Stichting Maatschappij en Politie (1991). *Veiligheid en politie: een beheersbare zaak*. Arnhem: Gouda Quint/Kluwer.
Strooper, M. N. (Ed.) (1994). *Politiewet 1993: tekstuitgave zoals deze luidt op de datum van inwerktreding (1 april 1994) met bijbehorende besluiten*. Den Haag: VUGA.
Terlouw, G.J., Kruissink, M., & Wiebrens, C. J. (1994). *Measuring police performances*. The Hague: Ministry of Justice, Dutch penal law and policy, Research and Documentation Centre (WODC), issue 9.
Van de Meeberg, D. (1996). Kwaliteit. *Algemeen Politie Blad, Jaargang 145*, Nr. 10, 7.
Van der Heijden, A. W. M. (1986). Nadere analyse van onrustgevoelens in 1985. In Centraal Bureau voor Statistiek, *Maandstatistiek rechtsbescherming en veiligheid*. Voorburg/Heerlen: CBvS.
Van der Hecjden, A. W. M., Tham, H., & Nierop, N. M. (1995). *Landelijke inventarisatie criminele groeperingen*. Zoetermeer: Divisie Centrale Recherche Informatie.
Van Traa, M. (1996). *Inzake opsporing*. Den Haag: SDU Uitgevers.

Van Zijl, F. (1996, October 5). Processen-Verbaal zeggen weinig over misdadigheid. *de Volkskrant*, p. 7.
Van Zwol, C. (1996, July 3). Nederland wil drugsbeleid niet opofferen aan Europa. *NRC Handelsblad*, p. 3.
Volkskrant, de (1996, May 31). *Dieven vangen in het donker, en graag keurig*, p. 7.
Willemse, H. M. (1993). *De belaagde toerist. SEC, Tijdschrift over samenleving en Criminaliteit, 7,* 18.

Additional Readings

Aronowitz, A. A. (1997). The Netherlands. Part of *The International Fact Book of Criminal Justice Systems*, Bureau of Justice Statistics, on the Internet at: http://www.ojp.usdoj.gov/pub/bjs/ascii/wfbcjnet.txt

de Groot-van Leeuwen, L., & Kester, J. G. C. (1991, June). *Occupational backgrounds and views of the Dutch judiciary*. University of Leiden, Department of Sociology, Working Paper No. 7. Paper prepared for the Joint Meetings of the International Conference of Law and Society, Amsterdam.

Department for Statistical Information and Policy Analysis. (1992). *Registered crime in the Netherlands 1965–1991*. The Hague: SIBa, Ministry of Justice/CDWO.

Fiselier, L., Moor, G., & Tak, P. (Eds.). (1992). *De staat van justitie*. Nijmegen: Sun.

Hoyng, W., & Schlingmann, F. (1992). The Netherlands. In M. Sheridan & J. Cameron (Eds.), *EC legal systems: An introductory guide* (pp. 1- 42). London: Butterworth and Co., Ltd.

Kroes, L., Meiberg, L., & Bruinsma, G. J. N. (1994). *Vernieuwingen in politiezorg*. Enschede: Internationaal Politie Instituut Twente, Universiteit Twente.

Kroes, L. & Scholtens, H. N. (1996). *Politie en de zorg voor veiligheid*. Enschede: Internationaal Politie Instituut Twente, Universiteit Twente.

Kruize, P. & Wijmer, D. J. (1991). The use of force in daily police procedure in The Hague. *Police Studies, 14,* Fall, 121–126.

Kuijvenhoven, A. (1992, December). The Police in the Netherlands. In *Focus on police research and development*. London: Police Research Group, Home Office Police Department, Issue Number 1, 42–44.

Lensing, H. & Rayar, L. (1992, September). Notes on criminal procedure in the Netherlands. *The Criminal Law Review*, 623–631.

Leuw, E. (1991). Drugs and drug policy in the Netherlands. In *Dutch penal law and policy: Notes on criminological research from the Research and Documentation Centre*. The Hague: Ministry of Justice.

Ministry of Justice. (1990). *Law in motion: a policy plan for justice in the years ahead*. The Hague.

Ministry of Justice (n.d.). *The Kingdom of the Netherlands: Facts and figures*. The Hague: Information Department.

Ministry of Justice, various publications available from the Information Department. P.O. Box 20301, 2500 EH The Hague, Netherlands.

The Netherlands. (1990, March–April). *International Criminal Police Review (Interpol)*, no. 423, 18–20.

Oudhof, J. (1992). *Fourth United Nations survey of crime trends and operations of criminal justice systems*. Voorburg/Heerlen: Central Bureau of Statistics.

Pease, K. & Hukkila, K. (Eds.). (1990). *Criminal justice systems in Europe and North America*. Helsinki: Helsinki Institute for Crime Prevention and Control, Publication Series No. 17.

Tak, P. J. P. (1993). *Criminal justice systems in Europe: The Netherlands*. Helsinki: HEUNI (European Institute for Crime Prevention and Control) and Deventer: Kluwer Law and Taxation Publishers.

Van Kalmthout, A. & Tak, P. J. P. (1992). The Netherlands. In *Sanctions-Systems in the member-states of the Council of Europe, Part II* (pp. 663–807). Deventer: Kluwer Law and Taxation Publisher.

Section III
Reflections on Challenges of Policing Democracies

12

Democracy, Democratization, Democratic Policing

OTWIN MARENIN

Any discussion of how to tackle the challenges of policing outlined in earlier chapters requires an understanding of the nature of democracy, of policing, and of processes of democratization. Societal changes and reforms, to be labeled democratization, must lead to particular kinds of innovations in politics, policy and policing (as well as other aspects of social life). In turn, democratic policing can only be defined by standards embodied in democratic theory and practice. At the same time, policing always occurs in specific societal contexts which shape its forms and functions. Democratic policing, then, is a response to two pressures: it represents the adaptation and incorporation of universal standards of democratic behavior to local policing conditions and needs.

In functioning democratic polities, policing and political norms will be congruent, hence little dissensus on the fundamental nature and meaning of either arises or, if disagreements occur, they can be resolved by established and legitimated political procedures. In emerging and mixed democracies, in contrast, both policing and politics are in transition; hence, the procedural means to reconcile new forms of policing with emerging political norms and practices must themselves be developed and become legitimated as part of the process of change. New, democratic policing in emerging democracies, hence, requires a double institutionalization—of policing itself and of the context which gives shape to and supports and constrains policing; not an easy task.

Democracy is here conceptualized as a dynamic set of balances between conflicting yet equally legitimate interests, demands and values. A fundamental tension and need for balance between populistic demands and personal integrity are often described as the most profound personal choice facing elected delegates in any democratic system. Should elected representatives merely be mouthpieces for the demands of communities they represent (and possibly violate deeply held personal values) or should they exercise their expertise, intelligence and conscience to arrive at decisions which they consider will contribute to the public good (and possible deviate from the expressed wishes of the citizens they represent)?

This dilemma exists for the police as the choice between obedience to conscience and law (doing what is right) versus responsiveness to community demands (doing what is expected)—all legitimate values and aspirations.[1] In practice, the balance is unresolvable on a permanent basis, by choosing one polar position or the other, and its changing and temporary resolutions in policy preferences need to be continually justified and be justifiable within democratic theory. The capacity to balance competing values, in turn, implies and demands discretion and a degree of autonomy for those faced with the choice—politicians, citizens and the police; that necessary autonomy of the police, in turn, must be legitimated and consented to by citizens and the state.

Policing is here conceptualized as a universal social function, a form of social control and order maintenance. Democratic policing is a particular version of how to conduct that function. Policing faces the same tension which confronts all democratic institutions. The police need to balance their capacity for discretion against demands by the state and public that they exercise their authority and capacity for services in specific ways. Every policing arrangement is problematic for policing and must balance coercion and consent, autonomy and responsiveness (Brogden, 1982), the desire for general order versus the wish for the protection of particular interests (Marenin, 1982), and the need for control of behavior and the desire by citizens to be left alone (except when they desire assistance). Every imbalance and swing of policing to one side or the other of the demand-autonomy balance benefits and penalizes groups and values differentially. No form of policing permanently solves either public, police or state concerns and conflicts, and every policing system embodies contingent resolutions (or balances) of conflicting claims.[2]

Policing is also work done for pay by trained craftsmen/women who have the authority to use specific powers, including (deadly) force. More specifically, the police face common issues which arise from the nature of their work (e.g., the proper use of force or when to lie and deceive) which can only be circumscribed by stated standards and principles (e.g., use force "whenever lawful, necessary or appropriate for the task") but cannot be embodied in precisely stated rules or by the preference of one polar value or the other (e.g., to use force whenever the state or citizens demand or to never use force even when force might be legal, useful and necessary).

Democratic ideals (the practical balancing of competing demands on the police) become effective democratic policing only if translated into operational policies, specifically as rules for working or guidelines for two sets of decisions: what to pay attention to, and when and how to exercise police powers if a situation merits attention. Rules for working link abstract principles to practical work, and condition the impacts of police power (and perceived justice) on individuals and groups in society. Rules for working are the operational definition of democratic policing.

Rules have multiple sources—law, professional standards, organizational norms, public demands, peer culture, etc.—each clamoring for a particular exercise of discretion, order maintenance or the enforcement of laws and expectations. Rules of working can develop in response to many pressures. Only particular substantive rules for working and only responsiveness by the police to particular patterns of pressures can be considered democratic policing. The substantive content of rules needs to reflect the norms being elaborated in emergent democratic policing regimes. Patterns of responsiveness to pressures and demands will reflect the adaptation by the police to the changing, democratizing conditions of their societies.

The chapters in this book exemplify the difficulties of creating new rules for working when functional needs change (e.g., domestic crime and violence explode, borders become open to organized crime and migration, publics become restless and demanding, and requests for international cooperation pour in), democratic values become the touchstone of acceptable behavior (in new police laws, in political rhetoric and practice) while rules for working continue to reflect the power of prior pressures on the police (the continuity of old police cultures, habits of stereotyping suspects and groups, the absence of new laws and legal cultures which enable and constrain the police, or the absence of trained police managers to enforce new sets of demands and patterns of responsiveness) (see also Shelley & Vigh, 1995).

Efforts to democratize policing, hence, require the development of rules for working (their specific substantive and procedural contents) which are effective responses to functional needs and also incorporate accepted and universal standards of democratic behavior. The conceptualization of democratic policing advocated here is based on the argument that the police cannot be made democratic or taught democracy in the abstract (even though these standards can be stated) or by exhortation to abide by human rights and procedural fairness, but the police can be taught, constrained and held accountable to proper behavior as stated in rules for working.

The challenge for democratic policing, I will argue, is not disagreement on what would constitute democratic practices. There is an emergent consensus on what democratic policing requires as goals and standards. The challenge, rather, is twofold. One aspect is to translate principles into rules of work which can be taught, enforced, rewarded if properly done and sanctioned if abused. Second, rules for work will only become operational and guide day-to-day decisions if they can be made to sink down and become part of the operative, formal and informal, police culture. In Reiner's (1992, p. 216) words, discussing efforts in England to enhance the accountability of the police to community norms, "what has to be achieved is the incorporation within the operative police sub-culture of working procedures and norms which embody the universalistic respect for the rights of even the weak and unpopular minorities which the rhetoric of legality purports to represent." Reform efforts which ignore the power of police culture will fail or be counterproductive. Organizational, operational and professional responses to challenges, as described by Das (see chapter 1), in effect seek to create just such rules and the justifications and mechanisms for their enforcement and make them part of police culture, largely so far by formal rather than informal socialization processes.

This chapter is organized into four sections. Section one discusses theories of democratization and finds them remarkably unconcerned with the analysis of the police, the coercive agency of the state. Section two argues that, as seen in the growth of international regimes and in scholarly writings, a consensus on the meaning of democratic policing has begun to emerge, but that the translation of that consensus into policies remains diverse and contested. This problem is no different from implementing the concept of democracy. Its core values can be defined fairly easily but they can take many historical forms.[3] Section three, using examples from the chapters in this book, discusses the challenge of translating notions of democracy into practical rules of democratic police work and behavior. What would democratic policing have to look like given the dominant conceptions of democracy, and existing and changing patterns of order and deviance? Section four sketches some lessons drawn from historical experience on how to transform policing toward a democratic form, again with reference to the chapters in the book.

POLICING AND DEMOCRACY

It is truly remarkable how little attention is paid to policing in the literature on democracy and democratization. One can search practically all textbooks and analyses of democratic governance without finding much mention of that most distinct and visible embodiment of the coercive power of the state, the police. They are hardly ever mentioned, far less analyzed, even in discussions of violations of human rights and dignity, of the control and the accountability of coercion and political power, or typologies of political systems. It is as if state power and control were a natural, self-generated occurrence and not something done and achieved by personnel employed to create social order, control deviance and crime, and ensure the continuity of the state.

The literature on democracy does not take policing seriously as a theoretical issue or practical problem because policing is generally misconceptualized. Basically, policing is perceived as done by agents of the state, at the behest of those who control the state, with little choice or discretion—that is, as law enforcement. It is assumed that police actions and impacts are explained by analyzing the nature of the state and its relations to civic society. Explaining the state explains the police. The police are not accepted as historical actors nor granted the capacity to act or have agency or discretion. This view is much too simplistic (Marenin, 1982, 1985, 1990).

The theoretical situation is not much better in discussions of political development, changes in the state and its relations to civil society, and democratization. With rare exceptions (Bayley, 1969, 1975; Enloe, 1980; Hills, 1996; Potholm, 1967; Tilly, 1990), the police have not been considered in the literature on development, despite their obvious presence and potential impact on the nature and direction of change, the nature of group conflicts, or the prospects for democratization and the protection of human rights.[4] None of the political change literatures (modernization, underdevelopment, the reproduction of the state, the changing nature of civil society, or democratization) provides a consistently worked out and empirically sustainable depiction and analysis of the roles of the police or policing in such processes.

For example, Diamond (1995c), not an unknown writer in this field (1992, 1994, 1995a, 1995b; Diamond, Linz, & Lipset, 1995), in a study commissioned by the Carnegie Foundation (1995c) to review the state of knowledge on democracy and democratization, hardly mentions the police at all. He notes, incidentally, that the MacArthur Foundation supported research efforts to increase the accountability of the police and military in Latin America (p. 26); that civilian police participated in peacekeeping operations in Namibia (p. 31); and that international legal and judicial assistance involves more than training "judges, magistrates, clerks, prosecutors, public defenders, police, and investigators" (p. 43), but only discusses institutional developments in the legal and judicial arenas. In none of the discussions of "actors and instruments, issues and imperatives" for democracy do the police appear. Yet failures of democracy—corruption, abuses of authority, violations of law, disruptions of electoral competitions, denials of human rights or outright killings and disappearances—are the actions of the police. Diamond is not atypical in this disregard.

Discussions of the police in the development literature tend to see them uncritically and unreflectively. There is a failure to think through what coercion means in real life, what legitimates coercion, and how the police can be constituted in such a way that they properly balance coercion and consent, repression and protection. Human rights activists understand very well that the police are the people who do the nasty stuff, but they have little theoretical leverage on how to explain police behavior or suggest reforms, other than in simplistic terms and as appeals to make them stop. But one cannot change what one does not understand.[5] In this case, that requires getting close to the police, in theory and in their practical work.

Surrounding this neglect of policing, and supporting the lack of discussion of the police, are existing theoretical and ideological disagreements on the meaning of development or how to promote democratic forms of politics in developing or transitional (i.e., former socialist) societies. After reviewing the literature, Remmer (1995, p. 105) concludes that an extraordinarily wide variety of conjunctural conditions, socioeconomic structures, patterns of state-society relationships, transitional paths, and historical traditions have been associated with recent transitions to political democracy, invalidating old theory and complicating efforts at synthesis and generalization.

There is little consensus on what promotes and sustains democracy or what factors are linked to the political and social development labeled democratization, hence little attention is paid to how coercion plays out in such processes. Everything matters and nothing can be learned.

We are faced with one of the more extraordinary lacunae in social science theorizing. It is as if discussions of the military and their political roles were devoid of any effort to see the military from within—by its motives, interests, organizational dynamics or structural roles in society. That the other formally coercive agency of the state—the police—are so neglected reflects a real silence in the discourse of positivist social science and an inaccurate model of coercion in the views of most scholars writing on this topic.[6]

Social order is sustained by many forces, of which policing is only one and possibly not the most important. All the same, policing is an essential and powerful social institution which has to be understood as an organized activity done within societal contexts.

THE NATURE OF POLICING

Theoretical Changes

What is missing from discussions of democracy and democratization is a theorized understanding of the police and their relations to the state and civil society. Yet a critical dialogue is carried on in the police literature which would be of use to democratic theory. Most fundamentally, much of this discussion argues for a redefinition of policing as ordering structures rather than viewing policing more narrowly as the work of state agents employing legitimate force; this conception implies a rethinking of the goals and means of policing and views citizens as consumers (and co-producers) of policing. In this view,

> policing constitutes an intricate pattern of relations between sources of authority, a set of goals, the selected means and available resources for achieving them, and the environment in which such activity takes place. [It is a set of] particular interactions of interests, power and authority (Hebenton and Thomas, 1995, p. 205; also, Brewer, et al., 1988; Brogden, 1982; Johnston, 1992; Reiner, 1992; Shearing, 1992; Shearing and Stenning, 1987; Waddington, 1994).

Policing can exist, everywhere and in every time, in multi-layered and pluralistic (Bayley and Shearing, 1996) forms (statist, private, communitarian, transnational), within general societal and international networks of ordering structures. In Brogden and Shearing's (1993, p. 175, p. 172) words,

> policing should be understood as a product of a network of interrelated institutions operating at different levels and with different knowledges and resources .... It is now something that is "owned" by, and done by, a variety of entities.[7]

Policing is a set of order promoting policies only some of which are located in the state and characterized by a monopoly of legitimate force, though that is the dominant type and can be found in a vast variety of forms (Bayley, 1985; Brewer, Guelke, Hume, Moxon-Browne & Wilford, 1988; Das, 1991, 1994; Marenin, 1992; Mawby, 1990; Skolnick and Bayley, 1988). But alongside state police forces exist private and corporate types of security and order maintenance (Johnston, 1992; Shearing and Stenning, 1987; Shearing, 1992). Populist, community based, self-help policing has a long history, from American vigilantism to state sponsored, but little-controlled, populist ordering organizations (e.g., security brigades in Zambia or the revolutionary vigilantes in Nicaragua). Lastly, transnational policing not tied to a state has begun to emerge in regional and international settings, such as the policing organizations associated with the European Union (Anderson, 1989; Anderson and den Boer, 1994; Benyon, Turnbull, Willis, Woodward, & Beck, 1993; Fijnaut, 1993; Hebenton and Thomas, 1995; Sheptycki, 1995; Walker, 1993a, 1993b, 1994; 1996) or with international peacekeeping efforts which incorporate domestic policing forces and goals (e.g., in Cambodia, El Salvador, or the former Yugoslavia) (Holiday and Stanley, 1993; WOLA, 1995).

Policing in Action

The discussion below will focus on one form of policing, the state police. Common themes in this literature stress that state policing is political; it is overdetermined; it is problematic; it is work done by people who have a stake in it; and explaining and reforming the police must take these complexities into account.

Foremost, the police must be understood in their historically given contexts. The police are a major nexus between civil society and the state, and they are not neutral:

> to portray public order policing as somehow insulated from political decisions about the choice and mix of state strategies to deal with disorder is not only misleading, it also obscures the chain of relations connecting the state, police and society. By the same token, to treat the police as mere ciphers dutifully implementing whatever strategy has been arrived at by the state elite is equally prone to error and mystification ... . Their pivotal role at the junction of state-society relations leaves them with immense strategic significance (Brewer, et al., 1988, p. 233).

The police are political by definition and in their work, in at least three senses (see also Brodeur, 1983; Reiner, 1992; Shelley, 1990, 1995; Turk, 1982; Weitzer, 1995).

Most narrowly, policing is political because most policing is done or overseen by the state. The state police are a government agency—paid, trained, disciplined and used by the government in its efforts to provide services to the public and meet the demands of the powerful.

Viewed from a broader perspective, policing raises issues which are part of the ebb and flow of political argument and conflict. Their street behavior, their powers, their impartiality or deviance from law and public norms, their claims for autonomy—all are stakes in the normal political process, most visibly so in democratic political systems. The nature and forms of policing increasingly become issues in political debates during democratization, for policing lies at the heart of the debate on the meaning of democracy and democratization. This process is clearly apparent in every country discussed in this book, whether the discussion is about granting the police greater search and seizure powers or the authority to conduct electronic surveillance (in response to new forms of crime and migration), or how policing should be structured to ensure greater transparency, accountability and effectiveness.

Seen most broadly, the police are prizes and actors, objects and subjects in the continuous reproduction of political systems and state-civil society relations. What they do even when they try to act unpolitically, or seek to be neutral, will influence claims, perceptions and realities of legitimacy, effectiveness and justice. As Waddington (1993, p. 159; also 1994) remarks, when, for example, the police control demonstrations, "it is not simply how they go about defending the state from violent disorder that is at issue but the fact that they defend the state at all, even when they do so with utmost restraint." When they act directly political, as they can, they alter the stakes and shapes of political conflicts. It is the police who establish most directly the rights and obligations of citizens by the treatment they impose in encounters; it is the police whose work creates law and order; and it is the police who arrest, who make political dissidents and other challengers to the seats of power disappear, and who shore up the state or challenge its leadership, sometimes violently (Bayley, 1985, pp. 190–201).

In short, policing is part of governance in all its forms (Shearing, 1996).[8] As governance, the nature of policing is a window on the quality of life and rule extant in societies, including its democratic character. Stated differently, democracy is not possible unless the police behave democratically. This argument is most clearly understood in transitional and developing countries for there the roles of state police (e.g., civil, secret, counter-insurgency) and of state-sponsored or supported auxiliaries, e.g., the druzhiny in the former Soviet Union or part-time RUC police in Northern Ireland (Mapstone, 1994) in the protection of the state, its occupants and state-connected elites is obvious and emphasized in practice in often quite brutal ways. The police can exhibit a purely instrumental and repressive political role rarely observed so nakedly in developed democracies. In non-democratic contexts and conditions, the police tend to exercise their normal powers with little administrative oversight or have implicit delegations of authority and permission from the state or the public to repress and kill dangerous, deviant or outcast groups.[9]

In all their actions the police defend the interests of others, but they are also interested actors in their own behalf. The police will not stand by idly while their professional norms, political power and organizational priorities are challenged. They will seek to shape their job, mandate, policies, programs, image and ideologies to suit their own conceptions of what is the problem they are asked to deal with and what needs to be done (Reiner, 1992). Their unavoidable and necessary capacity for discretion and autonomy stems from many sources.

Police officers are street-level bureaucrats (Lipsky, 1980). They have varying levels of discretion in how, when, against whom and for what reasons they wish to exercise their authority and powers. Control of discretion by police officers on the street has been the major goal and preoccupation of police managers, but with varying success (Brewer, 1993; Skolnick & Fyfe, 1993; Steytler, 1990). Attempts to eliminate discretion have proved futile. Discretion can, at best, be controlled (Goldstein, 1977). Much of police research in the USA or England has sought to understand the determinants of discretionary behavior, and it is generally acknowledged that street-level discretion has multiple determinations, including law, organizational priorities and rules, professional norms and codes, direct and indirect public demands, police culture, encounter conditions and behavior, individual personalities and styles of policing. The implications of this argument for implementing change are that altering one or some of the factors which determine discretion is not enough. Change in one set of pressures, for example in legal norms, may be offset by changes in others, for example, by modification in work groups norms and personal styles, or may be overridden by encounter conditions.

Police organizations embody formal and informal rules. It has become conventional wisdom that informal rules, organizational cultures, work groups, or peer relations are as important in determining job behavior as are formal rules (Chan, 1995; Crank, 1997; Klinger, 1997; Tifft, 1975). Many police reform efforts in the US have foundered on not understanding either the complexities of organizational behaviors or the interplay of formal and informal rules. For example, team policing was waylaid by organizational arrangements of power and by informal rules of control (Sherman, Millon, & Kelly, 1973).[10]

Policing takes the specific forms it does for many reasons; the multiple determination of police work has practical implications for control and reform. To give an example, violations of human rights by the police are to be condemned but violations will not be rectified by telling the police to stop their practices. Abuses occur for specific reasons, not forgetfulness or neglect or insufficient training. Reform requires an understanding of the occupational norms of the job and its political nature and relations to society and the state. Merely changing the technical skills of the police (e.g., how to file warrants or use surveillance equipment or investigate a crime) so that they have less need for coercive techniques (e.g., to beat confessions out of suspects) will not work as desired because the reasons for using the new skills and technology are embedded in larger political contexts and in the organizational, occupational and professional norms of the police. Contexts and cultures will direct the uses of new skills and must be changed as well before reform can take hold.

In sum, policing is a complicated job, a set of government organizations and programs, a measure and symbol of human rights and universal values and an exercise of power and politics—all simultaneously. Improvements in policing, hence, need to be approached from different theoretical perspectives—from the knowledge of what they are, how they operate, and why they do what they do, that is from the inside and by a clear view of the nature and determinants of their work. They also need to be approached from their contexts, the societal constraints which shape their design and powers and within which they operate, that is from the outside, from their structured position, and through the analysis of their interconnections to civil society, state and the international system. Policing must be seen as organizational arrangements which link individual police discretion and state and public control and guidance. Lastly, to be judged democratic in their work requires the clarification and imposition of standards of evaluation of policing. Out of the interactions of these sometimes discordant sets of pressures (self-interest, external demands, organizational priorities, democratic norms) arise forms of policing, and such forms will only change in preferred directions if those pressures are properly delineated, understood and manipulated.

A major research effort to help promote and guide changes in policing in Northern Ireland (O'Rawe and Moore, 1997) combines these perspectives. The research team attempted to extract general principles of democratic policing and police reform from the experiences of established democracies (Canada, Belgium, the Netherlands, Australia) and of countries which had recently undergone massive societal and political changes (South Africa, El Salvador, Spain—Basque country and Catalonia, Palestine/Israel). The researchers "sought to explore how [democratic] values are being pursued, and what good practices exist, outside Northern Ireland, and whether this might have any application to the situation here," (p. 11) that is how "transition and change have been introduced elsewhere and what lessons are of interest to Northern Ireland" (p. 17). The report concludes that change is possible but difficult; that international "experience and expertise" can be "invaluable" in developing new forces (p. 247); and that change is dependent upon a combination of the right planning, a willingness to be open and accountable, good leadership, and the right structures in place to manage change ... . Innovation requires input from outside agencies, but police involvement in the process is vital, [as is] commitment to change from within the highest echelons of the police and government authorities ... . There must also be involvement in the process by middle management and by regular officers, who will have to implement the change ... . The process of change cannot be on-off; real progress requires that whatever arrangements are agreed are continually monitored, evaluated. and where necessary re-formulated over time (p. 249).

PRACTICES OF DEMOCRATIC POLICING

Conceptions of Democratic Policing

Democratic policing can be defined easily enough, in the abstract, but, given the importance of cultural, legal and political contexts, good policing is difficult to define operationally, that is by specific practices. It is easy enough, given the existing or emerging international consensus of what constitutes violations of human rights, to know what is bad policing (e.g., torture to extract a confession), but good policing can take many forms (there are many legitimate ways to reach a confession, supposing the accused is guilty). There is no one look to good policing, no one set of practices which are valid in all places and times. What can be specified are general principles and their adaptation to the realities of politics, policing and societal relations in each situation (Goldstein, 1977).[11]

The literature on democratic policing points to two general sources of norms—emergent universal regimes of policing and the writings of scholars—which yield a general consensus (though expressed in widely different jargon) on general principles for democratic policing.

International Regimes

Regimes are rules of behavior which people tend to abide by even though they lack the status of law and there are no formal authorities to enforce them. Regimes tend to be based on common-sense notions of what it takes to work together effectively on common problems and issues; they rise to the level of obligation when agreement, if not consensus, on how to deal with specific questions develops, when they receive "widespread acceptance by the international community" and "have an undeniable moral force, providing practical guidance to States in their conduct" (UN, 1994, p. 2). Regime norms develop informally, often through the interactions of subnational bureaucracies (e.g., Williams and Savona, 1996, pp. 80–107).

Existing international policing regime norms are found in a variety of UN resolutions, conventions, declarations and understandings (Clark, 1994) which find their justifications in international human rights thinking and agreements. These norms have been collected in short form in a booklet developed for the guidance of civilian police elements involved in peacekeeping operations (UN, 1994). The four essential principles are that "force is to be used only when strictly necessary and not excessively," firearms are to be used only exceptionally and as a last resort, arrests must be "made only on legal grounds," and "the detained must be treated humanely" (frontispiece). More specifically, the police must carry out the duties "imposed on them by law," serve the community, protect "all people against illegal acts," and act professionally. They must "respect and protect human dignity and uphold the human rights of all persons." They must not engage in corrupt acts (p. 4). And they must never torture or kill suspects (pp. 21–24). The standards also stress the importance of accurate written records and that the exercise of police powers (arrest, force, detention, investigation) must be constrained by law and regulations.[12]

The operational rules implied by the standards were spelled out in greater detail in the guidelines for the International Police Task Force in Bosnia (UN, 1996, 1).[13] The report states seven principles of democratic policing, and thirty-five ancillary standards derived from them, which are further elaborated in a separate "Notes" (UN, 1996, 2) section. The seven principles are "orientation to democratic principles; adherence to a code of conduct worthy of the public trust; protection of life; public service; a central focus on crimes against people and property; respect for human dignity; and non-discrimination" (p. 2). Officers must work proactively to achieve these standards and must accept the idea that the public, and not the state, is their master.

Bilateral police assistance programs offered by democracies incorporate and justify themselves by domestic and international regime norms. For example, US police aid is offered as part of US foreign policy, which currently has sustainable democracy as one of its major goals. The Agency for International Development (USAID) defines sustainable democracies by four standards: a strong rule of law and respect for human rights, genuine and competitive political processes, the development of a vital civic and politically active society, and transparent and accountable government institutions. Each of these program goals is divided into a number of program approaches. Democracy is achieved when societies exhibit vital and lively activity in each of the four arenas for democratic actions (US, AID, 1997).

Policing assistance is a fundamental part of aid which seeks to promote democracy because any "assistance to help countries democratize their system of government and justice **must** be based on an effective and legitimate law enforcement system" (McHugh, 1994, p. 4; emphasis in original). Without good policing democracy cannot exist. Policing assistance seeks to strengthen the rule of law, ensure legal protection for citizens' rights, enhance the fairness of the administration of justice, improve the timeliness of the administration of justice, and increase citizen pressure for conformity with international human rights standards.[14]

For example, the International Criminal Investigation Training Assistance Program (ICITAP), the agency through which passes much US international policing assistance which seeks long-term institutional development, has been working with Federation of Bosnia and Herzegovina, since February 1997, to implement democratic policing in that country based on a Strategic Plan developed in mutual consultation (Coxey, 1998). The plan incorporates forty-three standards of democratic policing, "virtually all of [which] have been addressed, worked on, met and/or exceeded in the eight months" (p. 18) following the drafting of the plan. Standards include the adoption of a mission statement asserting democratic goals, strict separation of the police from the military and secret police, transparency, written regulations and a citizen complaint system, but also include much more specific requirements, such as ethnic quotas in hiring (e.g., 70 percent Bosnian, 15 percent Croat), individual nametags with picture, or "mandatory reporting and investigation of any and all discharges of police firearms" (p. 16). ICITAP is also assisting in the development of a new six-months basic police training course.

International and bilateral policing regimes, in their substantive and procedural content, mirror conventional conceptions of democracy. The traits required of a democratic police force—respect for human rights; respect for domestic and international law and regulations; fairness in administering police practices; timeliness of services and protections; and recruitment, training and accountability policies which lead to a police culture and organization supportive of those values—are the practices typically emphasized and the values normally taught to police forces in democratic societies (even though they may not always adhere to them).[15]

Scholarly Reflections

There exists, as well, a small sprinkling of writings on democratic policing, some of these in comparative research efforts and settings (Bayley, 1985, 1993, 1994; Goldstein, 1977, 1990; Jones, Newburn, & Smith, 1996; Mawby, 1990; McLaughlin, 1992; Walker, 1993a, 1993b).

Sheptycki (1996, p. 64) defines a democratic polity by three aspects. Processes of political decision-making must be transparent; decision-makers must be accountable and can be removed for failure or malfeasance; decision-making must be rule-governed or oriented to established law. Democratic policing incorporates these three "crucial aspects" into its organization and practices.

Jones et al. (1996), based on their review of democratic theory and British policing, argue that seven thematic values must be institutionalized in democratic policing in some way: "equity, delivery of service, responsiveness, distribution of power, information, redress and participation" (p. 190). They rank these values from most to least important in the order in which they are listed. They conclude that "some form of highly localized forum(s) with powers and resources" (p. 197) is necessary if those values are to be sustained.

The seven values and characteristics listed by Das in his chapter in this book: "rule of law, accountability to the public, transparency of decision-making, popular participation in policing, minimum use of force, creating an organization that facilitates learning of civil and human rights, [and] internal democracy in the organization" (p. 5) reflect a similar notion of what democracy, in terms of general principles, implies for policing practices.

O'Rawe and Moore (1997), rather than focus on specific democratic values derived from the literature and the experience of democratic countries and international regime developments, thought "something more fundamental was required" (p. 11) for judging the nature of policing, namely how representative of society the police are, what training in general democratic principles recruits and practicing officers receive, what legal accountability mechanisms are in place, what democratic accountability mechanisms can be devised by society, and what structures work well. In sum, a democratic police is representative and inclusive demographically of society in the composition of its personnel, especially of female officers; is properly socialized in its working rules to democratic norms; accountable and responsive legally and politically to outside groups and agencies; internally democratic providing all echelons of the police a stake in the operation of the force; effective in the performance of its assigned functions; and committed and open to change.[16]

Marenin (1998) argues that the literature on democratic policing can be summed up in six principles of good policing: effectiveness, efficiency, accessibility, accountability, congruence and general order. The principles incorporate technical, political and justice standards. How these principles would look as operational policies or street behavior will depend on the ingenuity and knowledge of the local police, the state and societal groups. These six principles are mutuallyreinforcing and equally necessary. There may be a need for temporary trade-offs but all principles are the ultimate standards by which democratic performance can be judged.

Efficiency and effectiveness represent technical standards. The police should use the resources given to them by society and the state to achieve stated goals and do so efficiently, that is, without corruption, fraud, waste or by failing to design and implement appropriate administrative structures, policies and programs.

Accessibility and accountability assert the political nature of the police. Policing is a service that must be accessible to all citizens and not be available, by its distribution or priorities of work, to only a few. Policing decisions and justifications must be open and transparent, even though specific policy tasks may require (temporary) secrecy. Policing must be accountable and must be required to explain itself when challenged by political criteria, whether these emanate from publics, interests groups, or state agencies.

Congruence and general order refer to dominant conceptions of justice. Policing must have at least a rough fit with societal values, expectations and demands. Political culture and police culture should be congruent. General order refers to the notion that policing, especially the potential for force, must seek to achieve a public rather than a private good, however defined in a society. The police, and this is the greatest test of democratic policing, must be able to resist and reject unjust demands; they "must be prepared to refuse to apply laws which violate human rights" (O'Rawe and Moore, 1997, p. 223).

Operationalizing Democracy in Policing: Rules, Needs, Standards

The Origins of Rules for Working

The alternatives available for the police on the street are extensive. At the street level, the police must decide what individuals, groups and acts to pay attention to, and they must decide when, how and for how long to apply police power. In a modified form, the same set of decisions faces the organization as a whole. The dilemma is always to whom should the police listen, and what kind of responsiveness, what mix of sources and pressures, is democratic policing.

The sources for demands on police work and rules for working are numerous and include law, professional standards, occupational lore, cultural imagery, organizational regulations, community demands from below, political demands from above and the conscience of officers. All sources specify particular forms of behavior as appropriate, the demands from different sources will be in conflict, officers will be exposed to cross-pressures, individual officers and agencies must choose which demands (and which sources) to grant priority and which to reject as inappropriate, illegal or unjust.

Law provides the foundation for the authority of the police. Professional standards state the accepted norms of ethical and effective behavior. Occupational lore embodies the collective practical wisdom of experience passed on in formal teaching and on-the-job socialization. Cultural imagery extant in society in which the police partake often helps create stereotypes which label individuals and groups as suspect, dangerous and deviant; or as likely offenders, potential criminals and decent citizens. Organizational regulations prescribe approved conduct and the means to reward and sanction adherence or deviance from rules of work. Community and political demands reflect the interests, values and biases of groups, albeit with different access to resources. Conscience is the inner convictions of individual officers and their view of themselves. Cross-pressures are normally resolved in styles of policing, in patterned practices which allow individual officers to do their job safely, comfortably, effectively and with some sense of satisfaction and achievement.

Democratizing policing requires a shift in the mix of salient and legitimate sources of demands, and a change in the content of demands. Being responsive to occupational lore, cultural imagery, community and political demands, and conscience pulls discretion toward undemocratic forms of police actions. Being responsive to law, professional standards and organizational regulations signals a movement toward democratic policing. If done effectively (through teaching, recruitment, organizational management, and the supervision of work), the hope is that policing will shift from being responsive to less democratic sources of rules to becoming sensitive to more democratic sources, and that this shift in the mix of sources will affect the substantive content of occupational lore, change cultural imagery, depoliticize demands, subdue conscience and thereby create democratic policing.

A shift in the mix of sources and pressures should bring a shift in content. As occupational lore is replaced by the acceptance of legal and professional norms, or responsiveness to demands of the powerful by the organizational control of discretion, the culture of policing will begin to incorporate different values and operating rules, and thereby help bring into existence forms of democratic policing.

The general goals and processes in this regard are well described in the chapters in the book. All transitional countries have embarked on the following: the creation of new police laws (as a source of pressure); new socialization and teaching schemes which incorporate an emphasis on human rights and practices which support or do not violate them, techniques for the proper use of force, and the disparagement of corruption; and managerial approaches and skills which de-emphasize and depoliticize responsiveness to state (and elite) demands and community pressures, and increase the salience of organizational regulations, legal requirements and norms of conduct in police culture.

The process of creating salient priorities for the exercise of discretion is never-ending and succeeds only if established as a continual process of organizational and personal adjustment and renewal.

Needs and Situations

Policing serves interests and values under the guise of service, law and order. One can examine the dynamics of choices faced and made by the police by analyzing the operative rules for balancing attention and responsiveness in the exercise of police powers in five essential functional areas: the use of force, the definition of order, the extent of services, symbolic representation and politicization.

Force can be used for repression or protection. The police cannot avoid the use of force to protect, for that is part of their mandate, but they cannot abuse their authority either, for that would be repression. The use of force can be resorted to easily, be widely practiced and unrestricted by organizational or legal guidelines, or it can be extremely narrow in application, rarely used and tightly constrained by law and regulations. Democratic policing requires a balance between the discretionary and arbitrary use of force and the aversion of the community to coercive resolutions of conflicts. The democratic balance in policing is the use of force governed by rules which stress its limited use which may escalate only in response to situational needs for protection and effectiveness.

The chapters in the book make clear the crucial importance of transforming policing practices in which force served the interests of the authoritarian state and the police into a force which is constrained, e.g, in Estonia, Poland or Russia. That is not an easy transformation because force supported by the authority of role and status (symbolized in the uniform and title) is an easy habit to acquire and the ultimate temptation for corruption.

The police can reproduce a social order which reflects the consent of the policed or their work can impose an order on people which the public neither desires nor benefits from. Social order can reflect and incorporate the interests and demands of specific groups to be protected in their persons, properties and values, or order can be the demand for maximum individual and group liberty unconstrained by altruism or conformity. Order can be stultifying and enforce conformity to dominant norms or it can be the joyful celebration of individual and group deviance from societal norms. The balance of democratic policing requires attention to real problems and instabilities of routines which would undermine individual and collective security but also requires maximum attention to the possibility for freedom.

The inertia of police culture, especially if the police organization must retain members who worked as police during discredited regimes to have any continuity at all in the provision of policing (such as in Estonia, Hungary or Russia), tends toward the promotion of order rather than rights and freedoms. This professional outlook imbues all police forces anywhere. But in transitional societies, in which demands for rights and freedom (which the police denied citizens under previous regimes) were precisely the reasons for agitation to overthrow the existing regime, this professional tendency is much more visible, controversial and harder to overcome.

Associated with the inertia of police culture and supported by popular beliefs is the tendency of the police to stereotype marginal groups as sources of deviance, crime and disruption. This is as true in Austria or the Netherlands as it is in Hungary, Macedonia or Russia (e.g., Gypsies or the various ethnic mafias). The opening of borders, the influx of organized crime, drugs, culturally diverse groups, in- and through-migration (e.g., in Poland), all combine to create public demands to do something. The professional response prompted by the police culture is to identify and pursue distinct minorities, that is, to treat groups and individuals unequally under law and regulations, a profound violation of democratic norms and a reproduction of a discriminatory social order.

The extent of services can be non-existent, drawn narrowly and granted, as if a gift from masters to supplicants, or services may be widely and legitimately available, on demand, by the public from their state and its agencies. At one pole, the police do absolutely nothing for the public but only serve themselves and their state masters; at the other pole, policing is available and responds to the demands for protection, security and service of all. Democratic policing embodies a balancing of services and demands. The police must find a place between the master and servant role which does not undermine their power to insist on the limited range of the skills and services they can provide yet also does not draw the definition of their roles and functions too narrowly and mainly to suit themselves. The police need to meet as many demands by all as their resources, and the specific nature of their resources, allow them to.

This often requires a profound redefinition or limitation of policing powers. This is an ongoing discussion in politics and in policing. It can be seen in case law and police legislation (e.g., PACE in England, see Chapter 10 of this book by Gregory Durston) or the restructuring of police into fewer agencies (such as in the Netherlands, see chapter 11 by van de Meeberg and Aronowitz), but is most apparent in mixed and emerging democracies. The question "what is policing for?" and what powers, authority and resources does policing need, these are fundamental issues in redesigning effective and legitimate police structures and functions. The thrust to community policing, which appears in the rhetoric of the administrators of most changing policing systems (e.g., in Croatia or Hungary), is precisely the search for new roles and meanings for the police. If implemented as the rhetoric defines it, community policing would be a profoundly democratic form of policing.

Symbolic representation refers to the rhetoric, symbols and rituals used by the police to justify their actions and inactions (Manning, 1979). The police can present themselves as agents of the state, as the coercive arm of state authority, or they can present themselves as representatives of law and of the communities in which they work. Democratic policing is the symbolic representation of policing as a service to the state and to society, and this claim has been pursued largely by presenting policing as a professional occupation, requiring an extensive body of knowledge and autonomy from political interference but balanced by responsiveness to community demands and needs.

Politicization refers to the extent to which the police take their marching orders and tailor their work to serve the interests of a few or the many. Politicization defines whether the police serve particularistic or universalistic interests. The police can be controlled by political power and be made to serve at the discretion and in the interests of the powerful or the police can seek to achieve universalistic goals by the uses of their resources and power. Democratic policing reflects a balancing of particularistic and universalistic demands.

Standards and Autonomy

The minimum structural and cultural requirement for democratic policing is semi-autonomy. Some capacity for discretion based on expertise, skills and experience but constrained by responsiveness to law, state and civic society demands is required.

Second, the police must recognize and accept that their job requires being responsive to conflicting demands, including their own views of what needs to be done. They must see the choices that they have and make democratic ones. Most fundamentally, such beliefs and the skills to make democratic decisions must filter into formal and informal police cultures.

This last requirement imposes the most difficult burden on democratizing police forces. Having been just liberated from the yoke of state and elite control they must now voluntarily submit themselves to external control of a different form and see the limits of their autonomy, not an easy struggle.

LESSONS

The case studies in this book, changes in the nature of policing, the development of international regimes and the increasing theoretical expansion of the police literature all point to similar lessons on how to institutionalize democratic policing.

First and foremost is the need to teach the police how to be political. Since policing is not and cannot be apolitical, it must be political, but in an appropriate way. Policing cannot be made apolitical but it can be depoliticized. Policing can be removed from partisan and particularistic control by subservience to law, organizational regulations, professional norms and a democratic police culture.

Second, democratic policing can take many forms in street work organization and relations to governance. Policing is becoming increasingly pluralistic (Bayley & Shearing, 1996) in forms and roles. Which of these forms is democratic will depend on how policing is done but also on the political and cultural environment in which policing occurs. Policing supports democracy, but democracy, in turn, must constrain the tendency of the police toward excessive claims for autonomy, and all its potential negative consequences (e.g., abuse of force, corruption, apathy in work).

Third, as all the chapters in the book note, there will be much advice offered from outside the country—through international organizations, states, groups and individuals—on how best to proceed toward democratic policing. (As noted by Lever and van der Spuy in Chapter 8, South Africa alone is engaged in fifty-three foreign-assisted policing reform projects.) That advice will come whether a country wants it or not, and if accepted will come with strings reflecting the particular orientation toward democratizing society and policing of the donors and advisors. In cases of bilateral aid, finances will be tied to expected reforms. In the case of aid by international organizations, groups or individuals who support change for humanitarian or professional reasons, criticism will surely follow if advice is not accepted and implemented. In sum, the effort to democratize policing will be public and commented on within and without the country. The luxury of claiming secrecy or sovereignty to support inactions or reforms is gone. Aid and advice, even when desired by the recipient country, as is most often the case, does not come for free.

Fourth, training for democratic policing must be done within the particular contexts of policing and be practically oriented. One cannot teach abstract democracy to the police, but one can teach rules for working which are democratic within their given contexts.

Fifth, the practicalities of democratic policing have to be stated and measurable. Standards of democratic policing must be spelled out as operational rules and measures of accountability. Principles of policing embodied in regimes and scholarly reflections will mean little to police on the beat or in the station house if they are not stated as doable work for which officers will be held accountable. Reforms will mean little if they do not reach the police culture, and they cannot reach the culture unless sanctions and incentives for change are meaningful and salient within the police officer's working world.

Lastly, mechanism for accountability in practical democratic policing must be in place. Democratic policing is not self-enforcing. It can be implemented by standards, training and expectations of work, but it must also be supervised and disciplined to ensure that officers carry teaching into practice.[17]

NOTES

[1] This issue remains one of the most difficult (and unexamined) problems in implementing community policing—the police must abide by law but they also must be responsive to community preferences. The two demands may clash in US cities; for example, when the police are expected to closely observe black males walking through white neighborhoods, or vigilante groups drive out registered sex offenders from a community.

[2] Such ambiguities are by design. Chevigny (1995, p. 251), argues that role confusion continues "because it represents a basic tension in governance between order and liberty, a tension that government does not really wish to resolve, but rather wishes to leave ambiguous so that the police can make choices as the situation changes."

[3] The logic of adapting principles to democratic policing is no different than the logic of democratization. For example, the principle that decision-makers be accountable supports the policy of elections as a form of public judgment and expression of preferences which, in turn, requires policies on the conduct of elections, party formation, voting procedures, and counting schemes. It would be difficult to argue, though, and impossible to persuade, that democracy, accountability and elections require two parties only, secret machine ballots, limited campaign periods, public financing, and that anything else (more than two parties, open voting, longer campaigns, and private financing) could not be democratic

politics. The principle of accountability remains the same, even though France, Mexico and India differ in how such principles are institutionalized.

4   For example, Seidman and Seidman (1994), in their interesting and often innovative analysis of the role of law and the state in promoting development and the failures of policy efforts, only mention the police as incidental *accouterments* of state power despite the fact that their argument centers on law and its enforcement. In their scheme and explanation, the bureaucratic bourgeoisie diverts resources and power to itself, almost as if there were no impact on its fortunes by how the coercive instruments of ruling are used.

5   There are some notable exceptions. For example, the ongoing work of the Washington Office on Latin America (WOLA), in studies of the police in Central America and Haiti, have shown an increasingly sophisticated grasp of the nature of policing, as well of traditional human rights rhetoric and policy (e.g., WOLA, 1995, 1998).

6   It is a historical oddity how infrequently the police have attempted to stage coups and take over a state. In some countries, such as in Latin America where the police and the military are closely interwoven, distinguishing police and military actions and coups may be impossible.

7   This argument supports King's assertion (1986, p. 82) that examining the roles of the police in political development or transitions to democracy may require a rethinking of Weber's conception of the state as the locus of legitimized coercion. He argues, in critiquing Offe and Habermas, that Weber's conception (which is the standard conventional definition of the state in positivist political science)

> may overestimate the necessity for legitimation and normative commitment to the existing order for it to function effectively. The instruments of surveillance and control available to the apparatuses of the 'secret state,' and the dependency of labour and the poor on either the labour market or the state's social services payments for daily comfort and sustenance, provide a compelling framework for individual compliance to state authority, largely irrespective of the levels of normative commitment to the existing order.

8   By governance Shearing means the relations of state and civic society, more specifically, the manner in which "state and non-state resources are networked," (1996, p. 286) including the various forms of social ordering labeled policing.

9   Chevigny (1995, pp. 145–179), in his discussion of police violence in Sao Paulo, Brazil, argues that much of the violence is a form of delegated vigilantism, the police being expected by the public and the state to use their discretion to deal with the violent and the deviant, for the formal criminal justice system will not punish the dangerous and will avoid the rich and powerful.

10 Yet middle-level managers can also be a source of innovation. For example, much of the community policing movement in the US, though intellectualized and advocated by reformers and progressive police chiefs, is designed, supported and carried out by all echelons in the police organizations. COP is a largely police-created innovation, strategy of policing and set of policies.

11 Two other forms of international learning and teaching are possible: the transfer of specific technical skills and the transmission of operational policies which are based on social science knowledge or practical experience. How to conduct a forensic investigation or run a buy-and-bust drug arrest are examples of the first; managerial policies to set up a personnel system or to establish an internal affairs unit (supposing that is the decision) are examples of the second. But in neither case is it easy to teach or transfer "proven" policies from one setting to another, unless the societal and cultural conditions which make policing effective in one situation or country are somewhat similar in the recipient country.

For example, the Japanese systems of *kobans* have been accepted in Singapore or in San Francisco (there in a modified form, mainly as an information location for tourists). The principle of dispersing police officers into the community to be available for contacts, which supports the *koban* institution, also underlies much of the decentralization thrust of community-oriented policing.

The main problem for teaching and learning through comparative and international examples is the contextual nature of policing, which will not be changed by reforms of particular aspects of policing alone.

[12] These principles are stated in a slightly different form from the nine admonitions in the Code of Conduct for Law Enforcement Officers adopted by the General Assembly in 1979, and include standards drawn from a number of other relevant UN human rights resolutions.

[13] The guidelines were written initially by a member of the Chicago Police Department, a lawyer on loan to the International Police Transitional Force (IPTF); elaborated by an American academic based on his observations of the policing situation in Bosnia; and spelled out as operational standards by a member of the Police Executive Research Forum (PERF) in Washington, DC (Interview, September 1997).

[14] The rule of law and respect for human rights have instrumental utility for the US as well. Both values undergird the effectiveness and legitimacy of all criminal justice institutions and are "essential to fighting terrorism, counterfeiting, drug trafficking, money-laundering, and refugee flows" (US, AID, 1997, pp. 2–9).

[15] Though one could argue that universalism in democratic or policing norms is not yet achieved or is biased towards "Western" views of the rights of individuals, communities, and states, there is more overlap than diversity. Even critics of Western democracy notions who stress the cultural specificity and distinctiveness of justice norms in non-Western civilizations would not argue that police torture, or arbitrariness, or corruption is condoned by culturally specific standards of justice. And countries which deviate, such as Israel, which formally use and legally justify mild forms of torture of prisoners as necessary for the state's and society's survival in a hostile world, are subject to universal censure.

[16] Reiner's (1992, p. 221) definition of democratic policing similarly stresses general principles rather than specific norms of performance.

> By [democratic policing] I mean that [the police] respect due process rights, do not discriminate unjustifiably in enforcement practices, and follow priorities which are in line with popular sentiment where this is clear, or which discreetly balance contending priorities in a divided community.

[17] This is the conclusion arrived at by officials in the International Criminal Investigative Training Assistance Program (ICITAP), a unit in the US Department of Justice, from their numerous efforts to help "stand up" (their jargon for establishing) or reform the policing institutions of other countries. Reforms which do not include a strong Office of Professional Responsibility will fail (Interviews, September 1997; also, ICITAP, 1997).

REFERENCES

Anderson, M. (1989). *Policing the world: Interpol and the politics of international police cooperation*. Oxford: Clarendon Press.

Anderson, M. & den Boer, M. (Eds.). (1994). *Policing across national boundaries*. London: Pinter Publishers.

Bayley, D. H. (1969). *The police and political development in India*. Princeton: Princeton University Press.

———. (1975). The police and political development in Europe. In C. Tilly (Ed.), *The formation of national states in Western Europe*. Princeton: Princeton University Press.

———. (1985). *Patterns of policing*. New Brunswick: Rutgers University Press.

———. (1993). A foreign policy for democratic policing. Paper, American Society of Criminology, Phoenix.

———. *(1994). Police for the future*. New York: Oxford University Press.

Bayley, D. H. & Shearing, C. D. (1996). The future of policing. *Law and Society Review*, 30, 585–606.

Benyon, J., Turnbull, L., Willis, A., Woodward, R., & Beck, A. (1993). *Police co-operation in Europe: An investigation*. Leicester: University of Leicester, Centre for the Study of Public Order.

Brewer, J. D. (1993). Re-educating the South African police: Comparative lessons. In M. L. Mathews, P. B. Heymann, & A.S. Mathews (Eds.). *Policing the conflict in South Africa* (pp. 194–209). Gainesville: University Press of Florida.

Brewer, J. D., Guelke, A., Hume, I., Moxon-Browne, E., & Wilford, R. (1988). *The police, public order and the state*. New York: St. Martin's Press.

Brodeur, J.-P. (1983). High policing and low policing: Remarks about the policing of political activities. *Social problems, 30*, 507–520

Brogden, M. (1982). *The police: Autonomy and consent*. New York: Academic Press.

Brogden, M. & Shearing, C. (1993). *Policing for a new South Africa*. London: Routledge.

Chan, J. B. L. (1997). *Changing police culture: Policing in a multicultural society*. Cambridge: Cambridge University Press.

Chevigny, P. (1995). *Edge of the knife: Police violence in the Americas*. New York: The New Press.

Clark, R. S. (1994). *The United Nations crime prevention and criminal justice program. Formulation of standards and efforts at their implementation*. Philadelphia: University of Pennsylvania Press.

Coxey, G. (1998). The new police of Sarajevo. *Crime & Justice International. 14*, 13–18.

Crank, J. (1997). *Understanding police culture*. Cincinnati: Anderson.

Das, D. K. (1991). Comparative police studies: An assessment. *Police Studies, 14*, 23–35.

———. (Ed.). (1994). *Police practices: An international review*. Metuchen, NJ: The Scarecrow Press.

Diamond, L. (1992). Promoting democracy. *Foreign Policy, 87*, 25–46.

———. (1994). Rethinking civil society: Towards democratic consolidation. *Journal of Democracy, 5*, 4–17.

———. (1995a). Democracy and economic reform: Tensions, compatibilities, and strategies of reconciliation. In E. Lazear (Ed.), *Economic transition in Eastern Europe and Russia: Realities of reform* (pp. 107–158). Stanford: Hoover Institution Press.

———. (1995b). Promoting democracy in Africa. In J. Harbeson & D. Rothchild (Eds.), *Africa in world politics* (2nd ed.). Boulder: Westview Press.

———. (1995c). *Promoting democracy in the 1990s: Actors and instruments, issues and imperatives. A Report to the Carnegie Commission on Preventing Deadly Conflict*. New York: Carnegie Corporation of New York.

Diamond, L., Linz, J., & Lipset, S. M. (Eds.). (1995). *Politics in developing countries: Comparing experiences with democracy* (2nd ed.). Boulder: Lynne Rienner Publishers.

Enloe, C. (1980). *Police, military and ethnicity: The foundations of state power*. New Brunswick: Transaction Books.

Fijnaut, C. (Ed.). (1993). *The internationalization of police co-operation in Western Europe*. Deventer: Kluwer Law and Taxation Publishers.

Goldstein, H. (1977). *Policing a free society*. Cambridge: Ballinger.

———. (1990). *Problem-oriented policing*. New York: McGraw-Hill.

Hebenton, B. & Thomas, T. (1995). *Policing Europe: Co-operation, conflict and control*. New York: St. Martin's Press.

Hills, A. (1996). Toward a critique of policing and national development in Africa. *The Journal of Modern African Studies, 34*, 271–291.

Holiday, D. & Stanley, W. (l993). Building the peace: Preliminary lessons from El Salvador. *Journal of International Affairs, 46*, 415–438.

ICITAP (1997). *Annual report of organizational development and training activities, 1996*. Mimeo. Washington, DC: ICITAP.

Johnston, L. (1992). *The rebirth of private policing*. London: Routledge.

Jones, T., Newburn, T., & Smith, T. J. (1996). Policing and the idea of democracy. *British Journal of Criminology, 36*, 182–198.

King, R. (1986). *The state in modern society*. Chatham, NJ: Chatham House Publishers.

Klinger, D. A. (1997). Negotiating order in patrol work: An ecological theory of police response to deviance. *Criminology, 35*, 277–306.

Lipsky, M. (1980). *Street-level bureaucracy*. New York: Russell Sage.

McHugh, H. S. (1994). *Key issues in police training: Lessons learned from USAID experience*. Washington, DC: USAID, Center for Development Information and Evaluation, Research and Reference Services Project, PN-ABY-304.

McLaughlin, E. (1992). The democratic deficit: European Union and the accountability of the British police. *British Journal of Criminology, 32*, 473–487.

Manning, P. K. (1979). *Police work: The social organization of policing*. Cambridge, MA: The MIT Press.

Mapstone, R. (1994). *Policing a divided society: A study of part-time policing in Northern Ireland.* Aldershot: Avebury.

Marenin, O. (1982). Parking tickets and class repression: The concept of policing in critical theories of criminal justice. *Contemporary Crises, 6,* 241–266.

———. (1985). Review essay: Police performance and state rule: Control and autonomy in the exercise of coercion. *Comparative Politics, 18,* 101–122.

———. (1990). The police and the coercive nature of the state. In E. S. Greenberg & T. F. Mayer (Eds.), *Changes in the state: Causes and consequences* (pp. 115–130). Newbury Park: Sage.

———. (1998). The goal of democracy in international police assistance programs. *Policing: An International Journal of Management and Policy.*

———. (1992). Policing the last frontier: Visions of social order and the development of the Village Public Safety Officer program in Alaska. *Policing and Society, 2,* 273–291.

Mawby, R. I. (1990). *Comparative policing issues.* London: Unwin Hyman.

O'Rawe, M. & Moore, L. (1997). *Human rights on duty. Principles for better policing—International lessons for Northern Ireland.* Belfast: Committee for the Administration of Justice.

Potholm, C. P. (1967). The multiple roles of the police as seen in the African context. *The Journal of Developing Areas, 3,* 139–158.

Reiner, R. (1992). *The politics of the police* (2nd ed.). Toronto: University of Toronto Press.

Remmer, K. L. (1995). New theoretical perspectives on democratization. *Comparative Politics, 28,* 103–122.

Seidman, A. & Seidman, R. (1994). *State and law in the development process: Problem-solving and institutional change in the Third World.* New York: St. Martin's Press.

Shearing, C. (1992). The relationship between public and private policing. In M. Tonry & N. Morris (Eds.). *Modern policing.* Chicago: University of Chicago Press.

———. (1996). Reinventing policing: Policing as governance. In O. Marenin (Ed.), *Policing change, changing police: International perspectives* (pp. 309–330). New York: Garland.

Shearing, C. D. & Stenning, P. C. (Eds.) (1987). *Private policing.* Newbury Park: Sage.

Shelley, L. I. (1990). The Soviet *militsiia:* Agents of political and social control. *Policing and Society, 1,* 39–56.

Shelley, L. I. (1995). *Policing Soviet society: The evolution of state control.* New York: Routledge.

Shelley, L. & Vigh, J. (Eds.). (1995). *Social changes, crime and the police.* Chur: Harwood Academic Publishers.

Sheptycki, J. W. E. (1995). Transnational policing and the makings of a postmodern state. *British Journal of Criminology, 35,* 613–635.

———. (1996). Law enforcement, justice and democracy in the transnational arena: Reflections on the war on drugs. *International Journal of the Sociology of Law, 24,* 61–75.

Sherman, L., Milton, C., & Kelly, T. (1973). *Team policing: Seven case studies.* Washington, DC: The Police Foundation.

Skolnick, J. & Bayley, D. H. (1988). *Community policing: Issues and practices around the world.* Cambridge, MA: ABT Associates.

Skolnick, J. H. & Fyfe, J. J. (1993). *Above the law: Police and the excessive use of force.* New York: The Free Press.

Steytler, N. (1990). Policing political opponents: Death squads and cop culture. In D. Hansson & D. van Zyl Smit (Eds.), *Towards justice/Crime and state control in South Africa* (pp. 106–134). Oxford: Oxford University Press.

Tifft, L. (1975). Control systems, social bases of power and power exercise in police organizations. *Journal of Police Science and Administration, 3,* 66–76.

Tilly, C. (1990). *Coercion, capital and European states, AD 990–1990.* Cambridge, MA: B. Blackwell.

Turk, A. T. (1982). *Political criminality: The defiance and defense of authority.* Beverly Hills: Sage.

United Nations (1994). *United Nations criminal justice standards for peace-keeping police.* Handbook prepared by the Crime Prevention and Criminal Justice Branch. Vienna: United Nations Office at Vienna.

United Nations (1996). *The United Nations and crime prevention: Seeking security and justice for all.* New York: UN, Sales No. E.96.IV.9.

US, Agency for International Development. (1997). *Agency performance report 1996.* Washington, DC: Author.

Waddington, P. A. J. (1993). Public order policing in Britain. In M. L. Mathews, P. B. Heymann & A. S. Mathews (Eds.), *Policing the conflict in South Africa* (pp. 133–160). Gainesville: University Press of Florida.

———. (1994). *Liberty and order. Public order policing in a capital city*. London: UCL Press.

Walker, N. (1993a). The accountability of European police institutions. *European Journal of Policy and Research, 1*, 34–52.

———. (1993b). The international dimension. In R. Reiner & S. Spencer (Eds.), *Accountable policing: Effectiveness, empowerment and equity* (pp. 113–171). London: Institute for Public Policy Research.

———. (1994). European integration and European policing: A complex relationship. In M. Anderson & M. den Boer (Eds.), *Policing across national boundaries* (pp. 22–55). London: Pinter.

———. (1996). Policing the European Union: The politics of transition. In O. Marenin (Ed.), *Policing change, changing police: International perspectives* (pp. 251–283). New York: Garland.

Washington Office on Latin America (WOLA). (1995). *Demilitarizing public order. The international community, police reform and human rights in Central America and Haiti*. Washington, DC: Author.

———. (1998). *Can Haiti's police reforms be sustained?* Washington, DC: National Coalition for Haitian Rights, Washington Office on Latin America.

Weitzer, R. (1995). *Policing under fire: Ethnic conflict and police-community relations in Northern Ireland*. Albany: State University of New York Press.

Williams, P. & Savona, E. U. (Eds.), (1996). *The United Nations and transnational organized crime*. London: Frank Cass.

Section IV
Country Studies

13

Policing Macedonia

DIME GUREV

BACKGROUND

The Republic of Macedonia was constituted as an independent sovereign unitary state in 1991. The country is located in the Central Balkans; its surface area encompasses 20,713 square kilometers and its population is 1,936,000 strong. The percentage of ethnic Macedonians is 66.5 percent; other ethnic groups include Albanians (22.9 %), Turks (4 %), Serbs (2 %), Gypsies and others. All groups enjoy the same human and civil rights and their equality is guaranteed by the Constitution of the country.

The Republic of Macedonia is a multiparty, parliamentary, democratic country, with a market-oriented economy. However, the potential worsening of inter-ethnic relations and the possible social disturbances that may result from the ongoing process of the privatization of the economy may lead to disruptive social tendencies and the destabilization of the country. In addition, the immediate international political environment of the country is not very favorable to stability and development. The unstable political situation in the Balkans is well known.

Worsening economic conditions are caused, to a large extent, by the Greek blockade on the southern border, as well as by UN sanctions against the republics of the former Yugoslavia. Both events have led to many difficulties in placing the products of the country on the world market. The quantity of industrial production has also decreased (for example, in 1994 there was a decrease of 10.5 percent from 1993).

Compared to the other post-communist countries from Central and Eastern Europe, the Republic of Macedonia provides a greater degree of protection for the life and property of its inhabitants. Despite a growing crime problem, the Macedonian police have succeeded in preventing the entrance of international criminal organizations into the Republic's territory. The number of police officers in the Republic of Macedonia is approximately at the level of Western European countries, that is about one police officer to every 400 inhabitants.

THE POLICE SYSTEM OF THE REPUBLIC OF MACEDONIA AND THE CURRENT PUBLIC SECURITY SITUATION

The enactment of the Constitution and the establishment of the Republic of Macedonia as an independent country had serious consequences for one of the most important responsibilities of the state, the protection of the constitutional order and of the property and personal safety of the citizens. The effort to strengthen the legal and democratic institutions of the state and the protection of human rights and freedoms immediately raised the issue of the transformation of the security agencies.

The process of transformation began in 1991 and was fully completed after the enactment of new laws on security in March 1995. The basic goals of the new laws were to begin the process of establishing a modern and efficient police force which would adopt a highly professional approach in the exercise of its powers and authority. Structural changes included separating the State Security Service, which is now established as an independent state agency—the Intelligence Agency—from the Ministry of the Interior.

The new laws divide security functions into two separate categories: the first, the protection of life, personal safety and property of citizens, and the protection of the constitutional order, fall under the competence of the Ministry of the Interior, and the second, the intelligence-gathering function, is directly coordinated and performed by the Intelligence Agency.

The competence of the Ministry of the Interior encompasses a wide range of activities related to protection of the life, personal safety and property of the citizens, the prevention of crime in general, the discovery and apprehension of perpetrators, the prevention of terrorism, illicit drug and arms traffic, the securing of public peace and order, safeguarding the state border, the regulation and control of road traffic, and other tasks of a predominantly police nature.

The protection of the constitutional order is to be performed by the Security and Counterintelligence Directorate which functions independently, but is incorporated into the Ministry of the Interior's organizational structure. The Directorate is authorized to perform the activities necessary for the protection of the state from espionage, terrorism and other activities directed toward the endangerment or violent destruction of the democratic institutions established in the Constitution, as well as the prevention and uncovering of serious forms of organized crime.

The new laws also established mechanisms and control procedures over the work of these agencies and classified the authority and competence of agencies with control powers. According to the Constitution, the administration is directly responsible for the supervision of the work of the Ministry of the Interior, while the Parliament indirectly controls the work of the Ministry through its supervision of the work of Government as a whole. The Director of the Security and Counterintelligence Agency is responsible to the Minister of the Interior and to the executive, that is to the organs which propose and confirm his or her appointment. Supervision over the work of the Directorate is also exercised by a committee appointed by the Parliament of the Republic of Macedonia.

Control over the work of the Counterintelligence Agency is similar to that exercised over the other security agencies. Control powers are held by three different bodies—the President of the Republic, the executive administration and the Parliament, each of which performs its control powers in accordance with its constitutional competence.

The Ministry of the Interior is responsible for the implementation of laws and regulations which concern internal security. Seven organizational units of the Ministry are responsible for public security matters.

The prevention and suppression of crime is the responsibility of the Criminalistic Police, which consists of four organizational units: economic crime, ordinary crime, illicit trafficking and corruption, and the unit for cooperation with Interpol.

The Uniformed Police are responsible for the maintenance of public peace and order, the protection of citizens, and the control and regulation of road traffic.

The territory of the Republic of Macedonia is divided into eleven departments, and at least one police station is responsible for the area of each municipal community. Depending to some degree on the nature of duties performed by each, police stations are responsible for the control of border crossings, the safety of road traffic, the protection of VIPs and the security of diplomatic and consular offices.

The Minister of the Interior is the head of the Ministry. He or she, according to the Constitution, must have been a civilian for at least three years prior to assignment at this post. The Minister of the Interior speaks for the Ministry, is responsible for the organization, supervision, and the lawful and efficient implementation of the appropriate laws and regulations which fall under the Ministry's authority and competence and is legally authorized to enact regulations related to the work of the Ministry.

The Minister is responsible to the executive and the Parliament for his or her work, for the functioning of the Ministry of the Interior as a whole, as well as for the security situation in the areas which are under the competence of the Ministry. The Minister is assisted by other high officials responsible for the coordination of the work of the separate organizational units within the Ministry.

THE COMPETENCE OF THE POLICE IN SITUATIONS OF BREACHES OF PUBLIC PEACE AND ORDER

The political and economic changes which occurred during the past few years, followed by extremely inconvenient circumstances and foreign influences, exerted a negative and destabilizing effect upon security conditions in the Republic of Macedonia.

The stability of public peace and order was particularly affected, in a negative sense, by manifestations of undesirable behavior at a number of public protests, strikes, demonstrations and similar expressions of discontent. These situations were often followed by provocative behavior, with the intent to cause conflicts, to provoke clashes with the police in order to use these circumstances for achieving political aims and to seek the destabilization of security in general.

In parallel with the process of self-transformation in order to adapt its organization and work to the ongoing social and political changes, the Ministry undertook a number of measures and activities to maintain public peace and order and to protect the security of the state, specifically by allocating human resources to specific security problems.

In the last couple of years there have been only a few serious breaches of public peace and order, mostly in 1992, and these came as a result of the organized resistance by the Albanian population (which mostly inhabits the western part of Macedonia) to state reforms and programs. Resistance was manifested by intentionally provocative behavior which sought to cause mass conflict, by resistance to measures and activities undertaken by the police and attempts to make the police physically unable to perform their duties.

To prevent further conflicts and their societal consequences, the Ministry of the Interior continuously undertakes coordinated efforts to keep the peace. At the same time, in common with the ongoing process of democratization in all areas of social life, the police are taking measures to transform themselves. The goals of this process include raising the ethical and professional standards of the police, eliminating any kind of political influence, avoiding the use of unnecessary force, strengthening the confidence of citizens in the police and ensuring that the performance of security duties is in accordance with constitutional and other legal regulations. These measures and the activities of the police are constantly being observed and analyzed to ensure that they are adapted to democratic standards of police practice.

The main focus of police activities is on the prevention of any serious disturbances of public peace and order and, subsequently, on the measures for reestablishing peace and order after any disturbances. In order to maintain control and prevent disturbances, the police gather intelligence about the likelihood of mass public gatherings, protest meetings, demonstrations and other manifestations of discontent, demonstrations, etc. The police also are present during these occasions to estimate their security aspects and to have a detailed picture of the existing conditions so that they can make adequate plans for the maintenance of public peace and order.

In conformity with such preventative tasks, the police provide complete physical and operative coverage of the area where the gatherings take place and ensure that traffic continues unimpeded. The police also contact the organizers of public gatherings to remind them of their duties to provide watchmen services.

In the case when a situation tends to become serious and complicated, the first steps of the police are directed toward identifying, calming and separating from the crowd suspected instigators and other participants likely to cause conflicts. If these measures are not sufficient and the possibility for more serious breaches of public peace and order is likely, preparations are made for the timely deployment of special intervention units.

During efforts to reestablish public peace and order, and depending on the behavior of the crowd, the police may utilize various means: making barriers and placing other obstacles in front of endangered objects and the use of physical force and other means of coercion in order to support the activities of intervention units in charge of extracting and arresting perpetrators. Depending on the seriousness of the disturbance and the aggressiveness of the participants, the application of the following legal means of coercion is available: use of physical force, nightsticks, water cannons and hoses, chemicals, trained dogs, armored vehicles, helicopters and firearms.

After the crowd is dispersed and public peace and order is restored, an analysis of the whole situation is undertaken to ensure that the experience and information gained from these events are used for the improvement of existing contingency plans.

The basic rules of police behavior and actions in circumstances of more serious disturbances of public peace and order are the following: to distinguish among citizens who take part in the protest meetings and those who do not, to cooperate with the organizers and the other participants in the protest gatherings, to engage in negotiations with calm spirits and restraint even in provocative circumstances, to differentiate among the participants and to isolate aggressive groups, and to determine the application of strategic and tactical countermeasures.

ANALYTICAL SURVEY OF THE MORE ORGANIZED AND AGGRESSIVE KINDS OF CRIMINAL BEHAVIOR

Ordinary Crime

Soon after the Republic of Macedonia was constituted as an independent state, crime in general increased enormously, a process very much similar to that noted in other Central and Eastern European countries undergoing a transition to new political and economic structures.

As a result of the economic situation in general, particularly the worsened living standard of the population and a high rate of unemployment, total crime in 1992 increased by 55 percent compared to the previous year. The largest increase (64 %)

occurred in crimes against property. The increase in crimes also affected the efficiency of detecting and investigating crime.

The analysis of crime trends in 1993 and 1994 confirms the positive effects of the activities of the Ministry of the Interior, which resulted in a decrease of total crime by 3.5 percent each for the years 1993 and 1994, mostly in ordinary crime. The analysis of ordinary crime leads to the conclusion that crime prevention methods adopted had their greatest impact on crimes against property. As a result of implementing surveillance measures to control known delinquents in the areas most often affected by this kind of crime, property crime decreased by 7.7 percent by 1993 and 12.4 percent by 1994 compared to 1992. Positive trends were also registered in respect to the percentage of minors as perpetrators of property crime, which fell from 12 percent in 1993 to 9.7 percent in 1994. It is estimated that this positive trend is mostly the result of the effective detection of this sort of crime by the police. According to this analysis, it is expected that a strong emphasis will continue to be placed on prevention methods, particularly by cooperating with educational and pedagogical institutions and by stressing the important role of the family as the basic structure of the society. All these efforts should lead to greater ethical, cultural and athletic achievements of the young population.

In contrast to the trends in property crime, the rate of violent crime has shown a significant increase, mostly because it is very difficult to devise effective methods of prevention. Therefore, the activities of the Ministry of the Interior are focused mainly on the detection of these criminal acts.

The rate of the violent crime which threatens the physical well-being of the population has experienced constant annual increases of 20 to 24 percent. The most common criminal acts in this category are bodily injuries, both minor and more serious ones. The rate of detection in regard to these crimes has risen to about 90 percent as a result of the Ministry's initiatives in this area.

The rate of criminal acts which threaten the people's dignity and morality has not shown any significant fluctuations, and their annual number ranges between 100 to 124. Rapes and lascivious misbehaviors are the most common offenses in this category, with an annual number of 38 to 43 rapes and 20 to 41 lascivious behaviors. The efficiency of detection of these kinds of criminal acts is very high and reaches up to 90 percent.

The crime of robbery has seen a constant increase. The number of robberies was 2.3 times greater in 1994 compared to 1991. The consequences of robberies were more serious as well, compared with the previous years. Considering the fact that this kind of crime causes serious anxiety among the public, the Ministry of the Interior has placed a strong emphasis on fighting these criminal acts, especially since the rate of detection is quite low.

The number of criminal acts of violent behavior had almost doubled by 1994 compared to 1991, but the efficiency of their detection is quite high.

Organized Crime

Changes in the social, political and economic system of the country, as well as the increase in private enterprises, have led to the phenomenon of organized groups who engage in extremely violent and aggressive behavior to enforce the payment of debts or to extort money from owners of private enterprises under the pretext of "protection." These activities resemble and are called, locally, "American racketeering."

These organized criminal groups use threats, blackmail, firearms, physical attacks and other means of extortion in order to force the owners of certain trading enterprises, discotheques, restaurants, etc., to pay them a percentage of their income. This is done usually through middlemen employed as bodyguards. The collection of money is also done through middlemen, while those in charge only appear on the scene in cases when the owners show "disobedience." Individuals who provide "racketeering services" have usually practiced boxing, karate, body building, etc., and their services normally amount to 20 to 30 percent of the total "payment."

Lately, since the phenomenon of not paying debts by owners of private enterprises has become quite common, new firms and detective agencies are being registered whose main occupation is the collection of payments. Their methods of work are very similar to the ones used by organized criminal groups.

The danger that the influence of organized criminal groups will spread to political parties and governmental institutions is very real, bearing in mind the amounts of money these agencies have at their disposal. Equally real is the possibility that these groups will become transformed into more organized types of criminal organizations, such as the "mafia," which then may threaten all sectors of social life. Therefore, in the last several years, the relevant units in the Ministry have paid special attention to the detection of these organized criminal groups, particularly their leaders and assistants, and have applied all available legal measures in order to prevent the spread of this phenomenon.

Illustrative examples of the activities of these criminal groups are the cases of five Macedonian citizens who were kidnapped, detained and blackmailed in order to collect debts. In two other cases, explosives were planted to collect unpaid debts owed by Bulgarian citizens.

According to the Ministry's information section, most of these criminal groups are found in the towns of Skopje, Ohrid and Strumica, while some of the owners of private enterprises located in the eastern part of Macedonia are most likely to maintain contacts with Bulgarian citizens (the so-called "fighters") for their "protection."

In the period from 1991 to 1994, the Ministry of the Interior instituted nineteen criminal charges (sixteen cases of using force for collecting payments and three cases of "protection") against fifty persons (six of whom were Bulgarian citizens). The criminal charges included twenty specific criminal acts (extortion, blackmail, kidnapping, illegal detention, unauthorized possession of firearms, usury, etc.). At the same time, charges were brought for misdemeanors, such as physical attacks and disturbances of the public peace and order.

In 1992, as a result of the Ministry's activities, an organized criminal group in Ohrid, led by a certain Vladimir Ruvinov, was detected and criminal charges were brought against the leaders and members of the group. This group of criminals had close contacts with influential individuals and political parties in Ohrid. The group controlled many tourist sites near Lake Ohrid which all together earned them illegal annual profits of approximately 500,000 German Marks (about $335,000).

A case in Skopje, in 1994, was quite similar. A violent group of forty persons caused significant material damage to the *Hard Rock* discotheque (which is a typical example of the methods of "American racketeering") because the owners refused to accept "protection." In this case, the leaders of the group (Roli Jakupovic and Luan Cavoli) did not succeed in "convincing" the owner to accept their conditions, which included: a count of tickets sold, the splitting of the profit of the discotheque, the leader of the group to be appointed as director of the discotheque, as well as the employment of all the members of the group. Because of this refusal, Jakupovic and Cavoli organized the demolition of the discotheque before it could be opened for the winter season. As in previous cases, criminal charges were brought by the Ministry against the leaders and members of the group.

Illegal drug trafficking in the Republic of Macedonia has experienced a constant increase in the last couple of years. The traffic in heroin is particularly disturbing because of the expansion of this phenomenon at the international level as a result of the interlinked activities of the narco-mafia in Europe and the world as a whole.

In the period from 1990 to 1995, the relevant units of the Ministry detected 231 cases of illicit drug production or trafficking. Criminal charges were brought against 419 suspected perpetrators and 279.5 kilograms of heroin, 64.4 kilograms of raw opium, 53.7 kilograms of marijuana, 39 kilograms of cocaine, 8.2 kilograms of morphine and smaller quantities of other drugs were seized. Compared to prior years, the rate of drug crimes in the last five-year period has increased several fold. There has been an enormous expansion of drug smuggling. This expansion is due to a number of factors: the geographic location of the Republic of Macedonia, the so-called "Balkan route," close family relations of Macedonian citizens with Turkish citizens, the significant number of temporarily employed Macedonian citizens in the Western European countries and the large profits which can be made by illicit drug trafficking. Drugs which are smuggled through the Balkan route to Western Europe originate from the Near and the Middle East. The "Balkan route" runs through Bulgaria, Macedonia, Albania and Italy toward Western Europe. A smaller quantity of heroin smuggled through the territory of the Republic of Macedonia goes to Kosovo province in the former Yugoslavia. Although many activities are being undertaken in order to stop the drug smuggling, perpetrators arrested are most often small-time criminals, the so called "transporters." The chief organizers of international smuggling, "the Bosses," remain free because they work outside Macedonian borders. The estimates are that only very small quantities of smuggled heroin remain in Macedonia, since it is not a consuming area and the number of drug abusers is small.

The illegal production of marijuana has increased in the last few years, by the illegal planting of the seeds of "cannabis sativa," since the climate in Macedonia is favorable for cultivation of these plants. Nevertheless, the estimates are that this illegal production is of local character and intended for domestic use. The same holds true for the traffic in raw opium. Much of this raw opium comes from old storage places where it has been kept, sometimes for a couple of decades, by owners who mostly live in rural areas.

The results achieved in the fight against illicit drug trafficking and abuse are considered quite good, correspond to the level of average European efficiency, and have been recognized as positive by the US Drug Enforcement Agency. In 1991, a special unit staffed with young and educated personnel was established whose main task is the coordination of counterdrug activities for the territory of the Republic of Macedonia as a whole. However, the level of the technical equipment is unsatisfactory and further investments to acquire modern technical means, specially equipped vehicles, communication devices, electronics, etc., are needed. Also, there is a lack of specially equipped garages for the mechanical search of cargo and passenger vehicles which, according to the Ministry's information, are most often used for illicit drug transports.

Keeping in mind the security aspects of the illicit drug trade which, together with illicit arms trafficking, are the most dangerous kinds of criminal activities on the international level, the Ministry of the Interior works intensively to increase its cooperative relations with police forces from all over the world to prevent and fight international crime by joint efforts.

14

Policing Democracy: The Slovenian Experience

JERNEJ VIDETIC

INTRODUCTION

The proper perspective is important if one intends to achieve a correct assessment of a particular subject. In our case, we are interested in the development of democracy in the Republic of Slovenia and the way in which its new political system is being policed in the new political, social and economic environment. When analyzing a newly founded state, an observer would certainly consider primarily its features and characteristics: geography, economy, demography, political structure, culture and traditions, and military-strategic situation.

Slovenia is a country of small size but with a strategic position in the center of Europe. The port of Koper on the Adriatic Sea provides good services for merchants from Asia seeking easy access to the increasingly lucrative markets of Eastern Europe. When the war in the Balkans ends, traffic routes to the Middle East will again open. Two neighboring countries, Austria and Italy, are members of the European Union. Another important factor is that many tourists and other travelers pass through Slovenia.

These economic and infrastructural factors influence the scope and types of crime in Slovenia in many ways. However, the development of crime is somewhat neutralized by the relatively even distribution of the population without any distinctly metropolitan areas. The areas most endangered by crime are the capital city of Ljubljana and the regional centers of Koper, Maribor and Celje.

SLOVENIAN INDEPENDENCE

Although the first Slovene words were already written in the tenth century and the Slovene nation consequently always had a strong cultural identity of its own, the historical periods which followed were marked by the presence of German, Italian, and Hungarian cultural and political influences. For many centuries, Slovenes lived under the rule of the Habsburg family, and compared to the time spent in the Yugoslav state after 1918 it could be said that the "Yugoslav experience" represents only a short era in Slovene national history.

In June of 1991, the Parliament of Slovenia declared Slovenia a sovereign and independent state. With this act, the Slovene people managed to acquire their long desired freedom, the ability to make their own political decisions and the opportunity to create their own destiny. But this was just the beginning of the long and arduous road to democracy, for after the proclamation of independence it was necessary to promulgate a new constitution that would introduce an open and transparent democratic system.

DEMOCRACY AND AUTHORITY

The transition from a socialist self-management system to a democratic society is a complex one. The goal is clear, but how to achieve it is not self-evident. It is similar to searching for something that many generations of scholars—from the period of the Greek philosophers to our own times—have been trying to discover without success.[1] A basic dilemma still remains, based on a paradox which the police must confront as well: "the permanent tension between individual freedom and authority" (Goodwin, 1992, p. 238).

Hannah Arendt in her essay "On Violence" offers a comprehensive definition of the word *authority* from which we can deduce that it is associated with two semantic derivatives, power and domination:

> Authority, which designates the most intangible of these phenomena and, because of this fact, is frequently the occasion for an abuse of language, can be attached to the individual—one can talk about personal authority, for example, among the relations between parents and children or professors and students—or it may even constitute an attribute of institutions, such, as for example, in the case of the senate of Rome (*auctoritas in senatu*) or the hierarchy of the Catholic church (a preacher in a drunken state can still, and validly, grant absolution). Its essential characteristic is that those who must give obedience must give it unconditionally; there is no need at all, in this case, for constraint or persuasion ... Authority can be maintained only as much as the institution or individual from which it emanates respects it. Condescension is the greatest enemy of authority, and laughter is its biggest threat.[2]

Policing is an agency established by the Ministry of the Interior for the enforcement of law and order. I proceed from Weber's definition of legal-rational-bureaucratic societies in which authority and power are attached to offices. Authority is deliberately transferred to the police for the sake of providing security and upholding law and order, or to paraphrase Uglow, "the State has delegated to the police the right to use force in civil-society—no other violence is regarded as legitimate" (Uglow, 1998, p. 11).

This statement unequivocally rejects the totalitarian aim—which is to suppress the civil society. There exist large variations in different political systems when talking about the relationships between "the use of force" and "the suppression of civil rights." In theory, force is used "legitimately" in both democratic and totalitarian systems, the difference being in the ideology that stands behind this coercive force or the who or what the police are protecting: the authoritarian regime or the people? But let us not misunderstand. Repression was applied in totalitarian regimes by the political police and as such they were distinct from the regular police who maintained public order. All the necessary power was conferred to the political branch of the police force for the protection of the authoritarian system, and consequently all other police organs had to follow its directives and thus act according to the commands conveyed. Once single-party rule dissolved and power returned to the people, who in turn elected their representatives in fair elections and a new mandate was given to the government, the foundations were laid for the consolidation of democracy. As Merkel (cited in Slovenian ..., 1994, p. 238) explains,

> The "founding elections" are considered to constitute an important *caesura* with which the phase of the transition to democracy ends, followed by a phase of democratic consolidation ... . How successful this first phase of democratic institutionalism is depends on the one hand upon the problem load inherited from the old authoritarian regime and, on the other hand, upon the immediate international contextual conditions.

Consequently, the real test of democratic rule is not in the executive power[3]—although it is important that old convictions are put aside and replaced by democratic beliefs—but in public control of the state's institutions and in legislation that cannot be misused for undemocratic actions through dubious interpretations of its articles. Only a political system in which the distribution of power is ensured by the constitution, and where a system of checks and balances prevents any one branch from abusing its power, provides the necessary political liberty and the expected level of democracy. Otherwise, a political system could acquire the characteristics of a totalitarian state so well described by Friedrich & Brzezinski (1965, pp. 21–22): (1) An official ideology incorporating a vision of the ideal state, belief in which is compulsory; unorthodoxy is punishable; (2) a single party which is bureaucratic and hierarchical, usually led by one man; (3) a terroristic police; (4) a monopoly of communications; (5) a monopoly of weapons; and, (6) a centrally directed economy.

THE CONFLICT OF VALUES

Words cannot convey the feeling of despair, deprivation, hopelessness and dissolution that affected people in a system in which fundamental human rights were violated and denied on a daily basis. The force behind this social apathy and resignation among the population was the political police who created a network of secret agents constantly eavesdropping, reporting conversations, or provoking persons to criticize the system. This is how fear was spread and the reason people stopped trusting each other. In the end everything converged into the "big lie": an empty promise of a heavenly future in the material world. The Marxist revolution slowly consumed itself until it simply collapsed because nothing could hold it together any longer. This shows how empty and unnatural Marxist ideology was.

Before we seriously tackle the subject of policing, it is important to see how the transformation of the political system affects past values and how difficult it is to promote new values or the values known to Western democracies which had existed in Slovenia before the Communist revolution. The values and beliefs of the Communist elite in Slovenia were Marxist. This is how Reynolds (1994, p. 46) defines the Marxist ideology:

> conflict is the dynamic of all development. In human society conflict occurs between classes, which are groups defined according to their relationship to ownership of the means of production. Human history is the record of the quantitative accumulation of conflicts between two classes to the point where a qualitative leap occurs by the revolutionary overthrow of the previously dominant by the previously subjected one. The victor in its turn becomes faced by a new class opponent. This historical evolution is inevitable. It is moral to act in accordance with, and promote the advance of, the inevitable course of history.

On the other hand, the values that are predominant in Slovenia derive from Western civilization, or as Reynolds would describe it, from the "past interaction of geographical, economic, demographic, racial, and sociopolitical influences" (1994, p. 45). The Slovene community thus historically draws its values from Greek, Judaic and Christian philosophy and ethics. Had the normal evolution of Slovene society not been cut short by a Marxist ideology alien to its tradition (or anyone's tradition),[4] it would certainly have developed, as in any other Western democracy, political institutions and procedures and social processes intended to maintain the inalienable rights and the inherent unique value of the individual human being. People in Slovenia were subjected to a perpetual revolution that was always looking for enemies of the system and was always producing them with the help of the political police. Show trials through which the citizens were induced to believe that the revolution was a never-ending process show the intention of the elite in power to create a revolution in perpetuity.

The consolidation of democracy today thus has to deal with the clash of values that must still come to terms with each other and find a way to coexist in tolerance. Unlike the Czech Republic, where members of the Communist Party were fired from their offices in public institutions after the fall of the authoritarian regime, a similar change of generations in Slovenia was only partial. A small number of Marxist doctrinaires in high-level posts referred to as the "nomenclature" were pensioned off administratively, while members of the middle, most numerous class kept their positions in all sectors of society and are today referred to by some sociologists as the "new elite." This fact influences the pace of democratization in many ways,[5] as well as the shift to a market economy, or as the Solvenian economist Mencinger correctly deduces when comparing economic changes in the new democracies of Eastern Europe:

> the specific means employed to bring about privatization can also in part be attributed to the faith of the new political elite in the supremacy of the market system; they are also dedicated to eliminating political competition by establishing control of the economy. This is confirmed by observing how technical approaches to privatization resemble or differ. The differences, and even more the similarities, indicate that genuine distinctions among the countries—political and social environment, existing institutional frameworks, degree of monetization of the economy, industrial structure, incorporation into the world market, and macroeconomic performance—have been of minor importance. The differences have instead reflected the specific distribution of political power in a particular country and have also been directly or indirectly influenced by the ideas of randomly chosen Western privatizers (cited in Slovenian ..., 1994, p. 412).

The transitional period therefore encounters two major difficulties: widespread social anomie or the absence of norms and moral standards, and a condition of minor disorganization caused by deficient legislation resulting from the newly enforced market-oriented economy and the process of consolidating the democracy.

As far as the pattern of criminality in the actual sociocultural matrix of Slovenian society is concerned, it is evident that the general conditions are both profitable and feasible for white-collar crime (fraud in trade and commerce, crime and corruption in the building trade, thefts in the industrial sector, crime in the ownership transformation of enterprises, tax evasion fraud and forms of internationally organized crime such as the illicit outflow of capital, fraud, corruption, smuggling and other crimes). However, it must be said that the legislation is being updated.

REFORMING THE POLICE

In an address to the parliamentary assembly of the Council of Europe in 1994, Slovenia's president Milan Kucan said, "freedom and the consolidation of democracy in a country undergoing transition are vitally linked to questions of the country's own national and international security." The president's speech was a reflection of the social, political and economic changes that swept across Slovenia, and raised questions about the internal security system and its obsoleteness and the way the police force had to reorganize itself to meet the new challenges created by the transition. To perform new tasks and to cope with new crime patterns, the police force had to undergo a process of renewal. The first measures included the depoliticization of the police (constitutionally prohibiting police officers from activity in any political party), the dismantling of the political police and the formation of an Intelligence Service independent of the Ministry of the Interior. Another positive element was the ability of the political parties to consensually agree that the primary service of the police is, and here I will paraphrase Uglow (1988, p. 5), to "do something about crime and crime rates."

With the imposition of legal limits on the police force, the prevention of direct political interference, and the search for political means and parliamentary supervision to control the extent of the penetration of civil society by the police, it has been possible so far for the police force to meet the expectations of society as it undergoes reconstruction. However, as in any other newly democratic country, the Slovenian police force must still determine its new relationship with the state and the civil society and discover where it stands, how to behave and how to communicate with the public in the new political order.

Confronted with the difficult tasks of meeting the high expectations of the public regarding the style of policing and delivery of police service and of achieving greater efficiency within the police service, the police force is also and simultaneously constantly confronted with the fear of having to cope with public dissatisfaction.[6] At the same time, it is not clear whether the Slovene police should assume a more preemptive role in its fight against crime, despite the opinion of Western police experts that this method does not reduce the number of offenses or affect the offenders. Nevertheless, the proactive or preemptive role of the police should be contemplated carefully, both in surveillance as well as in crime control. The same holds true for the availability and use of physical force (Uglow, 1988, p. 9).

The reorganization of the police force represents an important step towards its more transparent role in the eyes of the public.[7] Slovenia's police force is presently structured as follows: the Ministry at the national level, the Public Security Administration at the regional level, and the Police Station at the district level. The Ministry of the Interior performs police tasks in the domain of public security and tasks of national administration in the field of internal affairs.

In the annual report for the year 1994, the following information was a issued regarding the fight against crime in the Republic of Slovenia:

> there are some 8,000 employees in the Ministry of Internal Affairs, including 2,570 uniformed police officers and 490 investigators who deal directly with crime. An individual operative agent therefore deals with an average of 14.3 criminal cases per year. Some 83% of crimes are investigated independently by uniformed police officers, while the remaining criminal offenses, for which some special investigative knowledge is needed, are handled by investigators.
>
> In addition to dealing with tasks in the area of crime, uniformed police are concerned with the maintenance of public order and peace, they regulate and

control road traffic, and perform tasks related to the protection of national borders.

The Criminal Investigation Service is a specialized service of the Ministry of Internal Affairs in the battle against crime.

[Community policing has been adopted as a policy by the police.] Today in Slovenia, every police station has a certain number of community policing officers, depending on the population, the territorial size of the police station and the crime rate. In the Slovenian police, there are about 332 community policing officers, out of a total of 5,300 uniformed police officers and 700 crime investigators (Mikulan, Globocnik, Kos, & Podvršic, 1997, p. 55).

Due to constitutional changes, legislation had to be modified according to the new democratic standards that, as already stated, are primarily aimed at meeting expectations regarding human rights and the freedom of the individual. In conformity with the new political reality, the basic laws concerning police activities have been prepared and are currently under consideration in the parliament.

The draft proposal of the Public Safety Act or the "Police Law" will define the working domain of the Ministry of the Interior, the functioning of the police forces and the authority given to the police to perform their tasks. Through this law, public safety will become a commodity available to all. It will be formal, rational and under control. It will give citizens the necessary security in the event their rights are violated or the interests of the community are placed in jeopardy. The important feature of this draft law is that it clearly delimits the powers given to police officers.[8] Another new feature is that the means of coercion are clearly specified.[9]

Many voices have been raised lately in the public debate against the use of special methods and means in police work due to the fear that these powers could be misused for political or personal gain because of the inadequacy of public control mechanisms. The public concern is influenced by the memory of the past authoritarian system. This makes it hard for the police to explain that special surveillance is used only to fight crime or as Uglow (1988, p. 98) explains,

> the general justification for all surveillance techniques with their expensive equipment is crime prevention: they are said to be necessary both to combat the increasing sophistication of organized crime and of groups that use mass violence for political ends and to deal with crowd disorder.

The most controversial provision is Article 47 (wire-tapping and voice recording of a conversation or a statement in a closed or open space) which can be in conflict with the constitutional rights of personal dignity, privacy and the inviolability of dwellings. The Constitution, in Article 36, authorizes infringements on the inviolability of dwellings only in "exceptional circumstances" or "to apprehend a person who has committed a criminal offense or in order to protect persons or propert." The article on the Protection of Privacy of the Post and Other Means of Communication authorizes infringements only when,

> in accordance with statute, a court may authorize action on the privacy of the post or of other means of communication, or on the inviolability of individual privacy, where such actions are deemed necessary for the institution or continuance of criminal proceedings or for reasons of national security.

A more accurate description of special methods and means is found in the Criminal Procedure Law (Article 150) and in the draft of the Public Safety Act.[10] The draft also defines, in Article 46, those criminal acts, offenses, activities and threats for which the use of special methods and means can be authorized by the court.

CIRCUMSTANCES AFFECTING NATIONAL SECURITY

The security situation is influenced by many factors, both external and internal. The war in the Balkans has compelled many people in Bosnia and Croatia to flee their homes and find refuge in other countries. Slovenia has also been affected by this migration and has had to find means of coping with the refugee problem, which Solvenian authorities fear might evolve into a potential breeding ground for terrorist activity if the Balkan region does not stabilize in due time.

Another unstable region is already having an impact on the crime situation in Slovenia: the new states of the former Soviet Union. Slovenia is experiencing a growing influx of criminal activity from this part of Europe, especially in the fields of narcotics trafficking, weapon smuggling, car theft and the distribution of counterfeit money. The Slovenian police force is responding to these new challenges by reinforcing cooperation and improving contacts with other countries, especially in the fight against organized crime (ten agreements have been signed so far).

On the other hand, as we have already mentioned, due to the period of transition, the internal situation has an impact on social differentiation, the level of unemployment, the poverty line, the prospects for the younger generation and social uncertainty. The rise in the delinquency of minors is a confirmation that subculture behavioral patterns are changing. During recent years, the percentage of minors in the total number of convicted offenders has been growing steadily. The most disturbing fact is that criminal acts committed by juvenile delinquents are becoming more violent, are conceived in a more professional way, and are prepared in advance.

According to data from the Ministry of Internal Affairs, there were 43,635 criminal offenses recorded in Slovenia in 1994 or an average of 2,209.5 crimes per 100,000 citizens. In 1995, the crime rate stood at 1,933/100,000 (Mikulan, et al., 1997, p. 57). However, despite the spread of serious and organized forms of crime, the level of the safety of the population is still rather good, since Slovenia is still less endangered by crime than other countries in this part of Europe.

THE EUROPEAN PERSPECTIVE

The European Union has assumed a powerful role in Europe. Its structure can become the foundation for a new Europe; the possibility now exists for the EU to become the organizing framework for the whole of Europe.

Cooperation in police matters was established shortly after the proclamation of Slovenia's independence. It is occurring on several levels and is currently being arranged through Interpol, the UN, and the Council of Europe. The problem that Slovenia faces at the moment is the formation of the "Schengen space" with all its restrictions concerning immigration and the question of its impact on our borders with Austria, and later with Italy, once Slovenia fulfills the provisions of the Schengen Agreement.

One of Slovenia's foreign policy goals is admission to the European Union. An application has been presented to the European Union and to the European Commission, but talks have begun only recently. The Ministry of the Interior will certainly keep track of all changes the European Union adopts in the domain of domestic affairs. On March 26, 1995, the Schengen area finally began to exist, but still unresolved is the issue of the enforcement power of Europol and its relationship with Interpol and other law enforcement agencies such as the Europol Drugs Unit. The same concern applies for the External Borders Convention, which has many similarities to the Schengen Agreement (The Federal Trust, 1994).

Undoubtedly, the fight against international crime must be organized on a global level. Slovenia will need strong partners to cooperate with in the future to combat organized crime at home. Until Europol finally develops executive powers, Slovenia will work on strengthening ties with European intergovernmental agencies dealing with criminal matters and police cooperation. In any case, there should not be any conflicts regarding the exchange of data through information networks once a compatible system is installed throughout the European Community.

CONCLUSION

The search for security influences the policy of every state. This has been especially true of Europe since the end of Cold War bipolarity. The Old Continent must find a new stability that will ensure peaceful growth and prosperity.

The European Union will have to consider the security dilemma arising from the partial loss of national sovereignty by its members. With the removal of internal borders, the European Union will have to reconsider its policies toward third-country nationals, asylum-seekers, visas and illegal immigrants. As Slovenia touches the external borders of the European Union, there are fears that these problems might become a burden for the country. This new factor will certainly have to be dealt with by all sides.

Like the Partnership for Peace proposal in the area of defense, the issue of preserving democratic values should also be raised in the sphere of justice and cooperation in domestic affairs. This would be the best way to secure our common values and to police democracy at both the national and the European level.

NOTES

[1] Goodwin (1992) defines democracy, as it is classically understood, in the following way: (1) supremacy of the people; (2) the consent of the governed as the basis of legitimacy; (3) the rule of law: peaceful methods of conflict resolution; (4) the existence of a common good or public interest; (5) the value of the individual as a rational, moral active citizen, and (6) equal rights for all individuals.

[2] Autorite, qui designe le plus impalpable de ces phenomenes et qui de ce fait est frequemment l'occasion d'abus de langage, peut s'appliquer a la personne—on peut parler d'autorite personnell, par exemple dans les rapports entre parents et enfants, entre professeurs et eleves—ou encore elle peut constituer un attribut des institutions, comme, par exemple, dans le cas du senat romain (auctoritas in senatu) ou de la hierarchie de l'Eglise (un pretre en etat divresse peut valablement donner l'absolution). Sa caracteristique essentielle est que ceux dont l'obeissance est requise la reconnaissent inconditionnellement; il n'est dans ce cas nul besoin de contrainte ou de persuasion … L'autorite ne peut se maintenir qu'autant que l'institution ou la personne dont elle emane sont respectees. Le mepris est ainsi le plus rand ennemi de l'autorite, et le rire est pour elle la manace la plus redoutable (cited in Ricoeur, 1991, pp. 154–155).(Translation by the editors, with a little help from our friends.)

[3] The authors of *The Federalist* referred to the executive power as an agency designed to accomplish "extensive and arduous enterprises for the public benefit." Cited in Harvey C. Mansfield, *Taming the Prince: The Ambivalence of Modern Executive Power*, New York, Free Press, 1989, p. 247 and p. 270.

[4] Let us not forget that the Soviet Union encouraged socialist revolutions abroad and that many prominent Slovene Communists were trained in Moscow.

[5] For Fink Hafner (1994), the critical points for the long-term consolidation of democracy in Slovenia are "solving economic and social problems, development of the rule of law, development of a new, modern, political communication network, and a democratic political culture." (Cited in Slovenian ..., 1994, p. 403.)

[6] We cannot agree more with Feltes' (1994) viewpoint that political changes have an effect on police efficiency as well: "a changing system of values touches not only the society and the people living in it but also the police and the criminal justice system as a whole" (p. 40).

[7] The new public safety project deals particularly with changes at the district level, specifically police beats, police offices, and the specialization of work at police stations. These changes are necessary due to the latest administrative reform of the districts. Until now, a police station was situated in each district. The administrative reform produced a number of new, smaller districts, thus dictating a complex change in the organization and content of Slovenian police work. The reorganization of the Department of Internal Affairs was also implemented not long ago.

[8] "The duty of police officers is to protect the lives and personal safety of people; to prevent criminal acts and to apprehend their perpetrators; to maintain and restore public order at any time, even if the performance of the task could mean a danger to their lives and even if they are not at work, regardless of whether they were ordered to fulfill such a task or not. Whenever policemen perform their tasks in civilian clothes, they must identify themselves before an action. If or as circumstances allow, they must identify themselves orally as police officers" (Article 21).

[9] "1.—means for handcuffing and binding someone; 2.—gas spray; 3.—physical force; 4.—batons; 5.—gas and other means of pacification; 6.—jets and water; 7.—mounted police; 8.—special motor vehicles; 9.—police dogs; 10.—devices for coercive stopping of vehicles; 11.—fire arms" (Article 41).

[10] Special methods and means include undercover police activity, undercover cooperation, covert observation and tracking and picture-taking, fictitious purchase of objects, use of adjusted documents and identification signs, control of telephones and other communication means and devices, control of letters and other mail, as well as wire-tapping and voice recording (Article 45).

REFERENCES

Feltes, T. (1994, Summer). New Philosophies in policing. *Police Studies. The International Review of Police Development, 17.* 29–48.

Friedrich, C.J. & Brzezinski, Z. (1965). *Totalitarian dictatorship and autocracy* (2nd ed.). Cambridge, MA: Harvard University Press.

Goodwin, B. (1992). *Using political ideas.* New York: John Wiley & Sons.

The Federal Trust. (1994). *Maastricht and beyond—Building the European Union.* London: Routledge.

Mansfield, H. C. (1989). *Taming the prince: The ambivalence of modern executive power.* New York: Free Press.

Mikulan, M., Globocnik, M., Kos, D., & Podvrs[v]ic, A. (1997). The police experience in an emerging democracy. *The Police Chief*, June, 53ff.

Reynolds, P.A. (1994). *International relations.* London: Longman Group UK Limited.

Ricoeur, P. (1991). *Lectures 1—Autour du politique.* Paris: Editions du Seuil.

Slovenian Political Science Association. (1994). *Civil society, political society, democracy.* Ljubljana: Author.

Uglow, S. (1988). *Policing liberal society.* London: Oxford University Press.

15

The Impact of Human Rights on Identity Checks by the Police: a Swiss Perspective

LAURENT WALPEN

INTRODUCTION

The recognition, codification and protection of human rights is a recent development. In fact, codification is so recent that many police officers do not have even the faintest knowledge of the norms categorized under human rights. Older police officers in Switzerland (those over 50 years of age) have found it difficult to become accustomed to abiding by them. In their days, there was not much talk about safeguarding human rights. Methods were more direct and those who were charged with crimes and violations found this quite normal. Nowadays, though, a person held for the slightest misdemeanor will appeal to the Human Rights Court in Strasbourg. As the old adage says, "manners change with the times." (To place this discussion into historical context, a short description of the development of individual freedoms, culminating in the European convention for the Protection of Human Rights and Fundamental Freedoms, is included in the Appendix.)

The goal of this paper is to examine the relationship between human rights and a routine police procedure: identity checks. I will show how, even in such routine police work, adherence to democratic standards and the observation of human rights present a challenge to the police. I offer this explanation on the basis of Swiss institutions and my experience as a police officer in Geneva.

Identity checks are a procedure for establishing the identity of unknown persons or to check whether an individual is a wanted person. Checks may involve targeted individuals or passersby. When identity checks take the form of an organized and surprise police action to check the identities of a large number of people found in a particular place (station, park or public square), it is referred to as a police raid. Frequently, raids include a search for objects of a criminal nature or capable of being offered as evidence. It should be borne in mind that the rules governing identity checks do not include the search for objects as such. Objects are subject to rules governing searches or sequestration.

Since no federal directives exist in Switzerland in this matter, we must look for provisions in cantonal law. Few cantons have established an explicit legal basis for identity checks. In most of the cantons, the right to deprive a person of his or her liberty in order to check identity rests upon custom and is based on the legal concept of general police powers. It is evident, however, that police officers who operate on a clear legal basis will be much more at ease because unnecessary arguments with the public are more easily avoided. I shall examine, by way of example, the legal provisions prevailing in Geneva.

CONDUCTING IDENTITY CHECKS

The identity check may include eight steps: questioning, producing evidence of authority, requesting proof of identity, on-the-spot checking, questioning at the police station, the right of close relatives to be informed, searching, and identification measures such as fingerprinting.

Questioning

This must be carried out as discreetly as possible and with the utmost precaution in order to ensure the safety of the police officers.

Producing Evidence of Authority

The official who conducts a police check must always produce evidence of his or her authority. He or she must show that he or she is a member of the police force.

The only deviation from this rule is when exceptional circumstances militate against producing evidence of authority, for example, when this would make it impossible to intervene in the situation (e.g., for reasons of safety a plainclothes police officer must be able to disarm a threatening person without first providing a lawful explanation). For uniformed officers, the uniform itself is sufficient proof; but plainclothes officers must present their identity card.

Requesting Proof of Identity

Irrespective of the circumstances, requesting proof must be done in a polite manner. The maintenance of security and public order necessitates the police being able to interrogate, for purposes of establishing identity, not only persons suspected of having committed a felony or a misdemeanor, but also those persons who for any reason appear to have entered the country without legal authorization.

Legislation has therefore introduced the requirement that private persons must prove their identity when requested. But this obligation does not permit the police to interrogate any person under any circumstances they find convenient. In every case, the identity check must be supported by an objective reason. According to the Federal Tribunal, in Switzerland a certain number of conditions for instituting a check must be satisfied. These are:

(1) the existence of a troubling situation;
(2) the presence of the person checked within the vicinity of a place where an offense has just been committed;
(3) his or her resemblance to a wanted person; and
(4) his or her inclusion in a group of individuals concerning whom there are reasons to believe that one or some of them may be in an illegal situation.

If challenged about the justification for conducting a check, the police officer should always be in a position to provide legally supported reasons.

On-the-spot Checking

The identity check will normally take place on the spot and as quickly as possible. Proof of identity does not necessarily involve the production of a passport or identity card. Any other appropriate method will suffice. The Federal Tribunal, moreover, has expressly stated that

> the organs of the police must be flexible and individuals who are requested to identify themselves may do this in a variety of ways. The concept of identity documents is very broad, for it can include any document bearing a photograph and giving precise information about the person interviewed (such as a passport, identity card, driving license, cross-frontier permit, business pass, student card or diplomatic pass). In the absence of these documents, the police must question the person appropriately and, if necessary, verify his statements by employing the technical means at their disposal (radio contact with the police headquarters).

Identification at the Police Station

Generally speaking, an identity check must be carried out on the spot. If, however, the person involved is unable to establish his or her identity and a supplementary check becomes necessary, he or she may be taken to a police station for further identification. Taking a person to the police station for this purpose must remain exceptional and take place solely under the following circumstances:

(1) on-the-spot identification is impossible because the person concerned is unable for various reasons (drunkenness, sickness, unwillingness, etc.) to prove her/his identity;
(2) the behavior of the person is hostile;
(3) the police officer faces a hostile crowd;
(4) there is a grave danger of obstructing traffic;
(5) the documents presented appear dubious or fraudulent; or
(6) a radio check with police headquarters has revealed that the individual may be wanted by the police.

The fact that a person is unable to prove his or her identity does not, by itself, justify that he or she be taken to the police station; the need for a supplementary check must also be present. In making these two conditions cumulative, the legislature strictly applied the principle of proportionality. The restriction of individual liberty represented by the obligation for an individual to appear at a police station must be justified by special circumstances; and these circumstances may relate to the attitude of the person concerned, the locale or company in which he or she finds him- or herself, resemblance to a wanted person, or proximity to a place where an offense was just committed, etc.

If the person is taken to the police station for identification purposes, he or she is at this stage in no way suspected of having committed an offense. The restriction of individual liberty must therefore be strictly confined to the time necessary for establishing her/his identity. The law also prescribes that identification shall take place without delay. Once the formalities have been completed, the person shall immediately leave police premises. The process must not be interrupted under any circumstances. The necessary investigation must be continuous. It is not permissible for the person in question to be held while the officer performs other duties or while the official on duty goes for a rest or a meal. Sufficient time will, of course, be allotted for contact with other police forces or information centers. Legislation does not stipulate the maximum period a person may be held at the police station for identification purposes. This has been done to avoid the maximum permitted from becoming the rule.

The Right of Close Relatives to be Informed

The restriction of personal liberty connected with identification at a police state is clearly defined by its purposes. As the person involved is not under any definitive suspicion, he or she should not remain removed from the outside world. He or she therefore has the right to contact his or her near relatives immediately. Such contact, however, may not serve as an identification. This guarantee by the Federal Tribunal corresponds to that offered by Articles 8 and 10 of the European Convention on Human Rights.

Questioning for identification purposes must, of necessity, cease as soon as the identity of the person has been established. At this precise moment, the person is free to leave the police premises and cannot be detained any longer. This does not, of course, apply to a summons issued by a competent authority. In this case, the person is no longer being held at the police station for identification purposes but because he or she has been charged with a felony (a misdemeanor charge is insufficient).

It frequently happens that people who are invited or called for questioning come of their own free will to the police station for interrogation. As long as a summons has not been issued, they are not obliged to go to police premises or to stay there. The police have no right to hold them. Nevertheless, the competent authority may, when the appropriate conditions are met, issue a summons against a person who changes his or her mind and refuses to go to police premises or expresses the desire to leave them. The temporary forfeiture of liberty for the purposes of questioning, commonly termed "held for questioning," only exists within the parameters indicated above.

Searching

The search may be carried out for two reasons: to ensure security or to establish the identity of a person. The search involves a need for security when the person is suspected of being armed (even if an arrest is not envisioned) and irrespective of whether or not the carrying of arms was authorized. The Federal Tribunal has held that searches of persons held for identification purposes are permissible only for reasons of security. Yet a search may be necessary in order to establish the identity of a person who is unconscious, in a distressed state or deceased.

The legislature desired that the search should be adapted to the circumstances and as considerate as possible. In order to respect the personal domain, the persons must be searched only by police officers of the same sex, unless immediate security dictates otherwise.

It should be stressed that the demands for immediate security cannot justify complete strip or intimate searches. In effect, these provisions in the public interest are observed once the person under questioning has been disarmed or relieved of any object in his or her possession which could seriously be considered a danger to the life and bodily safety of other persons or of her/himself.

Identification Measures

When a person is not suspected of having committed a felony or misdemeanor offense, identification measures are authorized only if his or her identity is doubtful and cannot be established by any other means, especially when the declarations of the person appear to be inexact. The Federal Tribunal has stipulated that the taking of photographs and fingerprints must be seen as a "last resort," which must be considered only if the ordinary identification procedures have not yielded a satisfactory result. It should be remembered that if the person who has submitted to these identification measures is exonerated and the inquiry is over, he or she may demand the destruction of the material which has been assembled.

CONCLUSION

The need to check persons and, if necessary, to search them in order to prevent and detect offenses and maintain public order is one of the basic functions of the police. Nevertheless, the police officer who is experienced and suitably informed will be aware that the checking and searching of persons and vehicles can be conducted without unnecessary humiliation. Even a simple police activity such as identity checking must conform to the letter and spirit of human rights. This is not a question of formalistic or irksome obligations being imposed upon the police, but of fundamental guarantees to be enjoyed by citizens. Respect for such procedures does not come of its own accord. It is the responsibility of the higher police ranks to teach this respect to ensure that it exists at all levels of the organization. Freedom is priceless. Respect for human rights is certainly the price to be paid for acquiring and keeping it.

APPENDIX

A Short History of Human Rights

It is important to recall certain milestones in the history of freedom:

- 1215 Magna Carta
- 1679 Habeas Corpus Act
- 1789 Declaration of the Rights of Man and the Citizens
- 1948 Universal Declaration of Human Rights (United Nations)
- 1950 European Convention for the Protection of Human Rights and Fundamental Freedoms
- 1966 International Convention of Civil and Political Rights
- 1969 American Convention on Human Rights.

The Magna Carta was the first important text. Its sixty-three articles clearly defined reciprocal rights of the sovereign and the noble classes. The King was compelled to issue charters guaranteeing his vassals a certain number of privileges.

Strictly speaking, the Magna Carta was a catalogue of privileges rather than a declaration of rights, but its spirit inspired future generations. Article 39 provided that "no free man shall be taken, or imprisoned, or outlawed, or banished, or in any way injured ... except by the legal judgment of his peers, or by the law of the land. (John, King of England, Lord of Ireland, Duke of Normandy and Aquitaine, and Count of Anjou)."

The Habeas Corpus Act of 1679 is an essential part of English constitutional history, and the procedure it labelled is still effective in England and the Anglo-Saxon counties, including the United States. "Habeas corpus" is Latin expression meaning "(You must) have the body." It is applied to anyone who is holding a person under state of arrest. It is an order to produce that person before a court. The object of this rule is to guarantee individual liberty by removing the dangers of arbitrary arrest and confinement. The writ of habeas corpus may be defined as an order addressed by a judge to anyone detaining another person, commanding him or her to produce the detainee and furnish reasons for this detention.

The Declaration of the Rights of Man and of the Citizen states that "no man can be indicted, arrested or held in custody except for offenses legally defined and according to specified procedures" (Article 7). It is further stated that "everyone must be presumed innocent until he is pronounced guilty. If his arrest and detention are thought necessary, no more force may be used than is thought necessary to secure his person" (Article 9). And, finally, the Declaration holds that "the guarantee of the rights of man and of the citizen requires a police force; a force, therefore, which is instituted for the benefit of all, and not for the particular use of those to whom it is entrusted" (Article 11).

The Universal Declaration of Human Rights was unanimously adopted by the General Assembly of the United Nations in 1948. It expressly provides that "no one shall be subjected to arbitrary arrest" (Article 9) and that "everyone charged with a penal offense has the right to be presumed innocent until proved guilty according to law" (Article 11). The Universal Declaration of Human Rights is no more than an ideological and moral statement of principle and has no mandatory force. It is not a legally binding instrument and does not commit even those states who voted in its favor to abide by its provisions.

When states sign or ratify agreements, however, they undertake to guarantee the rights established therein and to introduce them into their own legal systems. These agreements are instrumental, therefore, in making it theoretically possible to compel states to account for any failure to comply with the obligations as defined within agreed conventions. For this reason, on the basis of the Universal Declaration of Human Rights, the United Nations Commission of Human Rights produced two bills, one on civil and political rights, the other on economic, social and cultural rights. In the international bill concerning political and civil rights, for example, one article states that "all persons deprived of their liberty shall be treated with humanity and with respect for the inherent dignity of the human person."

Certain agreements designed to safeguard human rights have been drawn up by groups of states linked by their geographic proximity, cultural traditions or ideological systems. In consequence, these regional conventions implement more precisely certain principles of the International Declaration of Human Rights. At the moment, there are two important regional conventions on human rights: the European Convention for the Protection of Human Rights and Fundamental Liberties, and the American Convention on Human Rights (the San Jose Agreement).

The European Convention on Human Rights was signed on November 4, 1950, by ministers of fifteen European countries meeting in Rome. The signing was a step of unprecedented significance and a milestone in the development of international law. The Convention came into force on September 3, 1953. Twenty-one members of the Council of Europe have now signed it, although not all of them have made the same commitments. The influence of this agreement has made itself felt not only throughout Europe but in all countries which have attempted to increase the protection of human rights. It has served, for instance, as a model for the Inter-American Convention of Human Rights which took effect in 1978.

Technically speaking, the European Convention is an international treaty, a type of contract by which states accept certain obligations. These obligations are notable in that they recognize certain individual rights. In addition, the Convention authorizes individuals who claim that their rights have been violated to proceed against the government whom they hold responsible.

Governments are under obligation to ensure that, in conformity with the agreement, the rights and liberties of all those under their jurisdiction are protected. This has compelled some states to modify internal legislation to bring it into line with the provisions of the Convention. Certain states have incorporated the agreement into their national law, thus enabling every citizen to bring a complaint or appeal, based on the provisions of the European act, before a national tribunal (or administrative authority). In countries where the Convention has not been integrated into the national law, the latter must not be in conflict with its provisions.

The Convention is not intended to replace national systems for the safeguarding of human rights, but to provide an international guarantee which supplements the right of appeal in these countries. Another essential principle of this Convention is that it does not apply solely to nationals of member states but to whoever is living in or visiting the country. Article 1 of the Convention clearly stipulates that the states must concede the rights and liberties defined in the text to all persons within their jurisdiction. Since the drafting of the Convention, new rights and obligations have been the subject of Additional Protocols, whose adoption is optional.

Specific rights protected by the European Convention include: "right to life," "right to liberty and security of the person," "right to a fair and public hearing ... by an independent and impartial tribunal established by law," "right (of a person) to respect for his private and family life, his home and his correspondence," "rights of freedom of thought, conscience and religion," "right to freedom of expression ... (and) to hold opinions," "right to freedom of peaceful association" and "right to marry and to found a family."

Rights guaranteed by the Protocols include: "right (of a person) to the peaceful enjoyment of his possessions," "(rights) in relations to education and to teaching," "right (of a person) to liberty of movement and freedom to choose his residence," and "right (of a person) to leave any country, including his own."

In addition, the spirit of the Convention and its Protocols forbids torture and inhumane or degrading treatment or punishment; slavery, servitude and forced labor; retroactive penal laws; discrimination in the enjoyment of the rights and freedoms guaranteed by the Convention; expulsion or repulsion by a state of its own nationals; and the collective expulsion of foreigners.

The Convention also, and correctly, recognizes that the majority of these rights cannot be unlimited in a democratic society, and for this reason admits that restrictions may become necessary on the grounds of: "public safety," "national security," "economic well-being of the country," "protection of health or morals," "protection of the rights and freedoms of others," "prevention of disorder," or "(prevention of) crime." The Convention also permits states to suspend their obligations in the case of war or other public dangers. But even in these cases of emergency, no state has the right to withdraw its obligation to respect the right to life, and torture, slavery, and retroactive penal laws are always forbidden.

Concerning the present subject—the Convention on Human Rights and the police—Article 5 of the Convention is of particular interest. This article envisions six situations for the deprivation of a person's liberty: "after conviction by a competent court," "to secure the fulfillment of an obligation prescribed by law," "on reasonable suspicion of having committed an offense or when it is reasonably considered necessary to prevent the committing of an offense or fleeing after having done so," "the detention of a minor … for purposes of educational supervision," "the lawful detention … for the prevention of the spreading of infectious diseases, of persons of unsound mind, drug addicts, alcoholics or vagrants," or "the lawful arrest or detention of a person to prevent his effecting an unauthorized entry into the country."

The Convention also lists four rights of a person under arrest: "everyone who is arrested shall be informed promptly … of the reasons for his arrest," "everyone who is arrested … shall be brought promptly before a judge," "everyone who is deprived of his liberty by arrest or detention … shall be entitled to take proceedings by which the lawfulness of his detention shall be decided speedily by a court," and "everyone who has been a victim of arrest or detention in contravention of the provisions of this Article (5) shall have an enforceable right to reparation."

Subject Index

Accountability, police,
    Austria, 235
    General, 16
    Croatia, 58, 65, 79–80
    England, 245–251, 277–278
    Estonia, 104
    Poland, 145
    Russia, 189
    The Netherlands, 298–300
Amnesty International, 138–139, 209, 231
Austria,
    Foreigners in, 216
    Police history, 216–217
    Traffic behavior, 221

Balkan Route, 74, 123, 154, 341
Broken windows thesis, 230–231

China, police, 26–27
Colonial policing, 32–33
Community policing,
    Austria, 234, 236
    England, 269–272
    South Africa, 198–199, 210
Cooperation, international,
    Austria, 233
    England, 272
    Estonia, 104–107
    General, 12–13, 319–321, 327
    Hungary, 126–128
    Poland, 154–157
    Russia, 187–188
    Slovenia, 349–350
    South Africa, 206–208
    The Netherlands, 12, 127, 295–297
Corruption, police,
    Austria, 232
    Croatia, 80
    England, 253–254
    General, 10, 16
    Poland, 164–165
Crime, causes of,
    Estonia, 110–111
    Hungary, 118–119
    Poland, 148
Crime, juvenile,
    Estonia, 97–98
    Poland, 148
Crime, normal,
    Austria, 223–228
    Croatia, 66–71
    England, 261–264
    Estonia, 95–97
    General, 7–8
    Hungary, 119–122
    Macedonia, 338–339
    Poland, 146–148
    Russia, 180–182
    Slovenia, 349
    South Africa, 200–202
    The Netherlands, 288
Crime, organized,
    General,
        Austria, 228–229
        Croatia, 71–74
        Macedonia, 339–341
        Poland, 149–150, 154–155
        Russia, 182–183
        The Netherlands, 289–290
    Drug smuggling, 156, 341
    Economic,
        Croatia, 74–75
        Poland, 148–149
    People smuggling,
        Poland, 151–154
Crime prevention,
    Croatia, 62
    Macedonia, 337–338

Crime prevention, (continued)
  The Netherlands, 290–291
Criminals, demographics,
  Estonia, 100
  The Netherlands, 288–289
Croatia,
  Elections, 46–47
  Police-military relations, 50–53
  Political history, 45–48
  Private security, 65
Culture, police,
  General, 6, 9–10, 322–325
  Croatia, 53–55, 76–77
  Russia, 177–178

Democratic policing, conceptions of,
  Austria, 219
  Croatia, 48–49
  England, 243–245, 279–281
  Estonia, 90–91
  General, 4–5, 7, 24, 81, 311–313, 319–322, 325–326
  Hungary, 115–117
  Poland, 144, 167–170
  Russia, 175–176
  Slovenia, 344–345
  South Africa, 196, 212
  The Netherlands, 286–287, 303–304
Drugs and crime,
  Russia, 184–185
  South Africa, laws on, 204

England,
  Constabulary independence, 248
  Fabrication of evidence, 256–257
  Holloway Road Transit Case, 254–255
  Lay Visitor Panels, 251
  Miners' Strike, 258–260
  Police, 29–30
  Police and Criminal Evidence Act 1984 (PACE), 250, 251, 274–277
  Police brutality, 254–256
  Special constabulary, 268

Estonia,
  Police history, 92–93
  Population, 88–89
EUROPOL, 91, 272–273, 296, 350

Fear of crime,
  Austria, 229–230, 235
  Croatia, 66
  Russia, 183–184
  The Netherlands, 290

Finland, police, 28–29

George Soros Fund, 105
Ghana, police, 33–34
GULAG, 174, 191, 193

Human rights, 357–360

ILEA (International Law Enforcement Academy), 127–128
INTERPOL, 79, 105–106, 126, 217, 296, 349
ICITAP (International Criminal Investigative Training Assistance Program), 320, 328

Legal changes,
  Austria, 235
  Croatia, 73
  England, 273–277
  Estonia, 107–109, 112–114
  General, 14–15
  Hungary, 130–135
  Poland, 125, 157–162
  Slovenia, 348–349
  Russia, 190–191
  South Africa, 204–206
  The Netherlands, 300–301
Los Angeles Police Department, 128

Media relations,
  England, 278–279
  Estonia, 109–110
  Hungary, 136
  Poland, 165–167
  Russia, 191–192
  The Netherlands, 301–303
MEPA (Middle European Police Academy), 127, 157, 188, 233
Migration,
  Austria, 231
  Estonia, 98–101
  England, 264–267
  General, 8–9
  Hungary, 122–124
  Poland, 150–154
  Russia, 184–187
  South Africa, 202–204
  The Netherlands, 293–295
Moral panics, 11, 202

Organization, police,
  Austria, 217–219, 233–234
  Centralization, 6–7, 27
  Croatia, 61–66

England, 245–249
Estonia, 94
Hungary, 117–118, 128–129
Macedonia, 335–337
Slovenia, 347–348
South Africa, 196–198
The Netherlands, 286–287
Organization, models of,
    Fragmented model, 28
    Coordinated model, 27–28
    General, 13–14

Poland,
    Police, 35–38
    Political history, 143
Police, abuse of power,
    Austria, 231–232
    Hungary, 125–126, 138–139
    Russia, 178–179
Police-community relations,
    Austria, citizen perceptions, 222–224
    England,
        Lay involvement, 267–268
        Local consultation, 250–251
        Police image, 260–261
    General, 25–27
Police equipment,
    Austria, 232–233
    Croatia, 75–76
Police functions,
    Austria, 219–221
    Estonia, 94–95
    General, 315–319
    Slovenia, 351
Police magazine,
    Gliny, Poland, 166
    Halo, Croatia, 58
    The Link, England, 278
    Zsaru, Poland, 136
Police personnel, Estonia, 102–103
Politics and policing,
    Democratization, 313–315
    England,
        public order policing, 257–260
        politicization, 267
    Hungary, 124–126, 132–135
    Poland, 146
    Politicization, 10–11
Polizei Fürhrungsakademie, 155
Professionalization, police,
    General, 25
    England, 252–253
    Poland, 162–163
    Russia, 188–189

    South Africa, 209–212
    The Netherlands, 297–298
propiska, internal passport, Russia, 177

Rechstaat, 130, 169
Refugees,
    Croatia, 47–48
    Hungary, 123
Royal Ulster Constabulary, 245
Rule of law, 34, 138
    Russia, 176–177
    The Netherlands, 292–293
Russia,
    Communist period, 173–175
    Police, 38–39
    Prisons, 190

SARPCCO (South African Regional Police
    Chiefs Co-ordinating Organization,
    208–209
Schengen Agreement, 231, 272–273, 296, 305, 350
Shadow economy, Estonia, 110–111
Slavonia, Eastern, 55–57
Socialism and public safety,
    Hungary, 120–122
    Slovenia, 345–346
South Africa,
    Political history, 195–196
    Non-state policing, 199
Stereotypes,
    Austria, foreigners, 226
    England, ethnic minorities, 264–267
    General, 11, 324
    Gypsies (Roma), 138–139
    Russia, migrants, 179, 187

The Netherlands,
    Detention, 291–292
    trias politica, 285
Training, police,
    Croatia, 51–52, 78–79
    England, 268–269
    Estonia, 101–102
    Hungary, 129–130
    Poland, 163–164
    South Africa, 210
Trevi agreement, 295–296

United States,
    police, 30–32
    racketeering, 8

War crimes, Croatia, 53, 64
Witdoeke, 213

Author Index

Abbott, D. J., 32
Abramkin, V., 190
Adamski, A., 158
Aiginger, K., 229
Alderson, J., 4, 116, 253, 280
Alfanasjev, V., 184
Allerbeck, K. R., 228
Amir, M., 5
Anderson, M., 316
Anschober, R., 232
Archer, R., 263
Arendt, H., 344
Aromaa, K., 187
Aronowitz, A. A., 296, 304, 324
Arthur, J. A., 33
Artner, D., 235
Arutjunjan, L., 175

Baker, K., 248
Baldwin, J., 10
Balokovic-Krklec, K., 49, 51, 52, 54, 67, 78
Bartunek, A., 215
Bayley, D. H., 7, 16, 17, 314, 315, 316, 317, 321, 325
Beck, A., 316
Bednarski, M., 149
Beidelt, I., 217
Bennett, T., 268
Bentkowski, A., 159
Benyon, J., 316
Berkley, G. E., 7, 116
Berman, H. J., 34
Bernheim, M., 235
Bittner, E., 4
Black, D., 19
Black, H. C., 25
Bögl, G., 217
Bojadziski, O., 58
Borisov, A., 175
Boross, P., 121
Bosiocic, G., 62

Bovenkerk, F., 294
Bracey, D. H., 26, 27
Brand, P., 167
Brewer, J. D., 204, 315, 316, 317
Brmbota-Devcic, T., 51
Brodeur, J.-P., 316
Brogden, M., 199, 312, 315
Brzezinski, Z., 345
Buble, N., 67
Bucqueroux, B., 16, 116
Burke, J., 274
Busch, H., 216
Butkovic, D., 66, 80, 81

Cajner, I., 66, 70
Campbell, D., 253
Carter, D. L., 19
Cebulak, W., 37
Cengic, S., 55
Chan, J. B. L., 318
Chervakov, V., 186
Chevigny, P., 326, 327
Chmielewski, R., 165
Cichorz, T., 165
Clark, R. S., 319
Clark, G. D. N., 246
Clinard, M. B., 32
Cohen, S., 11, 183
Coleman, J. W., 8
Concha, G., 117
Conquest, R., 175
Cornish, W. R., 246
Coxey, G., 320
Cullen, P., 272

Dai, Y. S., 27
Das, D., 7, 15, 17, 18, 24, 25, 34, 219, 220, 316
Daszkiewicz, W., 160, 162
Davy, B., 217
Davy, U. 217

de Haan, W., 29
Dearing, A., 226
DeKeseredy, W. S., 11, 15
Diamond, L. 314
Diederichs, O., 219
Dienaar, M., 206
Dikselius, A., 183
Djakov, S., 182
Djuric, M., 78
Doherty, M., 277
Dolgova, A., 180, 182
Dölling, D., 234
Doomen, J., 300
Douglas-Hamilton, D., 202
Dreher, G., 221
Dugard, J. 204
Dugin, A., 175
Dujmovic, Z., 67, 68, 70
Durston, G., 324
Dziemidowicz, Z., 149

East, R., 254, 255
Edelbacher, M., 228, 233
Egorshin, V., 177, 187
Eichwalder, R., 218
Eijken, A. W. M., 288, 289, 290, 291, 294, 305
Emerson, D., 5
Emsley, C., 253, 259, 264, 266, 270
Enloe, C., 314
Ettmayer, W., 228

Farmer, D., 19
Fassmann, H., 231
Fehervary, J., 234
Feltes, T., 221, 234, 351
Fiebig, J., 164
Fielding, N., 267
Fijnaut, C. J. C., 286, 316
Filar, M., 158
Findl, P., 216
Fischer-Kowalski, M., 222
Fogel, D., 19
Forrester, P., 7, 18, 24
Fosdick, R. B., 7, 117
Fox, W. F., 35
Friedman, S., 203, 204
Friedman, R. R., 4
Friedrich, C. J., 345
Fuchs, H., 219
Funk, B.-C., 219
Fyfe, J., 317

Gaines, L., 18
Galster, J., 145

Gartner, G., 263
Gataric, I., 64
Gatrell, V. A. C., 263
Gebhart, H., 217
Gibson, J. L., 14
Gilinskiy, Y., 8, 182, 184, 188, 189
Girtler, R., 216, 217
Gledec, Z., 59, 61, 64
Globocnik. M., 348
Golbert, V., 184
Goldsmith, A. J., 249, 270
Goldstein, H., 4, 16, 245, 317, 319, 321
Goodwin, B., 344, 350
Górniok, O., 148, 158
Graef, R., 252, 262, 263, 269
Guelke, A., 315, 316

Haberfeld, M., 81
Hain, P., 249
Halàsz, L., 216
Haller, B., 217, 234
Hanak, G., 221, 223, 226, 228
Hanausek, T., 161
Hanisch, E., 216
Harris, P. R., 18
Harrison, J., 16
Hauer, A., 219
Hebenton, B., 71, 72, 77, 315, 316
Hegel, G. W. F., 111
Heindl, W., 218
Henry, I., 262
Hesztera, G., 217
Heymann, P., 81
Hills, A., 314
Hirschfeld, A., 217
Hirst, P. Q., 18
Hobbs, D., 253
Hochenbichler, E., 217
Hofmański, P., 157, 160, 170
Holdaway, S., 17
Holiday, D., 316
Hołyst, B., 148, 149, 150, 158
Horvat, K., 71, 74, 75
Huang, C. Y., 27
Hume, I., 315, 316
Husain, S., 268

Iakolev, A., 34
Iljin, V., 187

Jäger, F., 217
Jakubski, K. J., 158
Jamróz, A., 167
Jansen, T. G., 297

Jarnjak, I., 66, 77, 79
Jaroch, W., 162
Jehle, J.-M., 229
Johnson, H. A., 31
Johnston, L., 315
Jones, T., 220, 321
Junger-Tas, J., 305
Jurina, M., 51
Juviler, P., 34, 175

Kappeler, V., 18
Karazman-Morawetz, I., 228, 230
Kelling, G., 281
Kelly, T., 318
Kemper, E., 228
Keplinger, R., 217, 219
King, R., 327
Kittel, B., 216
Klein, M., 305
Klinger, D. A., 318
Klockars, C. B., 54, 82
Kmety, K., 117
Knapp Commission Report, 10
Knoll, 35
Kobali, D., 49, 61, 72, 78
Kogler, R., 231, 233
Kojder, A., 165
Kołecki, H., 169
König, I., 217, 234
Konstantinov, A., 183
Korinek, L., 118, 124
Korunka, C., 218, 234
Korybut-Woroniecki, A., 151
Kos, D., 348
Kőszeg, F., 125, 126
Kovco, I., 68, 70
Krapac, D., 81
Kregar, J., 46, 47
Kruissink, M., 304
Kube, E., 149
Kudrelek, J., 161
Kuretic, Z., 58, 59, 61, 64, 65, 69, 80
Kutnjak Ivkovich, S., 54, 77, 82

Laagland, D. C. G., 296, 304
Lacqueur, E., 175
Laitinen, A., 28, 29
Lajic, I., 50
Lawton, Sir F., 252, 257
Lehti, M., 187
Leishman, F., 249, 266, 269, 280
Lenz, T., 228
Leps, A., 90, 94, 95, 99, 100, 101, 110, 111

Lernell, L., 159
Lesjak, K., 228
Lever, J., 326
Levin, B., 257
Levine, D. R., 18
Liit, A., 92
Lindmäe, H., 92, 93
Lintner, E., 3
Linz, J., 314
Lipset, S. M., 314
Lipsky, M., 317
Lisiecki, M., 161
Lobnikar, B., 78
Loef, C. J., 294
Los, M., 121
Louthan, W. C., 18
Louw, A., 200, 201
Loveday, B., 249, 266, 269, 280
Lovric, J., 59
Ludikowski, R. R., 35
Luneev, C., 186, 188, 189
Lustgarten, L., 280

Maguire, M., 262
Majer, P., 35, 40, 166
Makarevich, L., 175
Malygin, A., 175
Manning, P. K., 325
Mansfield, H. C., 350
Mapstone, R., 317
Marenin, O., 33, 198, 207, 312, 314, 316, 321
Markovic, L. J., 48, 70, 77
Martin, J. H., 7
Mason, K., 228
Mawby, R. I., 316, 321
Mayerhofer, C., 229
McCormack, T., 4
McEwan, J., 266, 280
McHugh, H. S., 320
McLaughlin, E., 246, 248, 321
McNee, Sir D., 261, 271
Meggeneder, O., 218, 223, 234
Merton, R. K., 3, 81
Mesle, F., 173
Michailovskaja, I., 188, 189
Miciñski, R., 165
Miedl, W., 217
Miksaj-Todorovic, L. J., 70
Mikulan, M., 348
Milton, C., 318
Milukov, S., 189
Mirosavljev, B., 50, 66, 67
Misiuk, A., 166

Monk, R. C., 19
Moore, L., 318, 321, 322
Morawetz, I., 226
More, H. W., 18
Moric, J., 47, 50, 51, 57, 58, 59, 61, 79
Morie, R., 216
Moxon-Brown, E., 315, 316
Mulac, V., 80
Münz, R., 231
Murck, M., 216
Murray, C., 281

Nadj, I., 72, 74
Nemeth, Z., 9
Newburn, T., 220, 321
Nierop, N. M., 289
Noll, A. J., 219
Nyíri, S., 126

O'Malley, P., 202
O'Rawe, M., 318, 321, 322
Ommens, H. C. D. M., 305
Osrecki, K., 77
Ovchinski, V., 182

Pagon, M., 78
Palka, P., 147
Paulides, G., 296, 304
Peberdy, S., 203
Pelikan, C., 226
Pelinka, A., 217, 234
Peneder, M., 229
Peric, I., 46, 47, 48
Perner, R. A., 230
Pichler, J. W., 222
Pilgram, A., 223, 226, 226, 228, 230
Pinkele, C. F., 19
Pinnock, D., 202
Piskor, M., 77
Plasser, F., 222, 224
Pływaczewski, E. W., 147, 150, 154, 156, 159, 164
Podvršic, A., 348
Potholm, C. P., 314
Pracki, H., 150
Pütter, N., 229

Rako, S., 51, 55, 76
Ratajczak, A., 158
Reiner, R., 7, 10, 245, 248, 249, 250, 258, 259, 260, 261, 263, 267, 277, 280, 313, 315, 316, 317, 328
Reinprecht, C., 216

Reismeller, J. G., 46
Reith, C., 116
Reitzes, M., 203, 204
Remmel, M. 90, 95
Remmer, K. L., 314
Reuss-Ianni, E., 218
Reynolds, P. A., 345, 346
Richards, M. 25, 30
Richardson, J. J., 29
Ricoeur, P., 350
Rider, B. 111
Rkman, I., 48
Roberts, D., 244
Robillard, St. J., 266, 280
Röglin, H.-C., 217
Romano, A. T., 7
Rombouts, R., 293
Rosenmayr, L., 222
Rubin, F., 137
Rzepliñski, A., 179

Sacic, Z., 72
Sandgruber, R., 229
Savage, S., 249, 266, 269, 280
Savona, E. U., 319
Scarman Report, p. 265, 270
Scheingold, S. A., 19
Schulte, R., 149, 155, 156, 216
Schulz, W., 232
Schwartz, M. D., 11, 15
Schwarzenegger,, C., 272, 273
Seegers, A., 199
Serdakowski, J., 150
Serrins, A. S., 19
Seyrl, H., 217
Shapiro, V., 186
Shaw, M. 200, 201, 202, 206
Shearing, C., 199, 315, 316, 317, 325
Shelley, L. I., 4, 175, 216, 233, 313, 316
Sheptycki, J. W. E., 316, 321
Sheregi, F., 186
Sherman, L., 318
Shkolnikov, V., 173
Shusta, R. M., 18
Siemaszko, A., 147, 149, 151, 169
Sigler, R. T., 34
Silverman, J., 278
Simic, J., 63
Simonovic, I., 46, 47
Singer, M., 66, 67, 70, 71
Sklepkowski, L., 151, 152, 154
Skolnick, J. H., 4, 16, 116, 316, 317
Smerdel, B., 46, 47

Smith, T. J., 220, 321
Smith, G., 254
Southgate, P., 270
Spencer, S., 245, 247, 248, 258, 277
Spencer, J., 71, 72, 77
Spotowski, A., 164
Springer, S., 35
Stanczyk, J., 12
Stanley, W., 316
Stead, P. J., 4, 30
Steinert, H., 228
Stenning, P. C., 315
Steytler, N., 317
Stierschneider, C., 217
Stone, R., 275, 276
Strooper, M. N., 286
Szabó-Kovács, J., 123
Szikinger, I., 117
Szikinger, I., 9, 12, 18
Szymanski, W., 219
Szyszkowski, W., 145

Tàlos, E., 216
Tarnawski, M., 157
Tendler, S., 272
Terlouw, G. J., 304, 305
Terrill, R. J., 23, 27, 28, 34, 38
Theobold, R., 14
Thom, H., 289
Thomas, T., 315, 316
Tifft, L., 318
Tilly, C., 314
Tomasic, R., 18
Tomcsányi, M., 117
Triandis, R., 4
Trojanowicz, R., 16, 116
Turk, 316
Turnbull, L., 316

Uglow, S., 244, 251, 274, 344, 347, 348
Ulram, P. A., 222, 224
Urquhart, F., 256
Urvaste, H., 98

Vallely, P. 265
Vallin, G., 173
van Wijngaarden, J. J., 305
van Sluis, V., 297

van der Spuy, E., 204, 326
van den Boer, M., 316
Van der Heijden, A. W. M., 289, 292, 294
Van Zijl, F., 292, 293
Van Zwol, C., 304
Van de Meeberg, D., 298, 324
Van Traa, M., 289, 301
Vardanian, R., 184, 186
Vaughn, J., 18
Veic, P., 51, 77
Vigh, J., 233, 313
Virjent-Novak, B., 78
Vitoskaya, G., 186

Wachowski, I., 166
Waddington, P. A. J., 18, 244, 245, 259, 260, 315, 317
Walker, N., 316, 321
Walker, D. B., 30
Waltoś, S., 161
Ward, R. H., 137
Wasek, A., 160
Wasik, Z., 145
Weitzer, R., 209, 316
Westley, W. A., 116
Wiebrens, C. J., 304
Wilford, R., 315, 316
Willemse, H. M., 305
Williams, P., 319
Williams, J. L., 19
Willis, A., 316
Wilson, J. Q., 281
Witkowski, Z., 145
Wójcik, D., 148
Wolchover, D., 257
Wolf, C., 218, 234
Wong, H. Z., 18
Woodward, R., 316
Woźniak, D., 151

Yang, C., 27

Zaslavskaja, T., 175
Zemskov, V., 174
Zima, H., 217
Zvekić, U., 71